DIVERSITY OF LIFE

CECIE STARR / RALPH TAGGART

BIOLOGY
THE UNITY AND DIVERSITY OF LIFE
EIGHTH EDITION

WADSWORTH PUBLISHING COMPANY

I(T)P® AN INTERNATIONAL THOMSON PUBLISHING COMPANY

Belmont, CA • Albany, NY • Bonn • Boston
Cincinnati • Detroit • Johannesburg • London • Madrid
Melbourne • Mexico City • New York
Paris • San Francisco • Singapore
Tokyo • Toronto • Washington

BIOLOGY PUBLISHER: Jack C. Carey

ASSISTANT EDITOR: Kristin Milotich

MEDIA PROJECT MANAGER: Pat Waldo

MARKETING MANAGER: Halee Dinsey

DEVELOPMENTAL EDITOR: Mary Arbogast

PROJECT EDITOR: Sandra Craig

EDITORIAL ASSISTANT: Michael Burgreen

PRINT BUYER: Karen Hunt

PRODUCTION: Mary Douglas, Rogue Valley Publications

TEXT AND COVER DESIGN, ART DIRECTION: Gary Head, Gary Head Design

ART COORDINATOR: Myrna Engler-Forkner

EDITORIAL PRODUCTION: Mary Roybal, Karen Stough, Susan Gall

PRIMARY ARTISTS: Raychel Ciemma; Precision Graphics (Jan Troutt, J.C. Morgan)

ADDITIONAL ARTISTS: Robert Demarest, Darwen Hennings, Vally Hennings, Betsy Palay, Nadine Sokol, Kevin Somerville, Lloyd Townsend

PHOTO RESEARCH, PERMISSIONS: Stephen Forsling, Roberta Broyer

COVER PHOTOGRAPH: *Minnehaha Falls*, © Richard Hamilton Smith

COMPOSITION: Precision Graphics (Jim Gallagher, Kirsten Dennison)

COLOR PROCESSING: H&S Graphics (Tom Anderson, Nancy Dean, John Deady, Rich Stanislawski)

PRINTING AND BINDING: World Color, Versailles

COPYRIGHT © 1998 by Wadsworth Publishing Company
A Division of International Thomson Publishing Inc.

The ITP logo is a registered trademark under license.

Printed in the United States of America
1 2 3 4 5 6 7 8 9 10

ISBN 0-534-53005-2

For more information, contact Wadsworth Publishing Company, 10 Davis Drive, Belmont, California 94002, or electronically at http://www.thomson.com/wadsworth.html

International Thomson Publishing Europe
Berkshire House 168-173, High Holborn
London, WC1V7AA, England

Thomas Nelson Australia
102 Dodds Street
South Melbourne 3205, Victoria, Australia

Nelson Canada
1120 Birchmount Road
Scarborough, Ontario, Canada M1K 5G4

International Thomson Editores
Campos Eliseos 385, Piso 7
Col. Polanco, 11560 México D.F. México

International Thomson Publishing GmbH
Königswinterer Strasse 418
53227 Bonn, Germany

International Thomson Publishing Asia
221 Henderson Road, #05-10 Henderson Building
Singapore 0315

International Thomson Publishing Japan
Hirakawacho Kyowa Building, 3F
2-2-1 Hirakawacho, Chiyoda-ku, Tokyo 102, Japan

International Thomson Publishing Southern Africa
Building 18, Constantia Park
240 Old Pretoria Road
Halfway House, 1685 South Africa

CONTENTS IN BRIEF

Highlighted chapters are included in DIVERSITY OF LIFE.

Highlighted chapters are included in DIVERSITY OF LIFE.

DETAILED CONTENTS

IV EVOLUTION AND DIVERSITY

MEERKATS AT SUNRISE

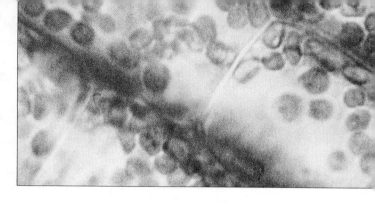

PREFACE

Not too long from now we will cross the threshold of a new millennium, a rite of passage that invites reflection on where biology has been and where it might be heading. About 500 years ago, during an age of global exploration, naturalists first started to systematically catalog and think about the staggering diversity of organisms all around the world. Less than 150 years ago, just before the start of a civil war that would shred the fabric of a new nation, the naturalist Charles Darwin shredded preconceived notions about life's diversity. It was only about 50 years ago that biologists caught their first glimpse of life's unity at the molecular level. Until that happened a biologist could still hope to be a generalist—someone who viewed life as Darwin did, without detailed knowledge of mechanisms that created it, that perpetuate it, that change it.

No more. Biology grew to encompass hundreds of specialized fields, each focused on one narrow aspect of life and yielding volumes of information about it. Twenty years ago I wondered whether introductory textbooks could possibly keep up with the rapid and divergent splintering of biological inquiry. James Bonner, a teacher and researcher at the California Institute of Technology, turned my thinking around on this. He foresaw that authors and instructors for introductory courses must become the new generalists, the ones who give each generation of students broad perspective on what we know about life and what we have yet to learn.

And we must do this, for the biological perspective remains one of the most powerful of education's gifts. With it, students who travel down specialized roads can sense intuitively that their research and its applications may have repercussions in unexpected places in the world of life. With that perspective, students in general might cut their own intellectual paths through social, medical, and environmental thickets. And they might come to understand the past and to predict possible futures for ourselves and all other organisms.

CONCERNING THE EIGHTH EDITION

Like earlier editions, this book starts with an overview of the basic concepts and scientific methods. Three units on the principles of biochemistry, inheritance, and evolution follow. The principles provide the conceptual background necessary for deeper probes into life's unity and diversity, starting with a richly illustrated evolutionary survey of each kingdom. Units on the comparative anatomy and physiology of plants, then animals, follow. The last unit focuses on the patterns and consequences of organisms interacting with one another and with their environment. Thus the organization parallels the levels of biological organization, from cells through the biosphere. We adhere to this traditional approach for good reason: it works.

As before, we identify and highlight the key concepts, current understandings, and research trends for the major fields of inquiry. Through examples of problem solving and experiments, we give ample evidence of "how we know what we know" and thus demonstrate the power of critical thinking. We explain the structure and functioning of a broad sampling of organisms in enough detail so that students can develop a working vocabulary about life's parts and processes. We also updated the glossary.

CONCEPT SPREADS

In the first chapter, an overview of the levels of biological organization kicks off a story that continues through the rest of the book. Telling such a big, complex story might be daunting unless you remind yourself of the question *"How do you eat an elephant?"* and its answer, *"One bite at a time."* We who have told the story again and again know how the parts fit together, but many students need help to keep the story line in focus within and between chapters. And they need to chew on concepts one at a time.

In every chapter we present each concept on its own table, so to speak. That is, we organize the descriptions, art, and supporting evidence for it on two facing pages, at most. Think of this as a concept spread, as in Figure A. Each starts with a numbered tab and ends with boldface statements to summarize the key points. Students can use these cues as reminders to digest one topic before starting on another. Well-crafted transitions between spreads help students focus on where topics fit in the larger story and gently discourage memorization for its own sake. The clear demarcation also gives instructors greater flexibility in assigning or skipping topics within a chapter.

By restricting the space available for each concept, we force ourselves to clear away the clutter of superfluous detail. Within each concept spread, we block out headings and subheadings to rank the importance of its various parts. Any good story has such a hierarchy of information, with background settings, major and minor characters, and high points and an ending where everything comes together. Without a hierarchy, a story has all the excitement, flow, and drama of an encyclopedia. Where details are useful as expansions of concepts, we integrate them into suitable illustrations to keep them from disrupting the text flow.

Not all students are biology majors, and many of them approach biology textbooks with apprehension. If the words don't engage them, they sometimes end up hating the book, and the subject. It comes down to line-by-line judgment calls. During twenty-two years of authorship, we developed a sense of when to leave core material alone and when to loosen it up to give students breathing room. Interrupting, say, an account of mitotic cell division with a distracting anecdote does no good. Plunking a humorous aside into a chapter that ties together the evolution of the Earth and life trivializes a magnificent story. Including an entertaining story is fine, provided that doing so reinforces a key concept. Thus, for example, we include the story of a misguided species introduction that resulted in wild European rabbits running amok through Australia.

BALANCING CONCEPTS WITH APPLICATIONS

Each chapter starts with a lively or sobering application that leads into an adjoining list of key concepts. The list is an advance organizer for the chapter as a whole. At strategic points, examples of applications parallel the core material—not so many as to be distracting, but enough to keep minds perking along with the conceptual development. Many brief applications are integrated in the text. Others are in *Focus* essays, which give more depth on medical, environmental, and social issues but do not interrupt the text flow.

FOUNDATIONS FOR CRITICAL THINKING

To help students develop a capacity for critical thinking, we walk them through experiments that yielded evidence in favor of or against hypotheses being discussed. The main index for the book will give you a sense of the number and types of experiments used (see the entry *Experiments*).

We use certain chapter introductions as well as entire chapters to show students some of the productive results of critical thinking. Among these are the introductions to the chapters on Mendelian genetics (11), DNA structure and function (14), speciation (19), immunology (40), and animal behavior (51).

Many *Focus on Science* essays provide more detailed, optional examples of how biologists apply critical thinking to problem solving. For example, one of these describes RFLP analysis (Section 16.3) and a few of its more jarring applications. Another essay (Section 21.4) helps convey to students that biology is not a closed book. Even when new research brings a sweeping story into sharp focus—in this case, the origin of the great prokaryotic and eukaryotic lineages—it also opens up new roads of inquiry.

This edition has *Critical Thinking* questions at the end of chapters. Katherine Denniston of Towson State University developed these thought-provoking questions. Chapters 11 and 12 also include a large selection of *Genetics Problems* that help students grasp the principles of inheritance.

To keep readers focused, we cover each concept on one or two facing pages, starting with a numbered tab . . .

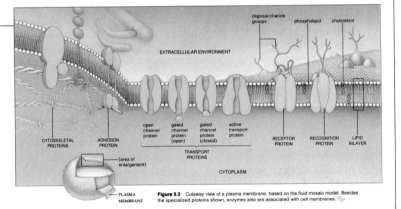

5.1 MEMBRANE STRUCTURE AND FUNCTION

Earlier chapters provided you with a brief look at the structure of cell membranes and the general functions of their component parts. Here, we incorporate some of the background information in a more detailed picture.

The Lipid Bilayer of Cell Membranes

Fluid bathes the two surfaces of a cell membrane and is vital for its functioning. The membrane, too, has a fluid quality; it is not a solid, static wall between cytoplasmic and extracellular fluids. For instance, puncture a cell with a fine needle, and its cytoplasm will not ooze out. The membrane will flow over the puncture site and seal it!

How does a fluid membrane remain distinct from its fluid surroundings? To arrive at the answer, start by reviewing what we have already learned about its most abundant components, the phospholipids. Recall that a **phospholipid** has a phosphate-containing head and two fatty acid tails attached to a glycerol backbone (Figure 5.2a). The head is hydrophilic; it easily dissolves in water. Its tails are hydrophobic; water repels them. Immerse a number of phospholipid molecules in water, and they will interact with water molecules and with one another until they spontaneously cluster in a sheet or film at the water's surface. Their jostlings may even force them to become organized in two layers, with all fatty acid tails sandwiched between all hydrophilic heads. This **lipid bilayer** arrangement, remember, is the structural basis of cell membranes (Section 4.1 and Figure 5.2c).

The organization of each lipid bilayer minimizes the total number of hydrophobic

groups exposed to water, so the fatty acid tails do not have to spend a lot of energy fighting water molecules, so to speak. A "punctured" membrane exhibits sealing behavior precisely because a puncture is energetically unfavorable. It leaves far too many hydrophobic groups exposed to the surrounding fluid.

Ordinarily, few cells get jabbed by fine needles. But the self-sealing behavior of membrane phospholipids is good for more than damage control. Among other things, it functions in vesicle formation. For example, as vesicles bud away from ER or Golgi membranes, phospholipids interact hydrophobically with cytoplasmic water. They get pushed together, and the rupture seals. You will read more about vesicle formation later in the chapter.

Fluid Mosaic Model of Membrane Structure

Figure 5.3 shows a bit of membrane that corresponds to the **fluid mosaic model**. By this model, cell membranes are a mixed composition—a "mosaic"—of phospholipids, glycolipids, sterols, and proteins. The phospholipid heads as well as the length and saturation of the tails are not all the same. (Recall that unsaturated fatty acids have one or more double bonds in their backbone and fully saturated ones have none.) The glycolipids are structurally similar to phospholipids, but their head incorporates one or more sugar monomers. In animal cell membranes, cholesterol is the most abundant sterol (Figure 5.2b). Phytosterols are their equivalent in plant cell membranes.

Also by this model, the membrane is "fluid" owing to the motions and interactions of its component parts.

Figure 5.2
(a) Structural formula of phosphatidylcholine, a phospholipid that is one of the most common components of the membranes of animal cells. *Orange* indicates its hydrophilic head; *yellow* indicates its hydrophobic tails.

(b) Structural formula of cholesterol, the major sterol in animal tissues.

(c) Diagram showing how lipids that are placed in liquid water may spontaneously organize themselves into a bilayer structure.

lipid bilayer

water — water

Overview of Membrane Proteins

The proteins embedded in a lipid bilayer or attached to one of its surfaces carry out most membrane functions. Many are enzyme components of metabolic machinery. Others are **transport proteins** that allow water-soluble substances to move through their interior, which spans the bilayer. They bind molecules or ions on one side of the membrane, then release them on the other side.

The **receptor proteins** bind extracellular substances, such as hormones, that trigger changes in cell activities.

Figure 5.3 Cutaway view of a plasma membrane, based on the fluid mosaic model. Besides the specialized proteins shown, enzymes also are associated with cell membranes.

The hydrophobic interactions that give rise to most of a membrane's structure are weaker than covalent bonds. This means most phospholipids and some proteins are free to drift sideways. Also, the phospholipids can spin about their long axis and flex their tails, which keeps neighboring molecules from packing together in a solid layer. Short or kinked (unsaturated) fatty acid tails also contribute to membrane fluidity.

The fluid mosaic model is a good starting point for exploring cell membranes. But bear in mind, membranes differ in the details of their molecular composition and arrangements, and they are not even the same on both surfaces of their bilayer. For example, oligosaccharides and other carbohydrates are covalently bonded to protein and lipid components of a plasma membrane, but only on its outward-facing surface (Figure 5.3). Moreover, they differ in number and kind from one species to the next, even among the different cells of the same individual.

For example, certain enzymes that crank up machinery for cell growth and division become switched on when somatotropin, a hormone, binds with receptors for it. Different cells have different combinations of receptors.

Diverse **recognition** proteins at the cell surface are like molecular fingerprints; their oligosaccharide chains identify a cell as being of a specific type. For example, "self" proteins pepper the plasma membrane of your cells. Certain white blood cells chemically recognize the proteins and leave your own cells alone, but they attack invading bacterial cells having "nonself" proteins at their surface. Finally, **adhesion proteins** of multicelled organisms help cells of the same type locate and stick to one another and stay positioned in the proper tissues. They are glycoproteins with oligosaccharides attached. After tissues form, the sites of adhesion may become a type of cell junction, as described earlier in Section 4.10.

A cell membrane has two layers composed mainly of lipids, phospholipids especially. This lipid bilayer is the structural foundation for the membrane and also serves as a barrier to water-soluble substances.

Hydrophilic heads of the phospholipids are dissolved in fluids that bathe the two outer surfaces of the bilayer. Their hydrophobic tails are sandwiched between the heads.

Proteins associated with the bilayer carry out most membrane functions. Many are enzymes, transporters of substances across the bilayer, or receptors for extracellular substances. Other types function in cell-to-cell recognition or adhesion.

. . . and ending with one or more summary statements.

FIGURE A   *A concept spread from this edition.*

VISUAL OVERVIEWS OF MAJOR CONCEPTS

While writing the text, we simultaneously develop the illustrations as inseparable parts of the same story. This integrative approach appeals to students who are visual learners. When they can first work their way through a visual overview of some process, then reading through the corresponding text becomes less intimidating. Over the years, students have repeatedly thanked us for our hundreds of overview illustrations, which contain step-by-step, written descriptions of biological parts and processes. We break down the information into a series of illustrated steps that are more inviting than a complex, "wordless" diagram. Figure *B* is a sample. Notice how simple descriptions, integrated with the art, take students through the stages by which mRNA transcripts become translated into polypeptide chains, one step at a time.

Similarly, we continue to create visual overviews for anatomical drawings. The illustrations integrate structure and function. Students need not jump back and forth from the text, to tables, to illustrations, and back again in order to comprehend how an organ system is put together and what its parts do. Even individual descriptions of parts are hierarchically arranged to reflect the structural and functional organization of that system.

COLOR CODING

In line illustrations, we consistently use the same colors for the same types of molecules and cell structures. Visual consistency makes it easier for students to track complex parts and processes. Figure C is the color coding chart.

ZOOM SEQUENCES

Many illustrations in the book progress from macroscopic to microscopic views of the same subject. Figure 7.2 is an example; this zoom sequence shows where the reactions of photosynthesis proceed, starting with a plant growing by a roadside. As another example, Figures 38.19 and 38.20 move down through levels of skeletal muscle contraction, starting with a muscle in the human arm.

ICONS

Within the text, small diagrams next to an illustration help relate the topic to the big picture. For instance, in Figure *A*, a simple representation of a cell subtly reminds students of the location of the plasma membrane relative to the cytoplasm. Other icons serve as reminders of the location of reactions and processes in cells and how they

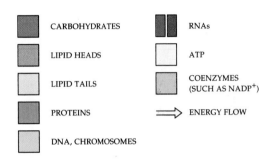

CARBOHYDRATES		RNAs
LIPID HEADS		ATP
LIPID TAILS		COENZYMES (SUCH AS NADP$^+$)
PROTEINS		ENERGY FLOW
DNA, CHROMOSOMES		

FIGURE C   *Color coding chart for the diagrams of biological molecules and cell structures.*

Step-by-step art with simple descriptions helps students visualize a process before reading text about it.

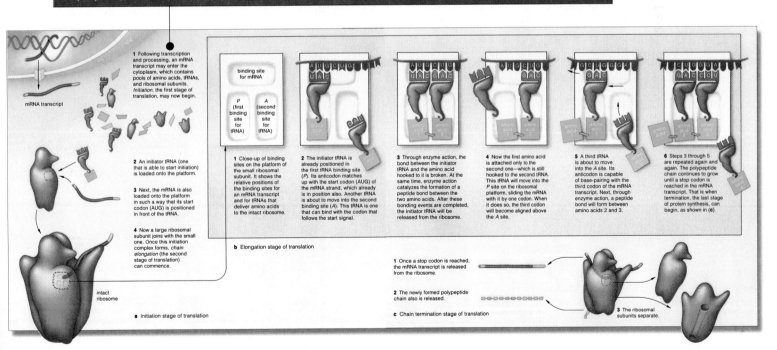

FIGURE B   *A visual overview from this edition.*

interrelate to one another. Still others remind students of the evolutionary relationships among groups of organisms, as in Chapter 26.

New to this edition are icons that invite students to use multimedia. One icon directs them to art in the CD-ROM enclosed with each student copy, another to supplemental material on the Web, and a third to InfoTrac:

CD-ROM
ICON:

WEB
ICON:

INFOTRAC
ICON:

END-OF-CHAPTER STUDY AIDS

The end-of-chapter study aids are designed to reinforce the key concepts. Each chapter ends with a summary in list form, review questions, a self-quiz, critical thinking questions, selected key terms, and a list of readings. Italicized page numbers tie the review questions and key terms to relevant text pages.

END-OF-BOOK STUDY AIDS

At the book's end, the detailed classification scheme in Appendix I is helpful for reference purposes. Appendix II includes metric-English conversion charts. Appendix III has answers to the self-quizzes at the end of each chapter.

A Glossary includes boldfaced terms from the text, with pronunciation guides and word origins to make the formidable words less so. The Appendixes as well as the Glossary are printed on paper that is tinted different colors to preclude frustrating searches for where one ends and the next begins. The Index is detailed enough to help readers find doors to the text more quickly.

CONTENT REVISIONS

Instructors who use *Biology: The Unity and Diversity of Life* may wish to evaluate this overview of key modifications. Overall, conceptual development is more integrated. The writing is still crisp but not too brief, because some topics can confuse students when presented in insufficient detail. New research called for some adjustments in the overall framework. For example, we subscribe to the six-kingdom classification scheme, which now has exceptional support from comparative biochemical studies. Even the end-of-chapter Critical Thinking questions incorporate current material.

UNIT IV. EVOLUTION AND DIVERSITY   Chapter 21 has a tree of life adapted to the six-kingdom scheme. Ediacaran and Cambrian forms are defined, and gymnosperm dominance is more clearly correlated with the Mesozoic. The global broiling theory is correlated with the K-T asteroid impact. Besides being a conceptual and chronological framework for the diversity chapters, this survey chapter may help students sense their place in nature. The greatly revised diversity chapters have crisp evolutionary story lines, a more balanced text-to-art ratio, and riveting applications (see Section 22.9 on infectious diseases). Chapter 23 states how opinions differ on which organisms are protistans. In keeping with current consensus, we treat chytrids, water molds, and algae in that chapter. We have a new red algal life cycle. Chapter 24 includes a better treatment of lichens. Chapter 25 includes more on the Carboniferous and new spreads on existing seed-bearing plants. In the Chapter 26 introduction, I shamelessly segue from my grandchild to the early invertebrates. The chapter has new sections on animal origins, molluscan evolution, and adaptations of arthropods. Chapter 27 is reorganized for clarity, as in the sections on vertebrate evolutionary trends, reptiles, and mammals. Chapter 28 integrates new fossil evidence.

THE ORIGIN AND EVOLUTION OF LIFE

In The Beginning . . .

Some clear evening, watch the moon as it rises from the horizon and think of the 380,000 kilometers between it and you. *Five billion trillion times* the distance between the moon and you are galaxies—systems of stars—at the boundary of the known universe. Wavelengths traveling through space move faster than anything else—millions of meters per second. Yet the long wavelengths that originated from faraway galaxies many billions of years ago are only now reaching the Earth.

By every known measure, all the near and distant galaxies suspended in the vast space of the universe are moving away from one another, which means that the universe is expanding. And the prevailing view of how the colossal expansion came about accounts for every bit of matter in the universe, in every living thing.

Think about how you rewind a videotape on a VCR, then imagine "rewinding" the universe. As you do this, the galaxies start moving back together. After 12 to 15

Figure 21.1 Part of the great Eagle nebula, a hotbed of star formation 7,000 light-years from Earth, in the constellation Serpens. (The Latin *nebula* means mist.) Not shown in this image are a few huge, young stars above the pillars. For the past few million years, intense ultraviolet radiation from the stars has been eroding the less dense surface of the pillars. (By analogy, visualize a strong prevailing wind blowing away sand in a desert and exposing rocks.) Globules of denser gases and dust that have resisted erosion are visible at the surface. Each one is wider than our solar system—more than 10 billion miles across! New stars are hatching from the protruding globules; some are shining brightly on the tips of gaseous streamers.

billion years of rewinding, all galaxies, all matter, and all of space are compressed into a hot, dense volume about the size of the sun. You have arrived at time zero.

That incredibly hot, dense state lasted only for an instant. What happened next is known as the **big bang,** a stupendous, nearly instantaneous distribution of all matter and energy everywhere, through all of the known universe. About a minute later, temperatures dropped to a billion degrees. Fusion reactions produced most of the light elements, including helium, which are still the most abundant elements throughout the universe. Radio telescopes have detected relics of the big bang in the form of cooled and diluted background radiation, left over from the beginning of time.

Over the next billion years, stars started forming as gaseous material contracted in response to the force of gravity. When stars became massive enough, nuclear reactions were ignited in their central region, and they gave off tremendous light and heat. As massive stars continued to contract, many became dense enough to promote the formation of heavier elements.

All stars have a life history, from birth to an often spectacularly explosive death. In what might be called the original stardust memories, the heavier elements released during the explosions became swept up during the gravitational contraction of new stars, and they became raw materials for the formation of even heavier elements. Even as you are reading this page, the Hubble space telescope is providing astounding glimpses of star-forming activity, as in the dust clouds of Orion, Serpens, and other constellations (Figure 21.1).

Now imagine a time long ago, when explosions of dying stars ripped through our galaxy and left behind a dense cloud of dust and gas that extended trillions of kilometers in space. As the cloud cooled, countless bits of matter gravitated toward one another. By 4.6 billion years ago, the cloud had flattened out into a slowly rotating disk. At the dense, hot center of that disk, the shining star of our solar system—the sun—was born.

The remainder of this chapter is a sweeping slice through time, one that cuts back to the formation of the Earth and the chemical origins of life. It is the starting point for the next four chapters, which will take us along lines of descent that led to the present range of species diversity. The story is not complete. Even so, all the available evidence, from many avenues of research, points to a principle that can help us organize separate bits of information about the past: *Life is a magnificent continuation of the physical and chemical evolution of the universe, of galaxies and stars, and of the planet Earth.*

KEY CONCEPTS

1. A great body of evidence suggests that life originated more than 3.8 billion years ago. Its origin and subsequent evolution have been linked to the physical and chemical evolution of the universe, the stars, and the planet Earth.

2. All of the inorganic and organic compounds necessary for self-replication, membrane assembly, and metabolism —that is, for the structure and functioning of living cells— could have formed spontaneously under conditions that existed on the early Earth.

3. The history of life, from its chemical beginnings to the present, spans five intervals of geologic time. It extends through two great eons—the Archean and the Proterozoic— and the Paleozoic, Mesozoic, and Cenozoic eras.

4. Not long after life originated, divergences led to two great prokaryotic lineages, called the archaebacteria and eubacteria. Shortly afterward, the ancestors of eukaryotes diverged from the archaebacterial lineage.

5. Archaebacteria and eubacteria dominated the Archean and Proterozoic eons. Eukaryotic cells originated late in the Proterozoic era and became spectacularly diverse. A theory of endosymbiosis helps explain the profusion of specialized organelles that arose in eukaryotic cells.

6. All six kingdoms of organisms are characterized by the persistences, extinctions, and radiations of many different lineages over time.

7. Throughout the history of life, asteroid impacts, drifting and colliding continents, and other environmental insults have had profound impact on the direction of evolution.

Origin of the Earth

Cloudlike regions of the universe, as shown in Figure 21.1, are mostly hydrogen gas. These clouds also contain water, iron, silicates, hydrogen cyanide, ammonia, methane, formaldehyde, and some other simple organic and inorganic substances. The contracting cloud from which our solar system evolved probably was similar in composition. Between 4.6 and 4.5 billion years ago, the periphery of the cloud cooled. Mineral grains and ice orbiting the sun started to clump together as a result of electrostatic attraction and the pull of gravity (Figure 21.2). In time, larger, faster clumps started colliding and shattering. Some became more massive by sweeping up asteroids, meteorites, and the other rocky remnants of collisions, and gradually they evolved into planets.

Figure 21.2   Representation of the cloud of dust, gases, and clumps of rock and ice around the early sun.

As the Earth was forming, much of its inner rocky material melted. Asteroid impacts as well as its own internal compression and radioactive decay of minerals could have generated the heat necessary to do this. As the rocks melted, nickel, iron, and other heavy materials moved to the Earth's interior; lighter ones floated to the surface. This process produced a crust, mantle, and core. The **crust** is an outer region of basalt, granite, and other low-density rocks. It envelops the intermediate-density rocks of the **mantle**, which wraps around a core of very high-density, partially molten nickel and iron.

Four billion years ago, the Earth was a thin-crusted inferno (Figure 21.3a). In less than 200 million years, life had originated on its surface! We have no record of the event. As far as we know, movements in the mantle and crust, volcanic activity, and erosion obliterated all traces of it. Still, we can put together a plausible explanation of how life originated by considering three questions:

First, *when life originated, what physical and chemical conditions prevailed on Earth?*

Second, *based on physical, chemical, and evolutionary principles, could large organic molecules have spontaneously formed and then evolved into molecular systems displaying the fundamental properties of life?*

Third, *can we devise experiments to test whether living systems could have emerged by chemical evolution?*

The First Atmosphere

When the first patches of crust were forming, hot gases blanketed the Earth. This first atmosphere probably was a mix of gaseous hydrogen (H_2), nitrogen (N_2), and carbon monoxide (CO), as well as carbon dioxide (CO_2). Was gaseous oxygen (O_2) also present? Probably not. Rocks release oxygen when they are subjected to the intense heat of volcanic eruptions, but not much. Besides, free oxygen would have reacted at once with other elements. At first, water at the Earth's molten surface must have evaporated into the atmosphere. After the crust cooled and solidified, rains from the clouds blanketing the Earth drenched the parched rocks. For millions of years the runoff eroded mineral salts and other compounds from the rocks. And salt-laden waters gradually collected in depressions in the crust and formed the early seas.

If the early atmosphere and seas had not been free of oxygen, then the organic compounds that became organized into the first cells might not have formed on their own. Oxygen would have attacked them, in the manner described in the introduction to Chapter 6. If liquid water had not accumulated, then cell membranes could not have formed. Cells are the basic units of life. *Each has a capacity to survive and reproduce on its own.*

Synthesis of Organic Compounds

Reduce a cell to its lowest common denominator and all that remains are proteins, complex carbohydrates and lipids, and nucleic acids. Existing cells assemble these molecules from smaller organic compounds: the simple sugars, fatty acids, amino acids, and nucleotides. Energy from the environment drives the synthesis reactions. Were small organic compounds also present on the early Earth? Were there sources of energy that spontaneously drove their assembly into the large molecules of life?

Mars, meteorites, the Earth's moon, and the Earth all formed at the same time, from the same cosmic cloud. Rocks collected from Mars, meteorites, and the moon contain precursors of biological molecules, so the same precursors must have been present on the Earth. If this were indeed the case, *then energy from sunlight, lightning, or even heat escaping from the crust could have been enough to drive their combination into organic molecules.*

Stanley Miller conducted the first experimental test of that prediction. First he mixed methane, hydrogen,

Figure 21.3 (a) Representation of the Earth during its formation, when the moon's orbit was much closer than it is today. If the Earth had condensed into a planet of smaller diameter, its gravitational mass would not have been great enough to hold onto an atmosphere. If it had settled into an orbit closer to the sun, water would have evaporated from its hot surface. If the Earth's orbit had been more distant from the sun, its surface would have been far colder, and water would have been locked up as ice. Without liquid water, life as we know it never would have originated on Earth.

(b) Stanley Miller's experimental apparatus, used to study the synthesis of organic compounds under conditions that presumably existed on the early Earth. The condenser cools circulating steam so that water droplets form.

ammonia, and water inside a reaction chamber of the sort depicted in Figure 21.3b. Then he kept the mixture circulating and bombarded it with a spark discharge to simulate lightning. In less than a week, amino acids and other small organic compounds had formed.

In other experiments that simulated conditions on the early Earth, glucose, ribose, deoxyribose, and other sugars formed spontaneously from formaldehyde, and adenine from hydrogen cyanide. Ribose and adenine are found in ATP, NAD, and other nucleotides vital to cells.

However, if *complex* organic compounds had formed directly in the seas, they would not have lasted long. The spontaneous direction of the necessary reactions would have been toward hydrolysis, not condensation, in water. How did more lasting bonds form?

By one hypothesis, clay in the rhythmically drained muck of tidal flats and estuaries served as templates (structural patterns) for the spontaneous assembly of proteins and other complex organic compounds. Clay consists of thin, stacked layers of aluminosilicates with metal ions at their surfaces, which attract amino acids. Expose amino acids to some clay, warm the clay with rays from the sun, then alternately moisten and dry it. Condensation reactions will proceed at its surfaces and yield proteins and other complex organic compounds.

By another hypothesis, complex organic compounds formed spontaneously near hydrothermal vents on the seafloor, where species of archaebacteria are thriving

today. As experimental tests by Sidney Fox and others show, when amino acids are placed in water and then heated, they spontaneously order themselves into small protein-like molecules, which Fox calls "protenoids."

However the first proteins formed, their molecular structure dictated how they would interact with other compounds. Suppose some proteins had the structure to function as weak enzymes. That is, they were able to hasten bond formation between amino acids. Enzyme-directed synthesis would have a selective advantage. In the chemical competition for available amino acids, the protein configurations that promoted reactions would win. Also, proteins have the capacity to bind metal ions and other carbon-based compounds. As you will see next, such chemical modification was the foundation for the evolution of metabolism. For now, simply reflect on the possibility that selection was at work before the origin of living cells, favoring the chemical evolution of enzymes and other complex organic compounds.

Many diverse experiments yield indirect evidence that the complex organic molecules characteristic of life could have formed under conditions that existed on the early Earth.

Origin of Agents of Metabolism

A defining characteristic of life is metabolism. The word refers to all the reactions by which cells harness energy and use it to drive their activities, such as biosynthesis. During the first 600 million years or so of Earth history, enzymes, ATP, and other organic compounds may have assembled spontaneously, and they may have done so in the same physical locations. If so, their close association would have naturally promoted chemical interactions and the beginning of metabolic pathways.

Imagine an ancient estuary, rich in clay deposits. It is a coastal region where seawater mixes with mineral-rich water being drained from the land. There, beneath the sun's rays, countless aggregations of organic molecules stick to the clay. At first there are quantities of an amino acid; call it D. Throughout the estuary, D molecules get incorporated into new proteins—until the supply of D dwindles. However, suppose an enzyme-like protein molecule also is present in the estuary. It can promote the formation of D by acting on an abundant, simpler substance—call it C. By chance, some aggregations of organic molecules include that particular enzyme-like protein, and so they have an advantage in the chemical competition for starting materials.

In time, C molecules become scarce. At that point, the advantage tilts to aggregations that can promote formation of C from even simpler organic substances B and A—say, from carbon dioxide and water. As you know, carbon dioxide and water occur in essentially unlimited amounts in the atmosphere and in the seas. Chemical selection has favored a synthetic pathway:

$$A + B \longrightarrow C \longrightarrow D$$

Finally, suppose some aggregations are better than others at absorbing and using energy. Which molecules could bestow such an advantage? Think of the energy-trapping pathway that now dominates the world of life: photosynthesis. It starts at pigments called chlorophylls. The light-absorbing and electron-donating portion of a chlorophyll molecule is called a porphyrin ring structure. Porphyrins are also present in cytochromes, which are part of electron transport systems in all photosynthetic and aerobically respiring cells. Porphyrin molecules can spontaneously assemble from formaldehyde—one of the molecular legacies of cosmic clouds (Figure 21.4). Was porphyrin an electron transporter of some of the early metabolic pathways? Perhaps.

Origin of Self-Replicating Systems

Another defining characteristic of life is a capacity for reproduction, which now starts with protein-building instructions in DNA. The DNA molecule is fairly stable,

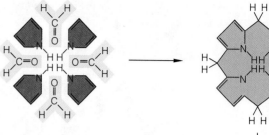

four pyrrole rings plus four
formaldehyde molecules

Figure 21.4 One hypothetical sequence by which formaldehyde, an organic compound, underwent chemical evolution into porphyrin. Formaldehyde was present when the Earth formed. Porphyrin is the light-trapping and electron-donating component of all existing chlorophyll molecules. It also is a component of cytochrome, which is a protein component of electron transport systems that are part of many metabolic pathways.

porphyrin ring system

chlorophyll *a*, one of the light-trapping pigments of photosynthetic plant cells

and it is easily replicated before each cell division. As you know from earlier chapters, arrays of enzymes and RNA molecules operate together to carry out DNA's encoded instructions.

Most existing enzymes are assisted by the small organic molecules or metal ions that are called coenzymes. Intriguingly, some types of coenzymes have a structure that is identical to that of the RNA nucleotides. Another clue: Heat nucleotide precursors and short chains of phosphate group together, and they will self-assemble into strands of RNA. On the early Earth, energy from the sun or from geothermal events would have been sufficient to drive the spontaneous formation of RNA molecules.

Very simple self-replicating systems of RNA, enzymes, and coenzymes have been created in some laboratories. Did RNA later become the information-storing templates for protein

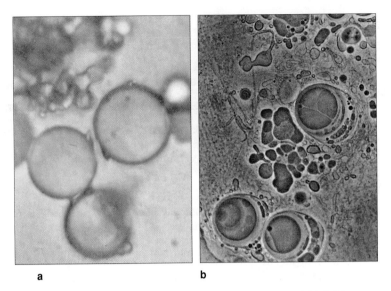

a   b

Figure 21.5 Microscopic spheres of (**a**) proteins and (**b**) lipids that self-assembled under abiotic conditions.

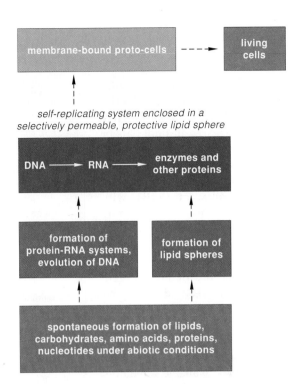

Figure 21.6 One possible sequence of events that led to the first self-replicating systems, then to the first living cells.

synthesis? Perhaps it initially did so, although existing RNA molecules are just too chemically fragile for such a role. Yet the use of RNA templates may have set up an **RNA world** that preceded DNA's dominance as the main informational molecule. Whatever the case, DNA eventually assumed this function, probably because it can form long nucleotide chains in more stable fashion.

We still don't know how DNA entered the picture. Until we identify the likely chemical ancestors of RNA and DNA, the story of life's origin will be incomplete. Filling in the details will require imaginative sleuthing. For instance, a few researchers ran a program on a rapid, advanced computer. The program included information about simple compounds, of the sort thought to have been present on the early Earth. The researchers asked the computer to subject the compounds that they had chosen to random chemical competition and to natural selection. They ran the program repeatedly. And the outcome was always the same: The simple precursors inevitably evolved to form interacting systems of large, complex molecules.

Origin of the First Plasma Membranes

Experimental tests are more revealing of the origin of the plasma membrane, the outermost component of every living cell. A plasma membrane consists of a lipid bilayer, studded with proteins that carry out diverse functions. Its main role is to control which substances move into and out of the cell. Without this control, cells cannot exist. Before cells, then, there must have been **proto-cells**. Maybe they were little more than membrane sacs that protected information-storing templates and

metabolic agents from the environment. We know that simple membrane sacs can form spontaneously.

For one experiment, Fox heated amino acids until they formed protein-like chains, which he then placed in hot water. After cooling, the chains assembled into small, stable spheres of the sort shown in Figure 21.5a. Like the membranes of cells, these proteinoid spheres were selectively permeable to a variety of substances. They picked up free lipids from the surroundings, and a lipid-protein film formed at their surface.

In still other experiments by David Deamer and his coworkers, fatty acids and glycerol combined to form long-tail lipid molecules under laboratory conditions that simulated evaporating tidepools. The lipids self-assembled into small, water-filled sacs. Many were like cell membranes (Figure 21.5b).

In short, there are major gaps in the story of life's origins. But there also is strong experimental evidence that chemical evolution probably led to the molecules and structures that are characteristic of life. Figure 21.6 summarizes the milestones in that chemical evolution, which preceded the first cells.

Although the story is not yet complete, many laboratory experiments and computer simulations indirectly show that chemical and molecular evolution gave rise to proto-cells.

21.3 ORIGIN OF PROKARYOTIC AND EUKARYOTIC CELLS

The first living cells originated in the **Archean** eon, which lasted from 3.9 billion to 2.5 billion years ago. Those cells emerged as molecular extensions of the evolving universe, of our solar system and Earth. They may have appeared in tidal flats or in muddy sediments of an ancient seafloor. Fossils indicate they were like existing bacteria. Specifically, they were **prokaryotic cells**, with no nucleus. At first they probably were no more than self-replicating, membrane-bound sacs of DNA and other complex organic molecules. Given the absence of free oxygen, they must have secured energy by anaerobic pathways—fermentation, most likely. Energy was plentiful. Geologic processes had enriched the seas with organic compounds. So "food" was available, predators were absent, and cellular structures were free from oxygen attacks.

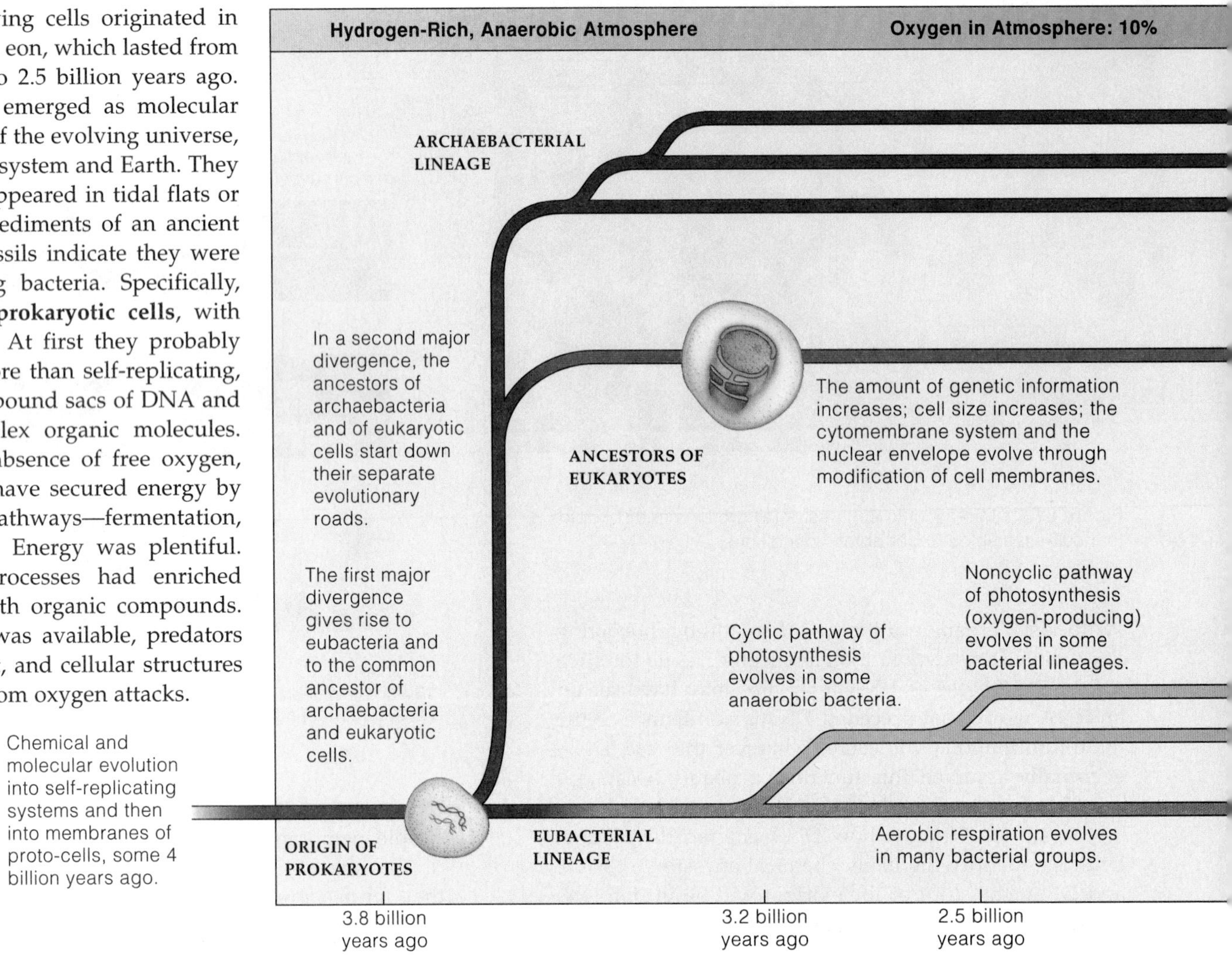

Hydrogen-Rich, Anaerobic Atmosphere   **Oxygen in Atmosphere: 10%**

ARCHAEBACTERIAL LINEAGE

In a second major divergence, the ancestors of archaebacteria and of eukaryotic cells start down their separate evolutionary roads.

The first major divergence gives rise to eubacteria and to the common ancestor of archaebacteria and eukaryotic cells.

Chemical and molecular evolution into self-replicating systems and then into membranes of proto-cells, some 4 billion years ago.

ANCESTORS OF EUKARYOTES

The amount of genetic information increases; cell size increases; the cytomembrane system and the nuclear envelope evolve through modification of cell membranes.

Cyclic pathway of photosynthesis evolves in some anaerobic bacteria.

Noncyclic pathway of photosynthesis (oxygen-producing) evolves in some bacterial lineages.

ORIGIN OF PROKARYOTES

EUBACTERIAL LINEAGE

Aerobic respiration evolves in many bacterial groups.

3.8 billion years ago

3.2 billion years ago

2.5 billion years ago

Some populations of those first prokaryotic cells apparently diverged in two major directions shortly after the time of origin. One evolutionary road led to the **eubacteria**. The other gave rise to the shared ancestors of the **archaebacteria** and **eukaryotic cells** (Figure 21.7).

Between 3.5 and 3.2 billion years ago, light-trapping pigments, electron transport systems, and other metabolic machinery evolved in some of the anaerobic eubacteria, which thereby became the first practitioners of the cyclic pathway of photosynthesis. (You read about this ATP-forming pathway in Section 7.4.) An unlimited source of energy—sunlight—had been tapped. For nearly 2 billion years, the photosynthetic descendants of those bacterial cells dominated the world of life. Their tiny but numerous populations formed very large mats in which sediments collected. The mats accumulated, one atop the other. Calcium deposits hardened and preserved the mats, which came to be called **stromatolites** (Figure 21.8).

Figure 21.7 An evolutionary tree of life that reflects mainstream thinking about the connections among major lineages. The diagram incorporates ideas about the origins of some eukaryotic organelles.

Figure 21.8 Stromatolites exposed at low tide in Western Australia's Shark Bay. These mounds started forming 2,000 years ago. They are identical in structure to stromatolites that formed more than 3 billion years ago.

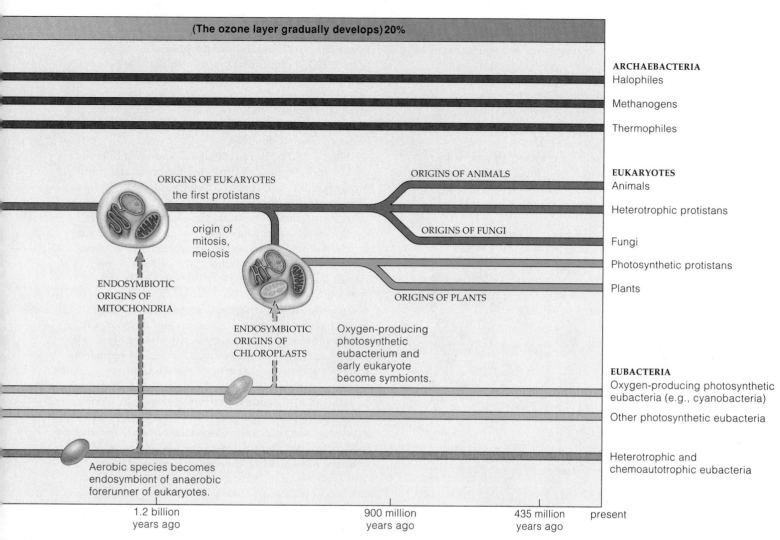

(The ozone layer gradually develops) 20%

ARCHAEBACTERIA
Halophiles

Methanogens

Thermophiles

EUKARYOTES
Animals

Heterotrophic protistans

Fungi

Photosynthetic protistans

Plants

ORIGINS OF EUKARYOTES
the first protistans

ORIGINS OF ANIMALS

origin of
mitosis,
meiosis

ORIGINS OF FUNGI

ENDOSYMBIOTIC
ORIGINS OF
MITOCHONDRIA

ENDOSYMBIOTIC
ORIGINS OF
CHLOROPLASTS

ORIGINS OF PLANTS

Oxygen-producing
photosynthetic
eubacterium and
early eukaryote
become symbionts.

EUBACTERIA
Oxygen-producing photosynthetic
eubacteria (e.g., cyanobacteria)

Other photosynthetic eubacteria

Heterotrophic and
chemoautotrophic eubacteria

Aerobic species becomes
endosymbiont of anaerobic
forerunner of eukaryotes.

1.2 billion years ago	900 million years ago	435 million years ago	present

By the dawn of the **Proterozoic** eon, 2.5 billion years ago, the photosynthetic machinery had become altered in some eubacterial species, and the noncyclic pathway of photosynthesis emerged. Oxygen, one of the pathway's by-products, started to accumulate. In time, this had two irreversible effects. First, *an oxygen-rich atmosphere stopped the further chemical origin of living cells.* Except in restricted anaerobic habitats, complex organic compounds could no longer form spontaneously and resist attack. Second, *aerobic respiration became the dominant energy-releasing pathway.* In many prokaryotic lineages, selection favored metabolic equipment that "neutralized" oxygen by using it as an electron acceptor. This innovation contributed to the rise of multicelled eukaryotes and their invasion of far-flung environments (Section 8.7).

Eukaryotic cells evolved in the Proterozoic, possibly before 1.2 billion years ago. We have fossils, 900 million years old, of well-developed red algae, green algae, fungi, and plant spores. Organelles, remember, are the hallmark of eukaryotic cells. *Where did they come from?* The next section describes a few plausible hypotheses.

About 800 million years ago, stromatolites began to decline dramatically. They had become a concentrated source of food for newly evolved, tiny bacteria-eating animals. About the same time, tiny soft-bodied animals were making tracks and digging burrows in seafloor sediments. They lived near the shores of Laurentia, an early supercontinent. About 570 million years ago, in "Precambrian" times, some of their descendants started the first adaptive radiation of animals.

The first cells evolved by about 3.8 billion years ago, during the Archean. All were prokaryotic cells, and most, if not all, probably made ATP by fermentation routes.

In the first major divergence, the ancestors of archaebacteria and of eukaryotic cells branched away from the road that would lead to modern eubacteria.

Oxygen-releasing photosynthetic bacteria evolved. In time, the oxygen-enriched atmosphere put an end to the further spontaneous chemical evolution of life. That atmosphere was a key selection pressure in the evolution of eukaryotic cells.

21.4 WHERE DID ORGANELLES COME FROM?

Thanks to Andrew Knoll, William Schopf, Jr., and other globe-hopping microfossil hunters, we have tantalizing evidence of early life. One fossil treasure is a strand of bacterial cells that lived 3.5 billion years ago, not long after the time that life originated. Others are from the Proterozoic. They were eukaryotic cells that contained a few membrane-bound organelles in the cytoplasm, as shown in Figure 21.9. Their living descendants have a profusion of organelles (Figure 21.10).

Where did eukaryotic organelles come from? Speculations abound. Some organelles probably evolved through gene mutations and natural selection. For others, researchers make a good case for evolution by way of endosymbiosis.

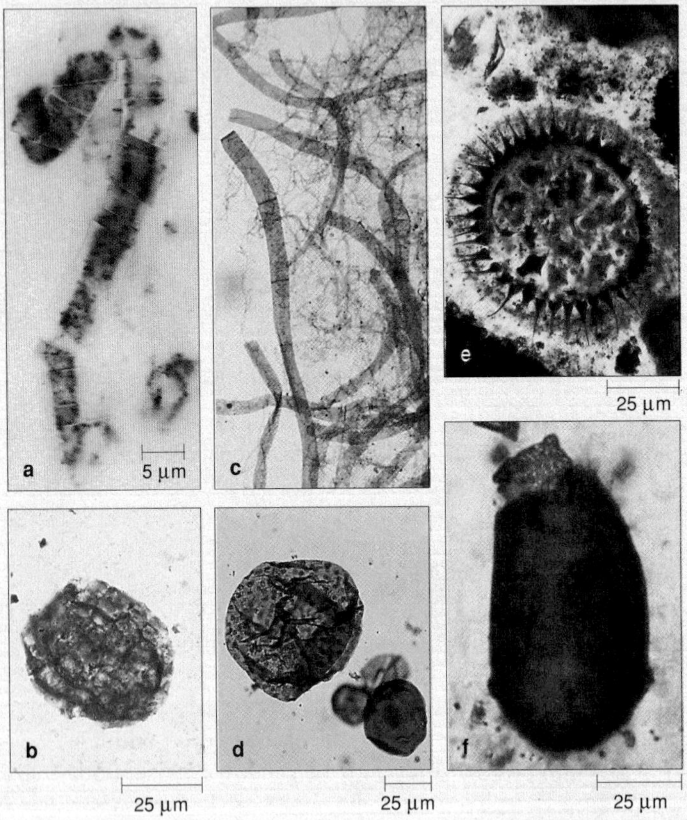

ORIGIN OF THE NUCLEUS AND ER Prokaryotic cells do not have an abundance of organelles, but some species have interesting infoldings of the plasma membrane (Figure 21.11). Embedded in that membrane are enzymes and other agents of metabolism. In the early forerunners of eukaryotic cells, similar infoldings may have extended into the cytoplasm and served as routes to the surface. They may have evolved into ER channels and into an envelope around the DNA.

What would be the advantage of such membranous enclosures? Maybe they protected the genes and protein products from "foreigners." Remember, bacterial species often transfer plasmid DNA among themselves. So do yeasts, which are very simple eukaryotic cells. At first, a nuclear envelope may have been favored because it got the cell's genes, replication enzymes, and transcription enzymes out of the cytoplasm. It would have allowed vital genetic messages to be copied and read, free of metabolic competition from what could become an unmanageable hodgepodge of foreign genes. Similarly, ER channels might have kept important proteins and other organic compounds away from metabolically hungry "guests"— foreign cells that one way or another became permanent residents inside the host cell, as described next.

A THEORY OF ENDOSYMBIOSIS It appears likely that accidental partnerships between a variety of prokaryotic species formed countless times on the evolutionary road to eukaryotic cells. Some partnerships possibly resulted in the origin of mitochondria, chloroplasts, and other organelles. This is a story of endosymbiosis, as developed in greatest

Figure 21.9 From Australia, (**a**) a strand of walled prokaryotic cells 3.5 billion years old and (**b**) one of the oldest known eukaryotes— a protistan 1.4 billion years old. From Siberia, (**c**) a multicelled alga 900 million to 1 billion years old and (**d**) eukaryotic microplankton 850 million years old. (**e**) From China, a splendid eukaryotic cell that lived 560 million years ago. (**f**) From Spitsbergen, Norway, a protistan that was alive 50 million years before the dawn of the Cambrian.

Figure 21.10 A fine example of the profusion of diverse organelles that are hallmarks of eukaryotic cells: *Euglena*, a single-celled protistan, sliced lengthwise for this micrograph. It also has a long flagellum, which could not fit in the image area at this magnification.

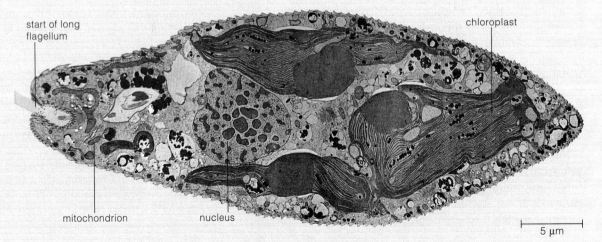

start of long flagellum

chloroplast

mitochondrion

nucleus

5 μm

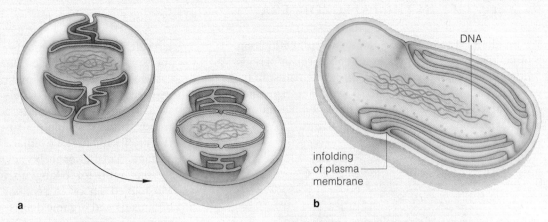

Figure 21.11 (**a**) One idea concerning the origin of endoplasmic reticulum and the nuclear envelope. In the prokaryotic ancestors of eukaryotic cells, infoldings of the plasma membrane may have given rise to these structures. (**b**) Such infoldings are present in the cytoplasm of many kinds of existing bacteria, including *Nitrobacter*, sketched here in cutaway view.

DNA

infolding of plasma membrane

detail by Lynn Margulis. *Endo-* means within; *symbiosis* means living together. In cases of **endosymbiosis**, one species (a guest) lives permanently inside another species (the host), and the interaction benefits both.

According to one theory, eukaryotic cells evolved by way of endosymbiosis long after the noncyclic pathway of photosynthesis had emerged and oxygen had accumulated to significant levels in the atmosphere. In some bacterial groups, certain electron transport systems in the plasma membrane had already expanded and now included "extra" cytochromes. Those cytochromes were able to donate electrons to oxygen. The bacteria could extract energy from organic compounds by aerobic respiration. By 1.2 billion years ago, and possibly much earlier, the forerunners of eukaryotes were engulfing aerobic bacteria. Maybe they were like existing soft-bodied, amoebalike cells that weakly tolerate free oxygen. If so, they would have trapped food by sending out cytoplasmic extensions from the cell body. Endocytic vesicles could form around food and deliver it to the cytoplasm for digestion.

A key point of the theory is that some aerobic bacteria resisted digestion. They actually thrived in the protected, nutrient-rich environment. In time, they were releasing extra ATP, which the host cells came to depend upon for growth, increased activity, and the assembly of hard body parts and other new structures. The guests were no longer duplicating metabolic functions that the hosts performed for them. The anaerobic and aerobic cells were incapable of independent life. The guests had become mitochondria, supreme suppliers of ATP.

EVIDENCE OF ENDOSYMBIOSIS Strong evidence favors Margulis's theory. There are plenty of examples of nature continuing to tinker with endosymbionts, including the cell in Figure 21.12. Its mitochondria are like bacteria in size and structure. The inner mitochondrial membrane is like a bacterial plasma membrane. Each mitochondrion replicates its own DNA and divides independently of the host cell's division. A few genetic code words in its DNA and mRNA have unique meanings. They are translated into a few proteins required for specialized mitochondrial tasks. Thus, compared to the genetic code of cells, the "mitochondrial code" has a few distinct differences.

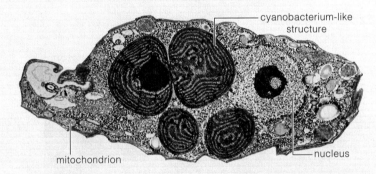

cyanobacterium-like structure

nucleus

mitochondrion

Figure 21.12 *Cyanophora paradoxa*, a protistan. Its mitochondria resemble bacteria. Its photosynthetic structures look like spherical cyanobacteria (which are photosynthetic) without the cell wall.

Also consider chloroplasts, which may be the stripped-down descendants of oxygen-evolving, photosynthetic bacteria. Perhaps predatory aerobic bacteria engulfed such photosynthetic cells, which escaped digestion, absorbed nutrients from their host's cytoplasm, and continued to function. By providing their respiring host with oxygen, their endosymbiotic existence would have been favored.

In their metabolism and overall nucleic acid sequence, chloroplasts resemble some eubacteria. Their DNA is self-replicating, and they divide independently of the cell's division. Chloroplasts vary in shape and in their array of light-absorbing pigments, just as various photosynthetic eubacteria do. They may have originated a number of times, in a number of different lineages. Adding to the intrigue are species of ciliated protistans, even marine slugs, that "enslave" chloroplasts! The slugs eat algae but retain the algal chloroplasts in their gut. The chloroplasts draw nutrients from the host tissues, and they continue to photosynthesize and release oxygen for weeks.

However they arose, new kinds of cells did appear on the evolutionary stage. They had become equipped with a nucleus, cytomembranes, and mitochondria, chloroplasts, or both. They were eukaryotic cells, the first **protistans**. With their efficient metabolic strategies, early protistans underwent rapid divergences and adaptive radiations. In no time at all, evolutionarily speaking, some of their descendants gave rise to the great kingdoms of plants, fungi, and animals, as sketched out earlier in Figure 21.7.

341

LIFE IN THE PALEOZOIC ERA

We divide the **Paleozoic** into the Cambrian, Ordovician, Silurian, Devonian, Carboniferous, and Permian periods. Before the dawn of that era, gradual rifting had split the supercontinent Laurentia apart. From Cambrian times on into the Silurian, its fragments straddled the equator, and warm, shallow seas lapped their margins:

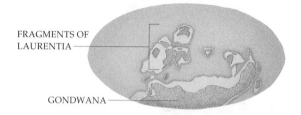

FRAGMENTS OF LAURENTIA

GONDWANA

The global conditions restricted pronounced seasonal changes in the prevailing winds, ocean currents, and the churning of nutrients upward from deep waters. As a result, supplies of nutrients along shorelines at or near the equator were stable but limited.

Most of the major animal phyla had evolved earlier, in Precambrian seas. Possibly some of their ancestors were among the **Ediacarans**, odd organisms shaped like fronds, disks, and blobs that nearly defy classification (Figure 21.13a,b). Like Ediacarans, the early Cambrian animals had flattened bodies, with a good surface-to-volume ratio for taking up nutrients (Figure 21.13c). Most lived on or in seafloor sediments, where dead organisms and organic debris settled. They ranged from sponges to simple vertebrates, and they were diverse.

How could so much diversity arise? Possibly genes governing early growth and development were far less intertwined than they are today, so there may not have been as much selection against mutant alleles and novel traits. Also, warm waters near vast new coasts afforded vacant adaptive zones, with splendid opportunities for new ways to secure food.

Entombed in sedimentary beds from the Cambrian are fossilized organisms that have punctures, missing chunks, and healed wounds. These are not artifacts of fossilization; the animals were injured while they were alive. Diverse predators and prey with armorlike shells, spines, mouths, and novel feeding structures evolved in short order. Things were starting to get lively!

Later in the Cambrian, temperatures in the shallow seas changed drastically. Trilobites (Figure 21.13d), one of the most common animals, almost vanished. In the Ordovician, the supercontinent Gondwana had been drifting south, and parts became submerged in shallow seas. Vast new marine environments opened up and promoted adaptive radiations. Many new reef organisms appeared. Among them were swift, shelled predators called nautiloids. Their surviving descendants include the chambered nautiluses, as shown in Section 49.12.

Later on, Gondwana straddled the South Pole, and vast glaciers formed across its surface. When enormous volumes of water were locked up as ice, shallow seas throughout the world were drained. This was the first ice age that we know about. It may have triggered the first global mass extinction. At the Ordovician-Silurian boundary, reef life everywhere collapsed.

Gondwana drifted north during the Silurian and on into the Devonian. This was a pivotal time of evolution. Reef communities recovered. Armor-plated fishes with massive jaws diversified. In the wet lowlands, small

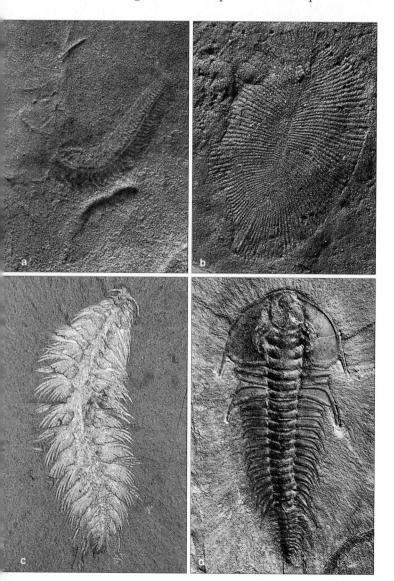

Figure 21.13 Representatives from the Precambrian and Cambrian seas. Two Ediacarans, about 600 million years old: (**a**) *Spriggina* and (**b**) *Dickensonia*. The oldest known Ediacarans lived 610 million years ago and the most recent in Cambrian times, 510 million years ago. (**c**) From the Burgess Shale of British Columbia, a fossilized marine worm. (**d**) A beautifully preserved fossil of one of the earliest trilobites.

a

b

Figure 21.14 (**a**) Life in the Silurian seas. Some of these shelled animals (nautiloids) were twelve feet long. (**b**) A Silurian swamp, dominated by the forerunners of modern ferns and club mosses. (**c**) Fossils of a Devonian plant (*Psilophyton*), possibly one of the earliest ancestors of seed-bearing plants. (**d**) Reptiles (*Dimetrodon*) of a largely hotter and drier time, the Permian. Fossils of these carnivores were found in Texas. Giant club mosses and horsetails had declined, and conifers, cycads, and other gymnosperms replaced them.

c

d

stalked plants appeared (Figure 21.14*b,c*). So did fungi and many invertebrates, such as segmented worms. In Devonian times, the fishes that would become ancestors to amphibians invaded the land. Those fishes had lobed fins, the forerunners of legs and other limbs. And they had simple lungs. As described in Section 27.6, lobed fins and lungs were key innovations—they would prove most advantageous for life out of water, in dry land habitats.

Then, as they say, something bad happened. At the Devonian-Carboniferous boundary, sea levels swung catastrophically, for unknown reasons, and triggered another mass extinction. Afterward, plants and insects embarked on adaptive radiations on land.

Throughout the Carboniferous, land masses were gradually submerged and drained many times. Organic debris piled up, then it was compacted and converted to coal, in the manner described in Section 25.4.

Insects, amphibians, and early reptiles flourished in vast swamp forests of Permian times (Figure 21.14*d*). Ancestors of the seed-producing plants called cycads, ginkgos, and conifers dominated those forests.

As the Permian drew to a close, the greatest of all mass extinctions took place. Nearly all known species perished. Pangea was forming at that time; all land masses were colliding together. The vast supercontinent eventually extended from pole to pole, and a single world ocean lapped its margins:

PANGEA

TETHYS SEA

As you will see, the new distribution of oceans, land masses, and land elevations had catastrophic effects on global climates—and on the course of life's evolution.

Early in the Paleozoic era, organisms of all six kingdoms were flourishing in the seas. By the end of the era, many lineages had successfully invaded the land, including the wet lowlands of the supercontinent Pangea.

LIFE IN THE MESOZOIC ERA

Speciation on a Grand Scale

We divide the **Mesozoic** into the Triassic, Jurassic, and Cretaceous periods. It lasted about 175 million years. Early on in the Cretaceous, the supercontinent Pangea started to break up. Its huge fragments slowly drifted apart, and we can assume that the resulting geographic isolation favored divergences and speciation:

This was an era of spectacular expansion in the range of global diversity. In the seas, invertebrates and fishes underwent adaptive radiations. On land, conifers and other seed-bearing **gymnosperms** as well as insects and reptiles became the most visibly dominant lineages. Flowering plants, or **angiosperms**, originated before the end of the era. Within a mere 30 to 40 million years, they would displace the conifers and related plants in nearly all environments (Figure 21.15 and Section 25.7).

Rise of the Ruling Reptiles

Early in the Triassic, the first **dinosaurs** evolved from a reptilian lineage. They were not much larger than a turkey. Possibly most species had high metabolic rates, and maybe they were warm-blooded. Many sprinted about on two legs. The dinosaurs weren't the dominant land animals then. Center stage belonged to *Lystrosaurus* and other plant-eating, mammal-like reptiles that were too large to be bothered by most predators.

In time, adaptive zones did open up for dinosaurs, possibly when an asteroid struck the Earth. In central Quebec, a crater about the size of Rhode Island shows how huge such an impact could have been. The blast wave and global firestorm, earthquakes, and lava flows would have been stupendous. Most of the animals that survived this time of mass extinction (and later ones) were smaller, equipped with higher metabolic rates, and less vulnerable than others to drastic changes in outside temperatures. Descendants of the surviving dinosaurs became the ruling reptiles; they endured for 140 million years. Some species reached monstrous proportions. Among them were the ultrasaurs, fifteen meters tall.

Many dinosaurs perished in another mass extinction at the end of the Jurassic, then in a pulse of extinctions in the Cretaceous. Perhaps plumes of molten material ruptured the crust and triggered changes in the global

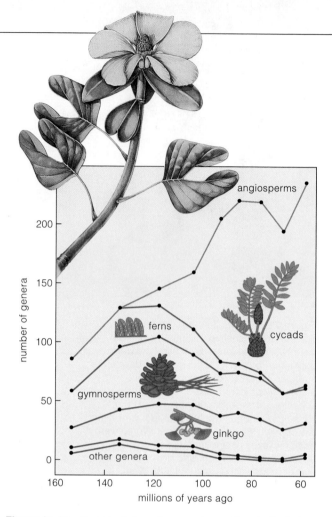

Figure 21.15 Range of diversity among vascular plants during the Jurassic and Cretaceous. Conifers and other gymnosperms were dominant before then. They were already declining when flowering plants began a spectacular radiation that continued into the Cenozoic. Also shown is a floral shoot of *Archaeanthus linnenbergeri*, a flowering plant of Cretaceous times. In many of its traits, this now-extinct species resembled living magnolias.

climate. Or perhaps a swarm of asteroids inflicted the blows. Whatever the causes, conditions changed for the dinosaurs and not all of them lived through it. Yet some lineages recovered and new ones evolved. Duckbilled dinosaurs appeared in forests and swamps. Tanklike *Triceratops* and other plant eaters flourished in open regions. They were prey for the fearsomely toothed, agile, and swift *Velociraptor* of motion picture fame.

About 120 million years ago, global temperatures shot up by 25 degrees. By one theory, plumes spread out beneath the crust and "greased" the crustal plates into moving twice as fast. A superplume or a rash of them broke through the crust. The crust in what is now the South Atlantic opened like a zipper. Basalt and lava poured from the fissures; volcanoes spewed nutrient-rich ashes. Simultaneously, the plumes released great amounts of carbon dioxide, one of the "greenhouse" gases that absorb some of the heat radiating from the Earth before it escapes into space. The nutrient-enriched

Figure 21.16 If dinosaurs of this sort had not disappeared at the end of the Cretaceous, would the then-tiny mammals ever have ventured out from under the shrubbery? Would *you* even be here today?

planet warmed up—and it stayed warm for 20 million years. On land and in the shallow seas, photosynthetic organisms flourished. Their remains were slowly buried and converted into the world's oil reserves.

About 65 million years ago, the last dinosaurs and many marine organisms vanished in a mass extinction (Figure 21.16). As described in the next section, their disappearance apparently coincided when an asteroid the size of Mount Everest slammed into the Earth. Over time, the impact site drifted into a position that we now call the northern Yucatán peninsula.

The Mesozoic was a time of major adaptive radiations and of a mass extinction in which the last dinosaurs and many marine organisms disappeared.

21.7 HORRENDOUS END TO DOMINANCE

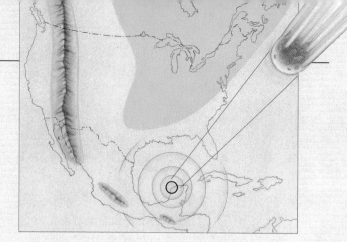

Figure 21.18 Artist's interpretation of what might have happened in the last few minutes of the Cretaceous.

It has only been about 50,000 years since the first fully modern humans walked the Earth. Since then, how many times have people puffed up with self-importance and set out to conquer neighbors, the land, and the seas? Think about it—then think about the dinosaurs. Were they good at reigning supreme? No question about it; their lineage dominated the land for 140 *million* years. In the end, did it matter? Not a bit. Sixty-five million years ago, at the Cretaceous-Tertiary (K-T) boundary, nearly all remaining members of their most excellent lineage perished. Why? Bad luck.

A thin layer of iridium-rich rock distributed around the world dates precisely to the K-T boundary. Iridium is rare on the Earth's surface but common in asteroids. **Asteroids** are rocky, metallic bodies, 1,000 kilometers to a few meters in diameter, that are hurtling through space. When the planets were forming, their gravitational pull swept most asteroids from the sky. At least 6,000 still orbit the sun in a belt between Mars and Jupiter (Figure 21.17). The orbits of dozens of others take them across Earth's orbit, like Russian roulette on a cosmic scale.

By analyzing iridium levels in soils, gravity maps, and other evidence, Walter Alvarez and Luis Alvarez hypothesized that an asteroid impact caused the K-T mass extinction. Later, researchers identified the impact site. Massive movements in the crust transported the site to what is now the northern Yucatán peninsula of Mexico (Figure 21.18). The impact crater is 9.6 kilometers deep and 300 kilometers across—wider than Connecticut. This crater as well as other evidence strongly supports what has become known as the **K-T asteroid impact theory**.

To make a crater that large, the asteroid had to hit the Earth at 160,000 kilometers (100,000 miles) per hour. At least 200,000 cubic kilometers of debris and dense gases were blasted skyward. The crust itself heaved violently. Monstrous waves, 120 meters high, raced across the ocean, obliterating life on islands and then slamming

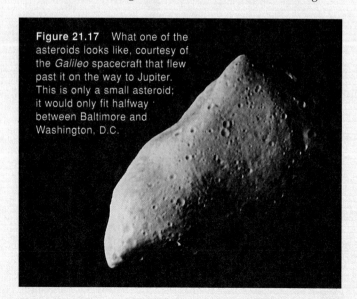

Figure 21.17 What one of the asteroids looks like, courtesy of the *Galileo* spacecraft that flew past it on the way to Jupiter. This is only a small asteroid; it would only fit halfway between Baltimore and Washington, D.C.

into the coasts of continents. Researchers long thought that atmospheric debris blocked out sunlight for months. If so, plants and other producers that sustained the web of life would have withered and died; animals and other consumers would have starved to death. The theory has problems. By some calculations, the volume of debris that was blasted aloft would not have been enough to have such significant consequences.

Then, in 1994, the comet Shoemaker-Levy 9 slammed into Jupiter. Particles blasted into the Jovian atmosphere caused intense heating over an area larger than the Earth. This outcome favors a **global broiling theory**, proposed first by H. J. Melosh and his colleagues. Energy released at the K-T impact site was equivalent to detonating 100 million nuclear bombs. Trillions of tons of vaporized debris rose in a colossal fireball, then rapidly condensed into particles the size of sand grains. Seconds later, a cooler fireball of steam, carbon dioxide, and unmelted rock formed. When the debris fell to the Earth, it raised the atmospheric temperature by thousands of degrees. The sky above the whole planet must have glowed with heat ten times more intense than the noonday sun above Death Valley in summer. In one horrific hour, nearly all plants erupted in flames and every animal out in the open was broiled alive.

Things haven't settled down much since the demise of the dinosaurs. For example, about 2.3 million years ago, a huge chunk of matter from space apparently hit the Pacific Ocean. At about the same time, vast ice sheets started forming abruptly in the Northern Hemisphere. Long-term shifts in climate may have been ushering in this most recent ice age, but the global impact would have accelerated it. Water vaporized by the impact could have contributed to a global cloud cover that limited the amount of sunlight reaching the surface. The ancestors of humans were around when this happened. The extreme climate shift surely tested their adaptability.

Almost certainly, severe environmental tests await all existing lineages. When we become too smitten with our importance in the world of life, we would do well to step back, from time to time, and reflect on what is going on above and beneath the Earth's surface. Asteroids and superplumes do have a way of leveling the playing field, as they did for the tiny mammals and gigantic dinosaurs.

21.8 LIFE IN THE CENOZOIC ERA

The breakup of Pangea triggered events that continued into the present era, the **Cenozoic**. At the dawn of the Cenozoic, land masses were on collision courses:

Coastlines fractured. The Cascades, Andes, Himalayas, and Alps rose through volcanic activity, uplifting, and other events at crustal rifts and plate boundaries. These geologic changes brought about major shifts in climate that influenced the further evolution of life.

During the Paleocene epoch, climates were warmer and wetter. Tropical and subtropical forests extended farther north and south than they do today. Woodlands and forests spread even into polar regions. Although their key traits developed before the dinosaurs left the scene, mammals now began their major radiation.

The global climate warmed even more in the Eocene epoch, and subtropical forests extended north into the polar regions. A variety of mammals, including primates, bats, rhinos, hippos, elephants, horses, and assorted carnivores, emerged. By the late Eocene, climates became cooler and drier, and seasonal changes became pronounced. Woodlands and semiarid grasslands now dominated vast tracts of land. Patterns of vegetational growth changed, and this drove many mammals to extinction.

From the Oligocene through the Pliocene, an abundance of grazing and browsing animals thrived in the woodlands and grasslands. Among them were camels and the "giraffe rhinoceros," along with some fearsome carnivores that stalked them (Figure 21.19).

Today the distribution of land masses favors species diversity. The richest ecosystems are the tropical forests of South America, Madagascar, and Southeast Asia, as well as the marine ecosystems of archipelagos in the tropical Pacific. Yet we are in the midst of what may be one of the greatest mass extinctions. About 50,000 years ago, nomadic humans followed migrating herds of wild animals around the Northern Hemisphere. Within a few

Figure 21.19 Representatives from Cenozoic times. (**a**) In the early Paleocene, in what is now Wyoming, diverse mammals lived in dense forests of sequoia trees, laurel, and other plants. On the ground is the raccoonlike *Chriacus*, facing a tree-climbing rodent (*Ptilodus*). Higher in the tree is *Peradectes*, a marsupial. (**b**) From the late Eocene to the early Miocene, *Indricotherium* browsed on woodland trees. This "giraffe rhinoceros," the largest land mammal known, weighed 15 tons and was 5.5 meters (18 feet) at the shoulder. (**c**) From the Pleistocene, a small horse and the saber-tooth cat (*Smilodon*). Both mammals are extinct. Fossils of them have been found in a pitch pool at Rancho LaBrea, California.

thousand years, major groups of mammals were extinct. The pace of extinction has since accelerated as humans hunt for food, fur, feathers, or fun, and as they destroy habitats to clear land for farm animals or crops. Chapter 50 focuses on the global repercussions.

Major geologic changes during the Cenozoic triggered shifts in climate. The great adaptive radiation of mammals began, first in tropical forests, then in woodlands and grasslands.

SUMMARY OF EARTH AND LIFE HISTORY

We conclude this overview chapter with an illustration that correlates milestones in the evolution of life with the evolution of the Earth. As you study Figure 21.20, keep in mind that it is only a generalized summary. For example, it charts five of the greatest mass extinctions, but there were many others in between. The diagram of the range of global diversity conveys the shrinking and expansion of species over time. However, the range represents *all* of the major groups on land and in the seas combined. Remember, each major group has its own history of persisting lineages, of radiations, and of extinctions.

And now, with these qualifications in mind, we are ready to turn to the next chapters in this unit. They will provide you with richer detail of the history and the current range of diversity for all six kingdoms of life.

Figure 21.20 Summary of major events in the evolution of the Earth and of life. As you read through the chapters to follow, you may wish to return to this illustration now and then to remind yourself of how the details fit into the greater evolutionary picture.

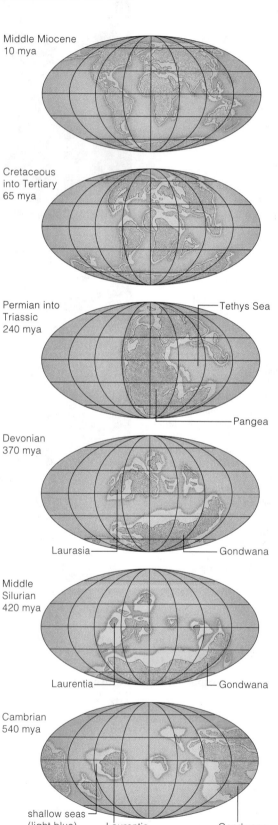

Middle Miocene
10 mya

Cretaceous
into Tertiary
65 mya

Permian into
Triassic
240 mya — Tethys Sea
— Pangea

Devonian
370 mya
Laurasia — — Gondwana

Middle
Silurian
420 mya
Laurentia — — Gondwana

Cambrian
540 mya
shallow seas
(light blue) — Laurentia — Gondwana

	Period	Epoch
CENOZOIC ERA	Quaternary	Recent
		Pleistocene
	Tertiary	Pliocene
		Miocene
		Oligocene
		Eocene
		Paleocene
MESOZOIC ERA	Cretaceous	Late
		Early
	Jurassic	
	Triassic	
PALEOZOIC ERA	Permian	
	Carboniferous	
	Devonian	
	Silurian	
	Ordovician	
	Cambrian	
PROTEROZOIC EON		
ARCHEAN EON AND EARLIER		

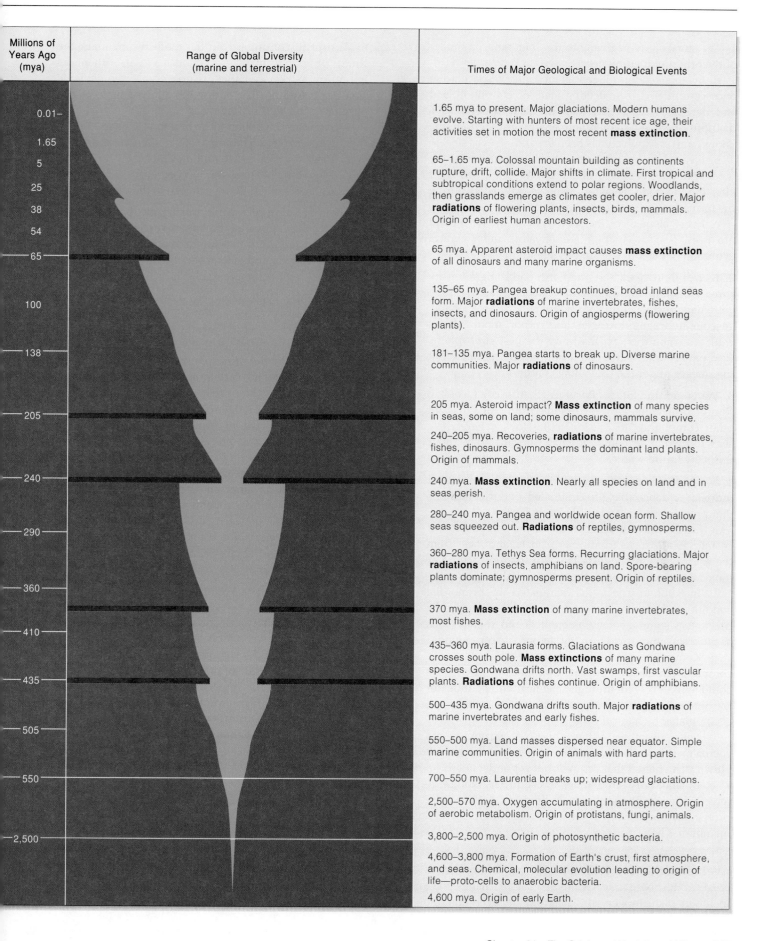

| Millions of Years Ago (mya) | Range of Global Diversity (marine and terrestrial) | Times of Major Geological and Biological Events |

1.65 mya to present. Major glaciations. Modern humans evolve. Starting with hunters of most recent ice age, their activities set in motion the most recent **mass extinction**.

65–1.65 mya. Colossal mountain building as continents rupture, drift, collide. Major shifts in climate. First tropical and subtropical conditions extend to polar regions. Woodlands, then grasslands emerge as climates get cooler, drier. Major **radiations** of flowering plants, insects, birds, mammals. Origin of earliest human ancestors.

65 mya. Apparent asteroid impact causes **mass extinction** of all dinosaurs and many marine organisms.

135–65 mya. Pangea breakup continues, broad inland seas form. Major **radiations** of marine invertebrates, fishes, insects, and dinosaurs. Origin of angiosperms (flowering plants).

181–135 mya. Pangea starts to break up. Diverse marine communities. Major **radiations** of dinosaurs.

205 mya. Asteroid impact? **Mass extinction** of many species in seas, some on land; some dinosaurs, mammals survive.

240–205 mya. Recoveries, **radiations** of marine invertebrates, fishes, dinosaurs. Gymnosperms the dominant land plants. Origin of mammals.

240 mya. Mass extinction. Nearly all species on land and in seas perish.

280–240 mya. Pangea and worldwide ocean form. Shallow seas squeezed out. **Radiations** of reptiles, gymnosperms.

360–280 mya. Tethys Sea forms. Recurring glaciations. Major **radiations** of insects, amphibians on land. Spore-bearing plants dominate; gymnosperms present. Origin of reptiles.

370 mya. Mass extinction of many marine invertebrates, most fishes.

435–360 mya. Laurasia forms. Glaciations as Gondwana crosses south pole. **Mass extinctions** of many marine species. Gondwana drifts north. Vast swamps, first vascular plants. **Radiations** of fishes continue. Origin of amphibians.

500–435 mya. Gondwana drifts south. Major **radiations** of marine invertebrates and early fishes.

550–500 mya. Land masses dispersed near equator. Simple marine communities. Origin of animals with hard parts.

700–550 mya. Laurentia breaks up; widespread glaciations.

2,500–570 mya. Oxygen accumulating in atmosphere. Origin of aerobic metabolism. Origin of protistans, fungi, animals.

3,800–2,500 mya. Origin of photosynthetic bacteria.

4,600–3,800 mya. Formation of Earth's crust, first atmosphere, and seas. Chemical, molecular evolution leading to origin of life—proto-cells to anaerobic bacteria.

4,600 mya. Origin of early Earth.

SUMMARY

1. The story of life begins with the "big bang," a model of the origin of the universe.

   a. By this model, all matter and all of space were once compressed in a fleeting state of enormous heat and density. Time began with the near-instantaneous distribution of all matter and energy throughout the known universe, which has been expanding ever since.

   b. Nearly all helium and other light elements, the most abundant elements of the universe, formed right after the big bang. Heavier elements originated during the formation, evolution, and death of stars.

   c. Every element of the solar system, the Earth, and life itself is a product of the physical and chemical evolution of the universe and its stars.

2. Four billion years ago, the Earth had a high-density core, a mantle of intermediate density, and a thin, extremely unstable crust of low-density rocks. Probably gaseous hydrogen, nitrogen, and carbon monoxide, as well as carbon dioxide, made up the first atmosphere. Free oxygen and water could not have accumulated at the surface under the prevailing conditions.

3. Water accumulated after the Earth's crust cooled. Runoff from rains carried dissolved mineral salts and other compounds to crustal depressions, where early seas formed. Life could not have originated without this salty liquid water.

4. Many diverse studies and experiments have yielded indirect evidence that life originated under conditions that presumably existed on the early Earth.

   a. Comparative investigations of the composition of cosmic clouds, rocks from other planets, and rocks from the Earth's moon suggests that precursors of complex molecules associated with life were available.

   b. In laboratory tests that simulated the primordial conditions, including the absence of free oxygen, the precursors spontaneously assembled into sugars (such as glucose), amino acids, and other organic compounds.

   c. Known chemical principles as well as advanced computer simulations indicate that metabolic pathways could have evolved through chemical competition for the limited supplies of organic molecules (which had accumulated by natural geologic processes in the seas).

   d. Self-replicating systems of RNA, enzymes, and coenzymes have been synthesized in the laboratory. How DNA entered the picture is not yet understood.

   e. In laboratory simulations of conditions thought to have existed on the early Earth, lipids as well as lipid-protein membranes having some of the properties of cell membranes have formed spontaneously.

5. Life originated by 3.8 billion years ago. Since then, it has been influenced by major changes in the Earth's crust, atmosphere, and oceans. Forces of change have included plate movements, asteroid impacts, and the activities of organisms (including oxygen-producing photosynthesizers and, currently, the human species).

6. Abrupt discontinuities in the fossil record mark the times of global mass extinctions. We use them as boundary markers for five great intervals in a geologic time scale. Radiometric dating has allowed us to assign absolute dates to this time scale:

   a. Archean: 3.9 billion to 2.5 billion years ago
   b. Proterozoic: 2.5 billion to 550 million years ago
   c. Paleozoic: 550 to 240 million years ago
   d. Mesozoic: 240 to 65 million years ago
   e. Cenozoic: 65 million years ago to the present

7. The first living cells were prokaryotic (bacteria). Not long after they appeared, the first divergence began that led to eubacteria, and to a common prokaryotic ancestor of archaebacteria and eukaryotes. Some eubacteria used a cyclic pathway of photosynthesis.

8. During the Proterozoic, the noncyclic pathway of photosynthesis evolved in some lineages of eubacteria. Oxygen, a by-product of the pathway, gradually started to accumulate in the atmosphere.

   a. Eventually, the atmospheric concentration of free oxygen prevented the further spontaneous formation of organic molecules. From that time on, the spontaneous origin of life was no longer possible on Earth.

   b. The abundance of atmospheric oxygen was a selective pressure that brought about the evolution of aerobic respiration. Aerobic respiration was a key step toward the origin of the first eukaryotic cells.

   c. Mitochondria and chloroplasts, two important eukaryotic organelles, probably evolved as an outcome of endosymbiosis between certain aerobic bacteria and the anaerobic bacterial forerunners of eukaryotes.

   d. The oxygen-rich atmosphere promoted formation of a layer of ozone (O_3). In time, that shield against destructive ultraviolet radiation allowed some lineages to move out of the seas, into low wetlands.

9. Early in the Paleozoic, diverse organisms of all six lineages had become established in the seas. By its end, the invasion of land was under way. From that time on, there have been pulses of mass extinctions and adaptive radiations. Asteroid impacts and other catastrophes triggered many of these events. So did plate movements that changed the distribution of oceans and land as well as the prevailing global and regional climates.

Review Questions

1. Compare the presumed chemical and physical conditions that are thought to have prevailed on the Earth 4 billion years ago with conditions that exist today. *21.1*

2. Describe examples of the kinds of experimental evidence for the spontaneous origin of (1) large organic molecules, (2) the self-assembly of proteins, and (3) the formation of organic membranes and spheres, under laboratory conditions similar to those of the early Earth. *21.1, 21.2*

3. Summarize the key points of the theory of endosymbiotic origins for mitochondria and chloroplasts. Cite evidence that favors this theory. *21.4*

4. Describe the prevailing conditions that probably favored the Cambrian "explosion" of diversity among marine animals, as evidenced by the fossil record. *21.5*

5. During which geologic time spans did plants, fungi, and insects invade the land? What kind of vertebrates first invaded the land, and when? *21.5*

6. What were global conditions like when gymnosperms and dinosaurs originated? *21.6*

7. Briefly explain how an asteroid impact and "global broiling" may have caused the mass extinctions at the K-T boundary. *21.7*

8. Would you expect the Paleozoic, Mesozoic, or Cenozoic to be called "the age of mammals"? As part of your answer, explain the differences between global conditions in each era. *21.8*

Self-Quiz *(Answers in Appendix IV)*

1. Life originated by _____ .
   a. 4.6 billion years ago   c. 3.8 billion years ago
   b. 2.8 million years ago   d. 3.8 million years ago

2. Through study of the geologic record, we know that the evolution of life has been profoundly influenced by _____ .
   a. tectonic movements of the Earth's crust
   b. bombardment of the Earth by celestial objects
   c. profound shifts in land masses, shorelines, and oceans
   d. physical and chemical evolution of the Earth
   e. all of the above

3. _____ was the first to obtain indirect evidence that organic molecules could have been formed on the early Earth.
   a. Darwin   c. Fox
   b. Miller   d. Margulis

4. An abundance of _____ was conspicuously absent from the Earth's atmosphere 4 billion years ago.
   a. hydrogen   c. carbon monoxide
   b. nitrogen   d. free oxygen

5. Which of the following statements is false?
   a. The first living cells were prokaryotes.
   b. The cyclic pathway of photosynthesis first appeared in some eubacterial species.
   c. Oxygen began accumulating in the atmosphere after the noncyclic pathway of photosynthesis evolved.
   d. In the Proterozoic, increasing levels of atmospheric oxygen enhanced the spontaneous formation of organic molecules.
   e. All are correct.

6. The first eukaryotic cells emerged during the _____ .
   a. Paleozoic   c. Archean   e. Cenozoic
   b. Mesozoic   d. Proterozoic

7. Match the geologic time interval with the events listed.
   ____ Archean   a. major radiations of dinosaurs, origin
   ____ Proterozoic   of flowering plants and mammals
   ____ Paleozoic   b. chemical evolution, origin of life
   ____ Mesozoic   c. major radiations of flowering plants,
   ____ Cenozoic   insects, birds, mammals; emergence of
                          human forms
                       d. oxygen present; origin of aerobic
                          metabolism, protistans, fungi, animals
                       e. rise of early plants, origin of
                          amphibians, origin of reptiles

Critical Thinking

1. Explain, in terms of hydrophilic and hydrophobic interactions, how proto-cells might have formed in water from aggregations of lipids, proteins, and nucleic acids.

2. The Atlantic Ocean is gradually widening, and the Pacific Ocean and Indian Ocean are closing. Many millions of years from now, the continents will collide and form a second Pangea. Write a short essay on what environmental conditions might be like on that future supercontinent and on what types of species might survive there.

3. By some estimates, there is a chance that about 10^{20} planets have formed in the universe that are capable of sustaining life—but only one chance at intelligent life per planet. Given your knowledge of molecular biology and evolutionary processes, speculate on why the odds are so low.

Selected Key Terms

angiosperm *21.6*
archaebacterium *21.3*
Archean *21.3*
asteroid *21.7*
big bang *CI*
Cenozoic *21.8*
crust, of Earth *21.1*
dinosaur *21.6*
Ediacaran *21.5*
endosymbiosis (theory of) *21.4*
eubacterium *21.3*
eukaryotic cell *21.3*
global broiling theory *21.7*
gymnosperm *21.6*
K-T asteroid impact theory *21.7*
mantle, of Earth *21.1*
Mesozoic *21.6*
Paleozoic *21.5*
prokaryotic cell *21.3*
Proterozoic *21.3*
protistan *21.4*
proto-cell *21.2*
RNA world *21.2*
stromatolite *21.3*

Readings

Bambach, R., C. Scotese, and A. Ziegler. 1980. "Before Pangea: The Geographies of the Paleozoic World." *American Scientist* 68(1): 26–38.

de Duve, C. September–October 1995. "The Beginnings of Life on Earth." *American Scientist* 83: 428–437.

————. April 1996. "The Birth of Complex Cells." *Scientific American* 274(4): 50–57.

Dobb, E. February 1992. "Hot Times in the Cretaceous." *Discover* 13: 11–13.

Dott, R., Jr., and R. Batten. 1988. *Evolution of the Earth.* Fourth edition. New York: McGraw-Hill.

Hartman, W., and Chris Impey. 1994. *Astronomy: The Cosmic Journey.* Fifth edition. Belmont, California: Wadsworth.

Horgan, J. February 1991. "Trends in Evolution: In the Beginning . . . ." *Scientific American* 264(2): 116–125.

Wright, K. March 1997. "When Life Was Odd." *Discover* 18(3): 52–61. Update on the bizarre Ediacarans.

Web Site See *http://www.wadsworth.com/biology* for practice quiz questions, hypercontents, BioUpdates, and critical thinking. The Wadsworth Biology Resource Center provides a wealth of information fully organized and integrated by chapter.

22 BACTERIA AND VIRUSES

The Unseen Multitudes

Did a friend ever mention that you are nearly 1/1,000 of a mile tall? Probably not. What would be the point of measuring people in units as big as miles? Even so, we think this way, in reverse, whenever we measure microorganisms. For the most part, **microorganisms** are single-celled organisms that are too small to be seen without the aid of a microscope.

The bacterial cells shown in Figure 22.1 are a case in point. To measure them, you would have to divide one meter into a thousand units, or millimeters. Next, you would have to divide one of the millimeters into a thousand smaller units, or micrometers. To give you a sense of how small that is, a single millimeter would be about as small as the dot of this "i." And a *thousand* bacteria would fit side by side on top of the dot!

To be sure, viruses are smaller still, as you might deduce after looking at Figure 22.2. We measure them in nanometers (billionths of a meter). But viruses are not alive. We consider them in this chapter because one kind or another infects just about every organism.

Bacteria generally are the smallest microorganisms, but they vastly outnumber the individuals in all other kingdoms combined. Their reproductive potential is staggering. Under ideal conditions, some divide about every twenty minutes. If that rate of reproduction were to hold constant, a single bacterium would have nearly a billion descendants in ten hours!

So why don't the unseen multitudes take over the world? Sooner or later, their burgeoning populations use up available nutrients and pollute the surroundings with their own metabolic wastes. In other words, they alter the very conditions that initially favored their reproduction. Besides this, certain kinds of viruses attack just about every species of bacterium and help keep their population sizes in check. So do seasonal changes in living conditions.

Of course, you probably do not find much comfort in this when you serve as host for one of the pathogenic types. **Pathogens** are infectious, disease-causing agents that invade target organisms and multiply inside or on

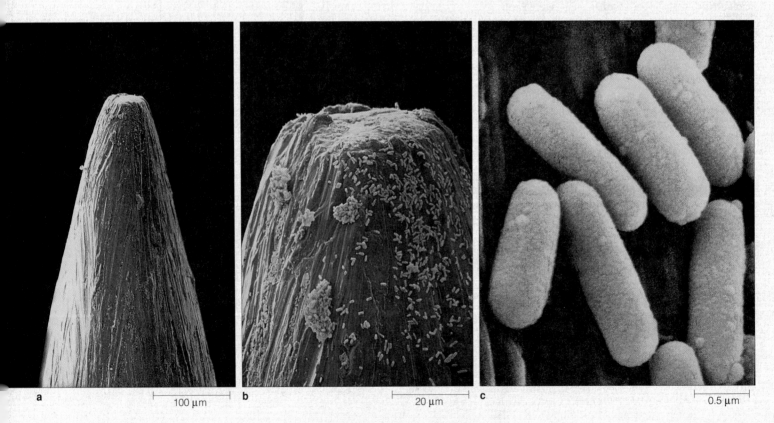

a   ⊢—— 100 μm ——⊣   b   ⊢—— 20 μm ——⊣   c   ⊢—— 0.5 μm ——⊣

Figure 22.1 (**a–c**) How small are bacteria? Shown here, *Bacillus* cells peppering the tip of a pin. The cells in (**c**) are magnified more than 16,600 times.

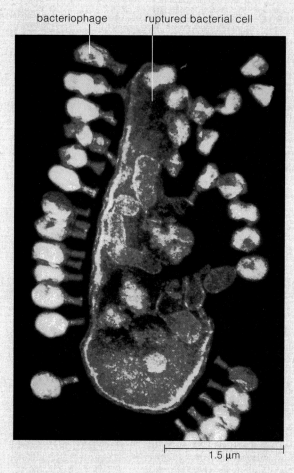

bacteriophage   ruptured bacterial cell

1.5 µm

Figure 22.2 How small are the viruses? Shown here, bacteriophage particles, each about 225 nanometers tall, that infected a bacterial cell. The cell has ruptured open.

them. Disease follows when the metabolic activities of pathogenic cells damage the tissues and interfere with the body's normal functioning.

Certain pathogens can indeed make you suffer, but they should not give every microorganism a bad name. For example, think back on the uncountable numbers of photosynthetic bacterial cells in the seas (Section 7.8). Collectively, they help provide food and oxygen for entire communities and have major roles in the global cycling of carbon. Or think about the kinds of bacterial species that feed on organic debris. Together with other decomposers, they help cycle nutrients that sustain entire communities.

From the human perspective, microorganisms are good or bad, even dangerous. Basically, however, they are simply surviving and reproducing like the rest of us, in ways that are the topics of this chapter.

KEY CONCEPTS

1. In structural terms, we find the simplest forms of life among the bacteria. Most of these are microscopically small. Smaller still are the viruses.

2. Bacteria alone are prokaryotic cells. They do not have a profusion of internal, membrane-bound organelles in their cytoplasm, as eukaryotic cells do. However, as a group, the bacteria show great metabolic diversity, and many species show complex behavior.

3. Most bacteria reproduce by prokaryotic fission. This cell division mechanism follows DNA replication and divides the parent cell into two genetically equivalent daughter cells.

4. Bacteria were the first living organisms on Earth. Not long after they originated, they diverged into two lineages. One lineage gave rise to eubacteria, and the other to the common ancestors of archaebacteria and eukaryotic cells.

5. A virus is a noncellular infectious particle. It consists of nucleic acid (either DNA or RNA), a protein coat, and sometimes an outer envelope. Viruses cannot be replicated without pirating the metabolic machinery of a specific type of host cell.

6. Nearly all viral multiplication cycles proceed through five steps: attachment to a host cell, penetration of its plasma membrane, replication of viral DNA or RNA and synthesis of viral proteins, then assembly of new viral particles, and finally release from the infected cell.

7. Most of us tend to judge microorganisms through the prism of human interests. Yet their lineages are the most ancient, their adaptations are astoundingly diverse, and they are simply surviving and reproducing like the rest of us.

Of all organisms, bacteria are the most abundant and far-flung. Many thousands of species live in places that range from deserts to hot springs, glaciers, and seas. The hardiest have been carrying on for millions of years 2,780 meters (9,600 feet) below the surface of the Earth! Billions of bacterial cells may occupy a handful of rich soil. The ones in your gut and on your skin outnumber your body cells. (Your cells are larger, so you are only "a few percent bacterial" by weight.) Bacteria also have the longest evolutionary history. Trace any lineage back far enough and you will find bacterial ancestors. From *Escherichia coli* to amoebas, elephants, clams, and coast redwoods, all living organisms interconnect, regardless of their differences in size, numbers, and evolutionary distance. Table 22.1 and Figure 22.3 introduce features that help characterize the remarkable bacteria.

Splendid Metabolic Diversity

All organisms take in energy and carbon to meet their nutritional requirements. Compared with other species, however, the bacteria show the greatest diversity in the means of securing these resources.

Like plants, *photoautotrophic* bacteria build organic compounds by photosynthesis; they are "self-feeders." They tap sunlight for energy and use carbon dioxide as their carbon source. Their plasma membrane houses the photosynthetic machinery. Some photosynthesizers use a noncyclic pathway, with electrons and hydrogen from water molecules being used for the synthesis reactions and oxygen being released as a by-product. Others are anaerobic; they cannot use oxygen or die in its presence. Those species use a cyclic pathway in which electrons and hydrogen are stripped from inorganic compounds, such as gaseous hydrogen and hydrogen sulfide.

Chemoautotrophic bacteria also are self-feeders. Most get carbon from carbon dioxide. For energy, some types use organic compounds as a source of electrons (and hydrogen). Others (the chemolithotrophs) use inorganic substances, such as gaseous hydrogen, sulfur, nitrogen compounds, and a form of iron.

By contrast, *photoheterotrophic* bacteria are not self-feeders. They can use sunlight for photosynthesis but cannot get carbon from carbon dioxide. Instead, their carbon sources are fatty acids, complex carbohydrates, and other compounds that other organisms produce.

Chemoheterotrophic bacteria are either parasites or saprobes, not self-feeders. The parasitic types live on or in a living host and draw glucose and other nutrients from it. One way or another, the saprobic types obtain nutrients from the organic products, wastes, or remains of other organisms.

Bacterial Sizes and Shapes

So far, you have a general sense of the microscopically small sizes of bacteria. Typically, the width or length of these cells falls between 1 and 10 micrometers.

Three basic shapes are common among bacteria. A spherical shape is a **coccus** (plural, cocci; from a word that means berries). A rod shape is a **bacillus** (plural, bacilli, meaning small staffs). A cell body with one or more twists is a **spirillum** (plural, spirilla):

coccus   bacillus   spirillum

Don't let these simple categories fool you. Cocci may also be oval or flattened. Bacilli may be skinny (like straws) or tapered (like cigars). Surface extensions give some bacteria a starlike shape. Even square bacteria live in salt ponds in Egypt. Besides this, after daughter cells divide, they may stay stuck together in chains, sheets,

Table 22.1 Characteristics of Bacterial Cells
1. All bacterial cells are prokaryotic; they do not have a membrane-bound nucleus.
2. Bacterial cells in general have a single chromosome (a circular DNA molecule); many species also have plasmids.
3. Most bacteria have a cell wall that is composed of peptidoglycan.
4. Most bacteria reproduce by prokaryotic fission.
5. Collectively, bacteria show great diversity in their modes of metabolism.

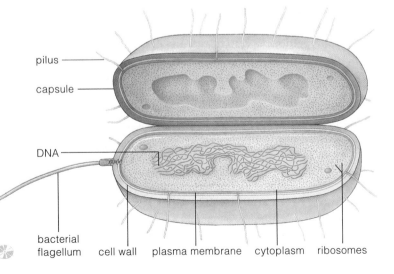

Figure 22.3 (*Right*) Generalized body plan of a bacterium.

pilus

capsule

DNA

bacterial flagellum   cell wall   plasma membrane   cytoplasm   ribosomes

Figure 22.4 (**a**) Surface view of numerous bacteria that have become attached, by means of a sticky mesh, to the surface of a human tooth; and doesn't this micrograph make you want to grab a toothbrush? (**b**) One example of pili, filamentous structures that project from the surface of many bacterial cells. This particular *Escherichia coli* cell is dividing in two.

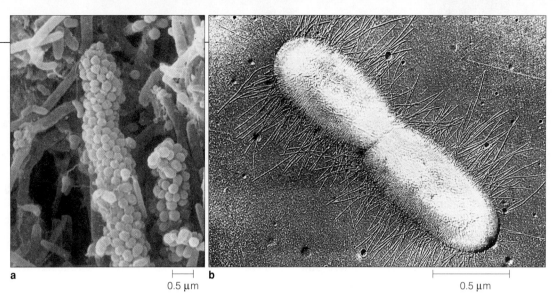

a   0.5 µm   b   0.5 µm

and other aggregations, as in Figure 22.4. Some spiral species are curved, like a comma, and others are flexible or like stiff corkscrews.

Structural Features

Bacteria alone are **prokaryotic cells**, meaning they were around before the evolution of nucleated cells (*pro-*, before; and *karyon*, nucleus). Few have membrane-bound compartments of any sort for isolating metabolic events. Reactions take place in the cytoplasm or at the plasma membrane. For example, protein synthesis proceeds at ribosomes that are distributed through the cytoplasm or attached to the plasma membrane. This does not mean bacteria are in some way inferior to the eukaryotic cells. Being tiny, fast reproducers, they do not require great internal complexity.

A wall usually surrounds the plasma membrane. A **cell wall** is a semirigid, permeable structure that helps

Figure 22.5 Gram staining. Cocci and rods smeared on a slide are stained with a purple dye (such as crystal violet), washed off, and then stained with iodine. All the cells are now purple. The slide is washed with alcohol, which renders the Gram-negative cells colorless. Now the slide is counterstained (with safranin), washed, and dried. Gram-positive cells (in this case, *Staphylococcus aureus*) remain purple. But the counterstain colors Gram-negative cells (in this case, *Escherichia coli*) pink.

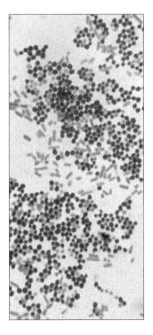

■ stain with purple dye
☐ stain with iodine
☐ wash with alcohol
■ counterstain with safranin

the cell maintain its shape and resist rupturing when internal fluid pressure increases (Section 5.4). Cell walls of eubacteria are composed of peptidoglycan molecules. In such molecules, peptide groups crosslink numerous polysaccharide strands to one another.

Clinicians can identify many bacterial species on the basis of their wall structure and composition. Consider the staining reaction called the **Gram stain**. A sample of unknown bacterial cells is exposed to a purple dye, then to iodine, an alcohol wash, and a counterstain. The cell wall of *Gram-positive* species remain purple. The wall of *Gram-negative* species loses color after the wash, but the counterstain turns it pink (Figure 22.5).

A sticky mesh, or **glycocalyx**, often surrounds the cell wall. It consists of polysaccharides, polypeptides, or both. When highly organized and attached firmly to the wall, we call the mesh a capsule. When less organized and loosely attached to the wall, we call it a slime layer. The mesh helps a bacterial cell attach to teeth, mucous membranes, rocks in streambeds, and other interesting surfaces (Figure 22.4*a*). It even helps some encapsulated species avoid being engulfed by phagocytic, infection-fighting cells of their host organism.

Some bacteria have one or more **bacterial flagella** (singular, flagellum), which are used in motility. These don't have the same structure as a eukaryotic flagellum, and they don't operate the same way. They move the cell by rotating like a propeller. Many species also have **pili** (singular, pilus). These short, filamentous proteins project above the cell wall, as in Figure 22.4*b*. Some pili help cells adhere to surfaces. Others help them attach to one another as a prelude to conjugation, an interaction that is described in the next section.

Bacteria are microscopic, prokaryotic cells. They generally have one circular bacterial chromosome, and often they have extra DNA in the form of plasmids.

Nearly all bacteria have a wall around the plasma membrane, and many have a capsule or slime layer around the cell wall.

The Nature of Bacterial Growth

Between cell divisions, bacteria grow through increases in their component parts. We measure the growth of a large, multicelled organism in terms of increases in size, but doing so for a microscopically small bacterium would be a bit pointless. Instead, we measure bacterial growth as an increase in the number of cells in a given population. Under ideal conditions, each cell divides in two, the division of two cells results in four, four result in eight, and so forth. Many types can divide every half hour; a few can do so every ten minutes. Such rates of increase lead to large population sizes in short order.

In nature, conditions that promote the growth of some bacterial populations might not sustain the growth of others. Most species could not establish themselves in the most extreme environments—say, Antarctica, the Negev Desert, or deep inside the Earth. There, the few species that do endure, inside rocks, rarely reproduce. And few can live in the highly acidic wastewater from mining operations. K-12, a strain of *E. coli* originally isolated from the human gut, has been cultivated for such a long time in the laboratory that it no longer is able to grow when reintroduced into the gut. Through microevolutionary processes, it has become adapted to the conditions in its artificial environment.

Prokaryotic Fission

When a bacterium has nearly doubled in size, it divides in two. Each daughter cell inherits a single **bacterial chromosome**, which is a circularized, double-stranded DNA molecule that has a few proteins attached to it. In some species, the daughter cell merely buds from the parent cell. Most often, however, bacteria reproduce by a division mechanism called **prokaryotic fission**.

As prokaryotic fission begins, a parent cell replicates its DNA (Figure 22.6). The two DNA molecules that result are attached to the plasma membrane at adjacent sites. The cell synthesizes lipid and protein molecules, which become incorporated in the membrane between the attachment sites. Membrane growth moves the two DNA molecules apart. New wall material is deposited above the membrane. The membrane and wall grow through the cell midsection and divide the cytoplasm. The result is two genetically equivalent daughter cells.

(Especially in microbiology, you may hear someone refer to this division mechanism as *binary* fission, but such usage can cause confusion. The same term applies to a form of asexual reproduction among flatworms and some other multicelled animals. It refers to growth by way of mitotic cell divisions, then division of the whole body into two parts of the same or different sizes.)

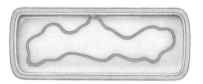

a Cutaway view of a bacterial cell before its DNA is replicated. The DNA molecule is attached to the plasma membrane.

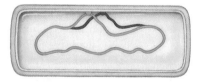

b Replication starts and proceeds in two directions, away from some point in the bacterial DNA molecule.

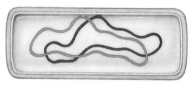

c The DNA copy is attached at a site on the plasma membrane, near the attachment site of the parent DNA molecule.

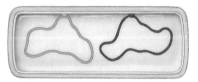

d New lipids and proteins are added to the plasma membrane between the two attachment sites; the membrane growth moves the two DNA molecules apart.

e New wall material is deposited on the outer surface of new membrane, and both start growing through the cell's midsection.

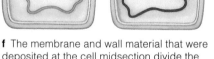

f The membrane and wall material that were deposited at the cell midsection divide the cytoplasm in two. When the two new walls part company, two separate daughter cells result.

Figure 22.6 Bacterial reproduction by way of prokaryotic fission, a cell division mechanism. The micrograph shows the division of the cytoplasm of *Bacillus cereus*, as brought about by the formation of new membrane and wall material. This cell has been magnified 13,000 times its actual size.

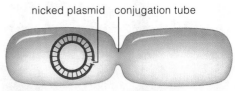

nicked plasmid   conjugation tube

a A conjugation tube has already formed between a donor and a recipient cell. An enzyme has nicked the donor's plasmid.

b DNA replication starts on the nicked plasmid. The displaced DNA strand moves through the tube and enters the recipient cell.

c In the recipient cell, replication starts on the transferred DNA.

d The cells separate from each other; the plasmids circularize.

Figure 22.7 A simplified sketch of bacterial conjugation. For the sake of clarity, the bacterial chromosome is not included in this sketch, and the size of the plasmid is greatly enlarged. (Compare Section 16.1.)

Bacterial Conjugation

Daughter bacterial cells also may inherit one or more plasmids. A **plasmid**, recall, is a small, self-replicating circle of extra DNA that has a few genes (Section 16.1). Some of the genes on the F (Fertility) plasmid confer the means to engage in **bacterial conjugation**. By this mechanism, a donor cell transfers plasmid DNA to a recipient cell. The transfers are known to occur among many types of bacteria, such as *Salmonella*, *Streptococcus*, and *E. coli*—even between *E. coli* and yeast cells, in the laboratory. The F plasmid carries genetic instructions for synthesizing a structure called a sex pilus. Sex pili at the surface of a donor cell can hook onto a recipient cell and pull it right next to the donor. Shortly after the two cells make contact, a conjugation tube develops between them. After this, plasmid DNA is transferred through the tube, in the manner shown in Figure 22.7.

Most bacteria reproduce by way of prokaryotic fission, a cell division mechanism that follows DNA replication. Each daughter cell inherits a single bacterial chromosome (one DNA molecule). Many species also transfer plasmid DNA.

Just a few decades ago, reconstructing the evolutionary history of bacteria appeared to be an impossible task. Except for stromatolites (Section 21.3), the most ancient groups of bacteria are not well represented in the fossil record. Most groups are not represented at all.

Given their elusive histories, the many thousands of known species of prokaryotic cells traditionally have been classified by **numerical taxonomy**. By this practice, traits of an unidentified cell are compared with those of a known bacterial group. Such traits typically include cell shape, motility, staining attributes of the cell wall, nutritional requirements, metabolic patterns, and the presence or absence of endospores. The greater the total number of traits that the cell has in common with the known group, the closer is their inferred relatedness.

Since the 1970s, nucleic acid hybridization studies, gene sequencing, and other methods of comparative biochemistry have been revealing compelling evidence of bacterial phylogenies (Section 20.5). Comparisons of ribosomal RNAs are notably useful. Remember, rRNAs are indispensable to protein synthesis, and they cannot undergo drastic changes in their overall base sequence without loss of function. The numerous small changes that accumulated in the rRNAs of different lineages can

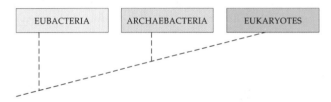

Figure 22.8 Evolutionary tree diagram showing the relationship of archaebacteria to eubacteria and eukaryotes, as suggested by evidence from comparative biochemistry.

be directly measured. Surprisingly, the measurements are uniting some groups that did not even appear to be related on the basis of other tests. For example, it now seems a key divergence began shortly after prokaryotic cells first appeared on Earth (Section 20.9). One branch led to the **eubacteria**, the most common prokaryotic cells. (Here, the *eu-* is meant to signify "typical.") The other branch led to the common ancestor of both the **archaebacteria** and the first eukaryotic cells. Figure 22.8 is a tree diagram of these evolutionary relationships.

Prokaryotic cells are classified by numerical taxonomy (the total percentage of observable traits they have in common with a known bacterial group). They also are classified more directly by comparisons at the biochemical level.

Prokaryotic cells are now assigned to one of two lineages, called the eubacteria and archaebacteria.

EUBACTERIA

More than 400 genera of prokaryotes are recognized, and by far, most are eubacteria. Unlike other bacterial cells, eubacteria have fatty acids incorporated into their plasma membrane. And when a species has a cell wall, as nearly all do, the wall incorporates peptidoglycan. We still do not know enough about the evolutionary histories of eubacteria to move much beyond taxonomic classification. Here we focus on modes of nutrition as a way to give you a sense of the diversity in this kingdom.

A Sampling of Diversity

PHOTOAUTOTROPHIC EUBACTERIA   The cyanobacteria, once called blue-green algae, are a good example of the photoautotrophic eubacteria. They are also among the most common photoautotrophs on Earth. Cyanobacteria are aerobic cells, and they themselves release oxygen when photosynthesizing. Most live in ponds and other freshwater habitats. They may grow as mucus-sheathed chains of cells, which can form dense, slimy mats near the surface of nutrient-enriched water.

Anabaena and other types also convert nitrogen gas (N_2) to ammonia, which they use in biosynthesis. When nitrogen compounds are scarce, some cells develop into **heterocysts**. These modified cells make a nitrogen-fixing enzyme (Figure 22.9). They produce and share nitrogen compounds with photosynthetic cells and they receive carbohydrates in return. They freely share substances by way of cytoplasmic junctions between cells.

Anaerobic photoautotrophs, such as green bacteria, get electrons from hydrogen sulfide or hydrogen gas, not from water. They may resemble anaerobic bacteria in which the cyclic pathway of photosynthesis evolved.

CHEMOAUTOTROPHIC EUBACTERIA   Many eubacteria in this category influence the global cycling of nitrogen, sulfur, and other nutrients. For example, as you know, nitrogen is a key building block for amino acids and proteins. Without it, there would be no life. Nitrifying bacteria in soil strip electrons from ammonia. Plants use the end product, nitrate, as a nitrogen source.

CHEMOHETEROTROPHIC EUBACTERIA   Most bacteria fall in this category. Many, including the pseudomonads, are decomposers; their enzymes break down organic compounds and even pesticides in soil. Other "good" species, at least by human standards, are *Lactobacillus* (employed in the manufacture of pickles, sauerkraut, buttermilk, and yogurt) and actinomycetes (sources of antibiotics). *E. coli* produces vitamin K and compounds that help humans digest fat, it helps newborns digest milk, and its metabolic activities help keep many food-borne pathogens from colonizing the gut. Sugarcane and corn benefit from a symbiont, the nitrogen-fixing

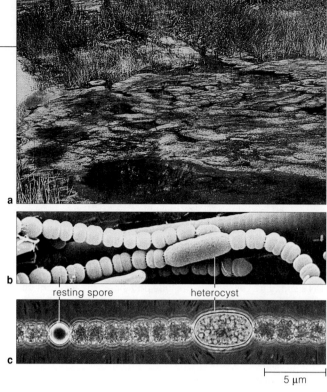

Figure 22.9 Cyanobacteria—common photoautotrophs. (**a**) A cyanobacterial population near the surface of a nutrient-enriched pond. (**b**,**c**) Resting spores form when conditions do not favor growth. A nitrogen-fixing heterocyst is also shown.

a

b
resting spore   heterocyst

c

5 µm

spirochete *Azospirillum*. The plants take up some of the nitrogen initially fixed by the bacterium, and they give up some sugars to it. Beans, peas, and other legumes are mutualists with *Rhizobium*, a symbiont in their roots.

Also in this category are most pathogenic bacteria. We admire pseudomonads as decomposers in soil, not when they grow on "our" soaps, antiseptics, and other carbon-rich goods. They are especially bad because they can transfer plasmids with antibiotic-resistance genes.

Some *E. coli* strains cause a form of diarrhea that is the main cause of infant death in developing countries. *Clostridium botulinum* can taint fermented grain as well as food in improperly sterilized or sealed cans and jars. Its toxins cause *botulism*, a form of poisoning that can disrupt breathing and lead to death. *C. tetani*, one of its relatives, causes the disease *tetanus* (Section 34.5).

Like many other bacteria that commonly live in soil, *C. tetani* can form an **endospore**. This resting structure forms inside the cell, around a copy of the bacterial chromosome and part of the cytoplasm (Figure 22.10). Endospores form when the depletion of nitrogen or some other nutrient stops cell growth. They are released

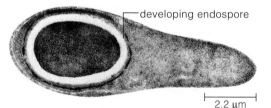

developing endospore

2.2 µm

Figure 22.10 An endospore developing in *Clostridium tetani*, one of the dangerous pathogens.

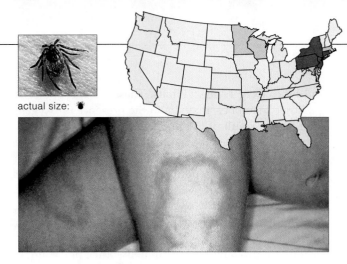

actual size: ●

Figure 22.11 An extreme reaction to an infection by *Borrelia burgdorferi*, a spirochete. The rash is a symptom of what is now the most common tick-borne disease in the United States: Lyme disease. Tick bites deliver the spirochete to new hosts. The map shows statewide cases reported in 1996 (*tan*, fewer than 10; *yellow*, 11 to 99; *gold*, 100 to 600; *red*, more than 2,000).

as free spores when the plasma membrane ruptures. They are exceptionally resistant to heat, and they also resist drying out, irradiation, acids, disinfectants, and boiling. Endospores can remain dormant, sometimes for many decades. When favorable conditions return, they become metabolically active and then develop into a single bacterial cell.

Many chemoautotrophic eubacteria taxi from host to host inside the gut of insects. For example, bites from blood-sucking ticks can transmit *Borrelia burgdorferi* from deer and some other wild animals to humans, who develop *Lyme disease*. A "bull's-eye" rash often develops around the bite (Figure 22.11). Severe headaches, backaches, chills, and fatigue follow. Without prompt treatment, the condition worsens.

Regarding the "Simple" Bacteria

Bacteria are small. Their insides are not elaborate. *But bacteria are not simple.* A brief look at their behavior will reinforce this point. Bacteria move toward nutrient-rich regions. Aerobes move toward oxygen; anaerobes avoid it. Photosynthetic types move into light and away from light that is too intense. Many species can tumble away from toxins. Such behaviors often start with membrane receptors that are stimulated by changes in chemical conditions or in the intensity of light coming in from a particular direction. Such a change triggers a shift in metabolic activities inside the cell, which might then lead to an adjustment in the direction of movement.

Magnetotactic bacteria contain a chain of magnetite particles that serves as a tiny compass (Figure 22.12*a*). The compass helps them sense which way is north and also down. These bacteria swim toward the bottom of a body of water, where oxygen concentrations are lower and therefore more suitable for their growth.

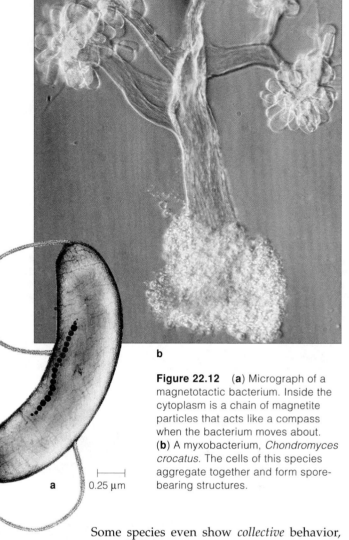

b

a   0.25 µm

Figure 22.12 (**a**) Micrograph of a magnetotactic bacterium. Inside the cytoplasm is a chain of magnetite particles that acts like a compass when the bacterium moves about. (**b**) A myxobacterium, *Chondromyces crocatus*. The cells of this species aggregate together and form spore-bearing structures.

Some species even show *collective* behavior, as when millions of *Myxococcus xanthus* cells form a "predatory" colony. The cells secrete enzymes that digest "prey," such as cyanobacteria, that get stuck to the colony. Then they absorb the breakdown products. What's more, the cells migrate, change direction, and move as a single unit toward what may be food!

Many myxobacteria colonies form **fruiting bodies** (spore-bearing structures). Under suitable conditions, some cells in the colony differentiate and form a slime stalk, others form branchings, and others form clusters of spores (Figure 22.12*b*). Spores disperse when a cluster bursts open; each may give rise to a new colony. As you will see in the next chapter, certain eukaryotic species also form spore-bearing structures.

Eubacteria are the most common and diverse prokaryotic cells. They are adapted to nearly all environments.

We divide the kingdom of archaebacteria into three major groups, called the methanogens, halophiles, and extreme thermophiles. In many respects, archaebacteria are unique in their composition, structure, metabolism, and nucleic acid sequences. They differ as much from other bacteria as they do from eukaryotic cells. Many investigators believe that the existing archaebacteria, which can withstand conditions as hostile as those on the early Earth, resemble the first living cells. Hence the name of the kingdom (*archae-* means beginning).

The **methanogens**, or "methane makers," are quite at home in swamps, sewage, stockyards, animal guts,

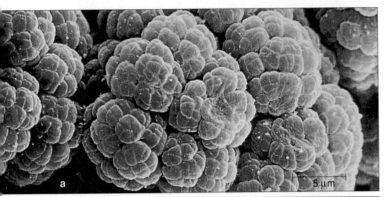

Figure 22.13 Archaebacteria. (a) Scanning electron micrograph of a dense population of *Methanosarcina* cells. (b) Transmission electron micrograph of *Methanococcus jannaschii*, which had its DNA fully sequenced (Section 20.9). (c) Pinkish, saline water in Great Salt Lake, Utah—a sign of colonies of halophilic bacteria and certain algae that contain pinkish to red-orange carotenoid pigments.

and other oxygen-free habitats. These prokaryotic cells are strict anaerobes; they die in the presence of oxygen. Mainly as a result of the findings from ribosomal RNA sequencing studies, the seventeen known genera have been assigned to seven categories. Figure 22.13 shows representatives. Although evolutionarily distant from humans, the methanogens still are closer relatives to you than are the eubacteria.

The methanogens make ATP by anaerobic electron transport, a pathway described in Section 8.5. Usually, they use hydrogen gas (H_2) as their source of electrons, although some groups get them from ethanol and other alcohols. Nearly all use carbon dioxide as their carbon source and as a final electron acceptor for the reactions, which end with the formation of methane (CH_4). As a group, the methanogens produce about 2 billion tons of methane every year. Collectively, their activities affect the levels of carbon dioxide in the atmosphere and the cycling of carbon through ecosystems.

Long ago, vast methane deposits accumulated on the seafloor. For example, geologists recently found 35 billion tons of it 400 kilometers off the South Carolina coast. If the ocean circulation were to change, as it has in the past, then all of that gas could move abruptly and explosively to the surface. The rapid release of so much methane would drastically change the atmosphere and therefore the global climate (Section 48.7).

The "salt lovers," or **halophiles**, live in brackish ponds, salt lakes, near volcanic fissures on the seafloor (hydrothermal vents), and other high-salinity settings (Figure 22.13c). Halophiles can spoil salted fish, animal hides, and commercially produced sea salt. Most make ATP by aerobic pathways. When oxygen levels are low, some also produce ATP by photosynthesis. A unique photosynthetic pigment (bacteriorhodopsin) forms in the plasma membrane and absorbs light energy. Absorption starts a metabolic process that leads to an increase in an H^+ gradient across the membrane. ATP forms when these ions flow down the gradient, through the interior of membrane proteins (compare Section 7.5).

The "heat lovers," or **extreme thermophiles**, live in such places as highly acidic soils, hot springs, even coal mine wastes. Some species are the basis of remarkable food webs in sediments around hydrothermal vents. Sometimes the water above the sediments is as hot as 110°C. Extreme thermophiles use the hydrogen sulfide spewing from the vents as a source of electrons for ATP formation. Their existence at vents is cited as evidence that life could have originated deep in the oceans.

Like the first cells on Earth, archaebacteria live in extremely inhospitable habitats. In their properties, they differ as much from eubacteria as from eukaryotic cells.

SUMMARY OF MAJOR BACTERIAL GROUPS

By now, you probably have sensed that the *species* is the basic unit in bacterial classification schemes. However, the definition of species that fits sexually reproducing organisms does not fit bacteria, which do not form reproductively isolated populations of interbreeding individuals. Each bacterial cell generally does its own thing; genetic recombination is infrequent. Besides this, bacteria do not show spectacular variation in traits, and variations that do occur are dictated by relatively few genes. If two types show only minor differences, one may be classified as a **strain**, not a new species.

These are just a few of the problems we face when attempting to bring order to our understanding of the bacterial kingdoms. Until evolutionary relationships are sorted out, bacteria will continue to be grouped mainly according to numerical taxonomy, as described earlier. Table 22.2 summarizes some of the major groupings that are touched upon in this book.

Table 22.2 Summary of Representative Eubacteria and Archaebacteria			
Some Major Groups	Main Habitats	Characteristics	Representatives
EUBACTERIA			
Photoautotrophs:			
Cyanobacteria, green sulfur bacteria, and purple sulfur bacteria	Mostly lakes, ponds; some marine, terrestrial habitats	Photosynthetic; use sunlight energy, carbon dioxide; cyanobacteria use oxygen-producing noncyclic pathway; some also use cyclic route	*Anabaena, Nostoc, Rhodopseudomonas, Chloroflexus*
Photoheterotrophs:			
Purple nonsulfur and green nonsulfur bacteria	Anaerobic, organically rich muddy soils, and sediments of aquatic habitats	Use sunlight energy; organic compounds as electron donors; some purple nonsulfur may also grow chemotrophically	*Rhodospirillum, Chlorobium*
Chemoautotrophs:			
Nitrifying, sulfur-oxidizing, and iron-oxidizing bacteria	Soil; freshwater, marine habitats	Use carbon dioxide, inorganic compounds as electron donors; influence crop yields, cycling of nutrients in ecosystems	*Nitrosomonas, Nitrobacter, Thiobacillus*
Chemoheterotrophs:			
Spirochetes	Aquatic habitats; parasites of animals	Helically coiled, motile; free-living and parasitic species; some major pathogens	*Spirochaeta, Treponema*
Gram-negative aerobic rods and cocci	Soil, aquatic habitats; parasites of animals, plants	Some major pathogens; some fix nitrogen (e.g., *Rhizobium*)	*Pseudomonas, Neisseria, Rhizobium, Agrobacterium*
Gram-negative facultative anaerobic rods	Soil, plants, animal gut	Many major pathogens; one bioluminescent (*Photobacterium*)	*Salmonella, Escherichia, Proteus, Photobacterium*
Rickettsias and chlamydias	Host cells of animals	Intracellular parasites; many pathogens	*Rickettsia, Chlamydia*
Myxobacteria	Decaying organic material; bark of living trees	Gliding, rod-shaped; aggregation and collective migration of cells	*Myxococcus*
Gram-positive cocci	Soil; skin and mucous membranes of animals	Some major pathogens	*Staphylococcus, Streptococcus*
Endospore-forming rods and cocci	Soil; animal gut	Some major pathogens	*Bacillus, Clostridium*
Gram-positive nonsporulating rods	Fermenting plant, animal material; gut, vaginal tract	Some important in dairy industry, others major contaminators of milk, cheese	*Lactobacillus, Listeria*
Actinomycetes	Soil; some aquatic habitats	Include anaerobes and strict aerobes; major producers of antibiotics	*Actinomyces, Streptomyces*
ARCHAEBACTERIA			
Methanogens	Anaerobic sediments of lakes, swamps; animal gut	Chemosynthetic; methane producers; used in sewage treatment facilities	*Methanobacterium*
Halophiles	Brines (extremely salty water)	Heterotrophic; also, unique photosynthetic pigments (bacteriorhodopsin) form in some	*Halobacterium*
Extreme thermophiles	Acidic soil, hot springs, hydrothermal vents	Heterotrophic or chemosynthetic; use inorganic substances as electron donors	*Sulfolobus, Thermoplasma*

22.7　THE VIRUSES

Defining Characteristics

In ancient Rome, *virus* meant "poison" or "venomous secretion." In the late 1800s, this rather nasty word was bestowed on newly discovered pathogens, smaller than the bacteria being studied by Louis Pasteur and others. Many viruses deserve the name. They attack humans, cats, cattle, birds, insects, plants, fungi, protistans, and bacteria. You name it, there are viruses that can infect it.

Today we define a **virus** as a noncellular infectious agent that has two characteristics. First, a viral particle consists of a protein coat surrounding a nucleic acid core—that is, around its genetic material. Second, a virus cannot reproduce itself. It can be reproduced only after its genetic material and a few enzymes enter a host cell and subvert the cell's biosynthetic machinery.

The genetic material of a virus is DNA *or* RNA. The coat consists of one or more types of protein subunits organized into a rodlike or many-sided shape (Figure 22.14). The coat protects the genetic material during the journey from one host cell to the next. It also contains proteins that can bind with specific receptors on host cells. Some coats are enclosed in an envelope of mostly membrane remnants from the previously infected cell. These envelopes bristle with glycoprotein spikes. Coats of complex viruses have sheaths, tail fibers, and other accessory structures attached.

The immune system of vertebrates can detect certain viral proteins. The problem is, the genes for many viral proteins mutate so frequently that a virus may elude the immune fighters. For example, people susceptible to lung infections get new "flu shots" each year because envelope spikes on influenza viruses keep changing.

Examples of Viruses

Each kind of virus can multiply only in certain hosts. It cannot be studied easily unless investigators culture living host cells. This is why much of our knowledge of viruses comes from **bacteriophages**, a group of viruses that infect bacterial cells. Unlike cells of humans and other complex, multicelled species, bacterial cells can be cultured easily and rapidly. This is also why bacteria and bacteriophages were used in early experiments to determine DNA function (Section 13.1). They are still used as research tools in genetic engineering.

Table 22.3 lists some major groups of animal viruses. They contain double- or single-stranded DNA or RNA, which is replicated in various ways. Animal viruses range in size from parvoviruses (18 nanometers) to the brick-shaped poxviruses (350 nanometers). Many cause diseases, including the common cold, certain cancers, warts, herpes, and influenza (Figure 22.15). One, HIV, is the trigger for *AIDS*. By attacking certain white blood

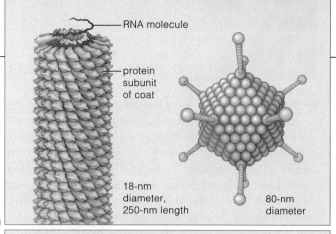

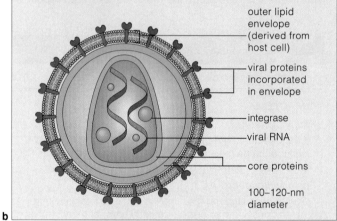

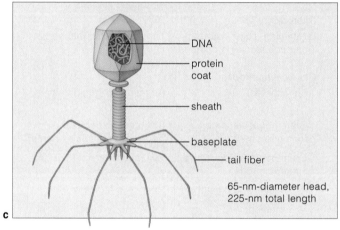

Figure 22.14　Body plans of viruses. (**a**) To the left, *helical* viruses have a rod-shaped coat of protein subunits, coiled helically around the nucleic acid. The upper subunits have been removed from this portion of a tobacco mosaic virus to reveal the RNA. To the right, *polyhedral* viruses, such as this adenovirus, have a many-sided coat. (**b**) *Enveloped* viruses, such as HIV, have an envelope around a helical or polyhedral coat. (**c**) *Complex* viruses, such as T-even bacteriophages, have additional structures attached to the coat.

cells, it weakens the immune system's ability to fight infections that might not otherwise be life threatening. Researchers attempting to develop drugs against HIV and other diseases use HeLa cells and other immortal cell lines for early experiments (Section 9.6). Later they must use laboratory animals, then human volunteers, to test a new drug for toxicity and effectiveness. Why? It takes a functioning immune system to test responses.

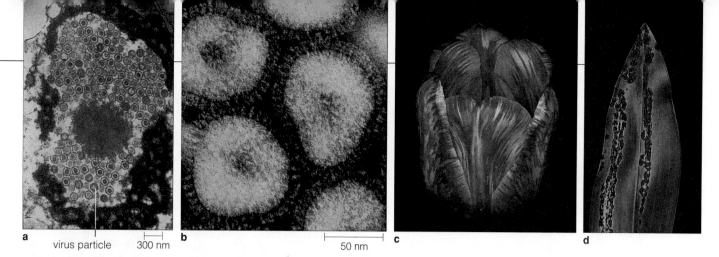

a virus particle |—| 300 nm b |—————| 50 nm c d

Table 22.3 Classification of Some Major Animal Viruses

DNA Viruses	Some Diseases
Parvoviruses	Gastroenteritis; roseola (fever, rash) in small children; aggravation of symptoms of sickle-cell anemia
Adenoviruses	Respiratory infections; some cause tumors in animals
Papovaviruses	Benign and malignant warts
Poxviruses	Smallpox, cowpox
Herpesviruses:	
H. simplex type I	Oral herpes, cold sores
H. simplex type II	Genital herpes
Varicella-zoster	Chicken pox, shingles
Epstein-Barr	Infectious mononucleosis, some forms of cancer
Hepadnavirus	Hepatitis B, severe liver damage

RNA Viruses	Some Diseases
Picornaviruses:	
Enteroviruses	Polio, hemorrhagic eye disease, hepatitis A (infectious hepatitis)
Rhinoviruses	Common cold
Hepatitis A virus	Inflammation of liver, kidneys, spleen
Togaviruses	Forms of encephalitis, rubella
Flaviviruses	Yellow fever (fever, chills, jaundice), dengue (fever, severe muscle pain)
Coronaviruses	Upper respiratory infections, colds
Rhabdoviruses	Rabies, other animal diseases
Filoviruses	Hemorrhagic fevers, as by *Ebola* virus
Paramyxoviruses	Measles, mumps
Orthomyxoviruses	Influenza
Bunyaviruses	Hemorrhagic fevers, as by hantaviruses
Arenaviruses	Hemorrhagic fevers
Retroviruses:	
HTLV I, II	Some cancers (leukemia)
HIV	AIDS
Reoviruses	Respiratory and intestinal infections

Figure 22.15 Some viruses and their effects. (**a**) Particles of a DNA virus that causes a herpes infection in humans. (**b**) Particles of an enveloped RNA virus that causes influenza in humans. Spikes project from the lipid envelope. (**c**) Streaking of a tulip blossom. A harmless virus infected pigment-forming cells in the colorless parts. (**d**) An orchid leaf infected by a rhabdovirus.

Plant viruses must breach plant cell walls to cause diseases. They typically hitch rides on the piercing or sucking devices of insects that feed on plant juices. Some RNA viruses infect tobacco plants (the tobacco mosaic virus), barley, potatoes, and other major crop plants. Certain DNA viruses infect such valuable crops as cauliflower and corn. Figure 22.15*c,d* shows visible effects of two viral infections.

Infectious Agents Tinier Than Viruses

Some infectious agents are more stripped down than viruses. **Prions** are small proteins linked to eight rare, fatal degenerative diseases of the nervous system. They are altered products of a gene that is present in normal as well as infected individuals. Unaltered and altered forms of the protein are found at the surface of neurons (communication cells of the nervous system). You may have heard of *kuru* and *Creutzfeldt-Jakob* diseases. They slowly destroy muscle coordination and brain function in humans (Section 22.9). *Scrapie*, a disease of sheep, is so named because infected animals rub against trees or posts until they scrape off most of their wool.

Viroids are tightly folded strands or circles of RNA, smaller than anything in viruses. They resemble introns (noncoding portions of eukaryotic DNA) and may have evolved from them. Viroids have no protein coat, but their tight folding may help protect them from a host's enzymes. These bits of "naked" RNA are known to cause plant diseases. Each year, they destroy millions of dollars' worth of potatoes, citrus, and other crop plants.

A virus is a nonliving infectious particle that consists of nucleic acid enclosed in a protein coat and sometimes an outer envelope. It cannot be replicated without pirating the metabolic machinery of a specific type of host cell.

VIRAL MULTIPLICATION CYCLES

Viruses multiply in a variety of ways. Even so, nearly all of their multiplication cycles proceed through five basic steps, as outlined here:

1. *Attachment.* The virus attaches to a host cell. Any cell is a suitable host if a virus can chemically recognize and lock onto specific molecular groups at the cell surface.

2. *Penetration.* Either the whole virus or its genetic material alone penetrates the cell's cytoplasm.

3. *Replication and Synthesis.* In an act of molecular piracy, the viral DNA or RNA directs the host cell into producing many copies of viral nucleic acids and proteins, including enzymes.

4. *Assembly.* The viral nucleic acids and viral proteins are put together to form new infectious particles.

5. *Release.* New virus particles are released from the cell.

Consider this list with respect to some bacteriophages. Lytic and lysogenic pathways are common among their replication cycles. In a **lytic pathway**, steps 1 through 4 proceed rapidly, and new particles are released when the host cell undergoes lysis (Figure 22.16). **Lysis** means the plasma membrane, cell wall, or both are damaged, so that cytoplasm leaks out and the cell dies. Late into most lytic pathways, a viral enzyme is synthesized that causes swift destruction of the bacterial cell wall.

In a **lysogenic pathway**, a latent period extends the cycle. The virus does not kill its host outright. Instead, a viral enzyme cuts a host chromosome, then integrates viral genes into it. When the infected cell prepares to divide, the recombinant molecule gets replicated. As a result of that single instance of genetic recombination, miniature time bombs get passed on to all of that cell's descendants. Later on, a molecular signal or some other stimulus may reactivate the cycle.

Latency occurs in the multiplication cycles of many viruses, not just among bacteriophages. Type I *Herpes simplex*, which causes *cold sores*, is an example. Nearly everybody harbors this virus. It remains latent inside a ganglion (a cluster of cell bodies of neurons) in the face. Stress factors, such as sunburn, can reactivate the virus. Then, virus particles move down the neurons to their tips near the skin. There they infect epithelial cells and cause painful skin eruptions.

Figure 22.16 Generalized multiplication cycle for some of the bacteriophages. New viral particles may be produced and then released by way of a lytic pathway alone. For certain viruses, the lytic pathway may be expanded upon to include a lysogenic pathway.

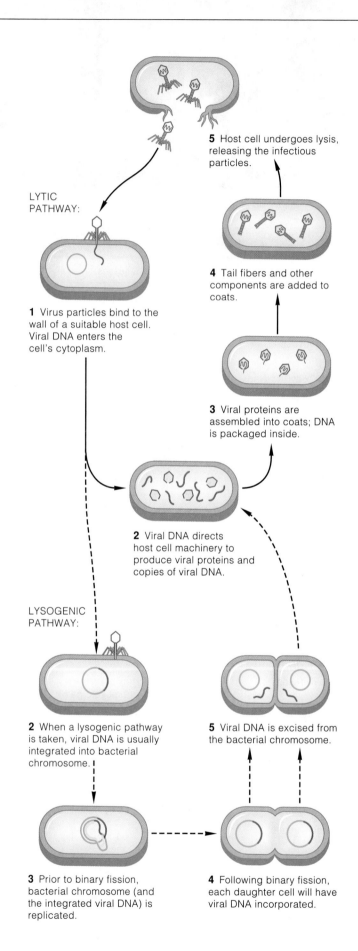

5 Host cell undergoes lysis, releasing the infectious particles.

LYTIC PATHWAY:

1 Virus particles bind to the wall of a suitable host cell. Viral DNA enters the cell's cytoplasm.

4 Tail fibers and other components are added to coats.

3 Viral proteins are assembled into coats; DNA is packaged inside.

2 Viral DNA directs host cell machinery to produce viral proteins and copies of viral DNA.

LYSOGENIC PATHWAY:

2 When a lysogenic pathway is taken, viral DNA is usually integrated into bacterial chromosome.

5 Viral DNA is excised from the bacterial chromosome.

3 Prior to binary fission, bacterial chromosome (and the integrated viral DNA) is replicated.

4 Following binary fission, each daughter cell will have viral DNA incorporated.

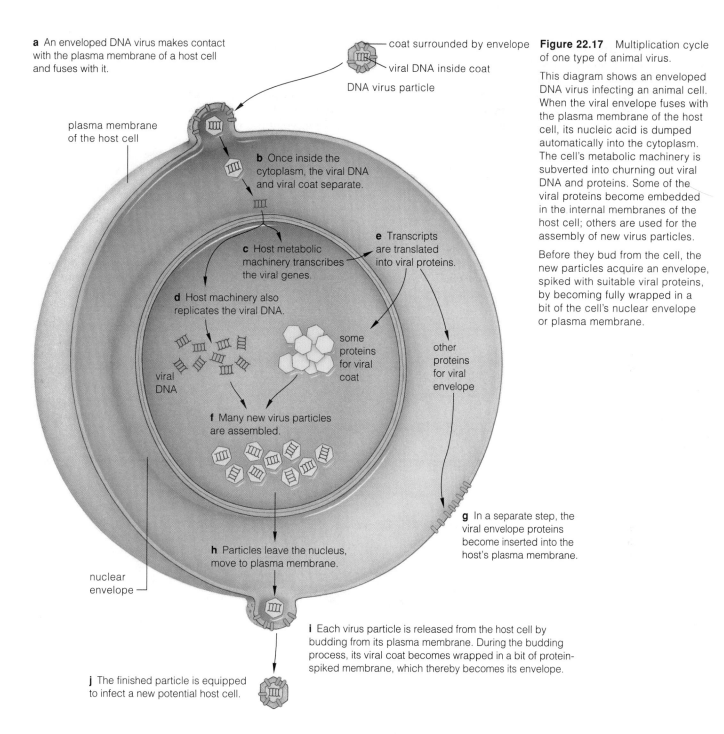

a An enveloped DNA virus makes contact with the plasma membrane of a host cell and fuses with it.

coat surrounded by envelope

viral DNA inside coat

DNA virus particle

plasma membrane of the host cell

b Once inside the cytoplasm, the viral DNA and viral coat separate.

c Host metabolic machinery transcribes the viral genes.

e Transcripts are translated into viral proteins.

d Host machinery also replicates the viral DNA.

viral DNA

some proteins for viral coat

other proteins for viral envelope

f Many new virus particles are assembled.

g In a separate step, the viral envelope proteins become inserted into the host's plasma membrane.

h Particles leave the nucleus, move to plasma membrane.

nuclear envelope

i Each virus particle is released from the host cell by budding from its plasma membrane. During the budding process, its viral coat becomes wrapped in a bit of protein-spiked membrane, which thereby becomes its envelope.

j The finished particle is equipped to infect a new potential host cell.

Figure 22.17 Multiplication cycle of one type of animal virus.

This diagram shows an enveloped DNA virus infecting an animal cell. When the viral envelope fuses with the plasma membrane of the host cell, its nucleic acid is dumped automatically into the cytoplasm. The cell's metabolic machinery is subverted into churning out viral DNA and proteins. Some of the viral proteins become embedded in the internal membranes of the host cell; others are used for the assembly of new virus particles.

Before they bud from the cell, the new particles acquire an envelope, spiked with suitable viral proteins, by becoming fully wrapped in a bit of the cell's nuclear envelope or plasma membrane.

Like other enveloped viruses, the herpesviruses can enter a host cell by their own version of endocytosis, then leave it by budding from the plasma membrane. Figure 22.17 shows how they accomplish this.

The multiplication cycle of the RNA viruses has an interesting twist to it. In the host cell's cytoplasm, their RNA serves as a template for synthesizing either DNA or mRNA. For example, HIV, a retrovirus, carts its own enzymes into cells. These assemble DNA on viral RNA by reverse transcription (Sections 16.4 and 40.11).

The multiplication cycles of nearly all viruses include five basic steps: attachment to a suitable host cell, penetration into the cell, viral DNA or RNA replication and synthesis of viral proteins, assembly of new viral particles, and then release from the infected cell.

Bacteriophage replication cycles commonly follow a rapid, lytic pathway and an extended, lysogenic pathway.

Replication cycles of RNA viruses involve the use of viral RNA as a template for synthesizing DNA or mRNA.

THE NATURE OF INFECTIOUS DISEASES

CATEGORIES OF DISEASES   Just by being human, you are a potential host for a great many pathogenic bacteria, viruses, fungi, protozoans, and parasitic worms. When a pathogen has invaded your body and is multiplying in host cells and tissues, this is an **infection**. Its outcome, **disease**, results if the body's defenses cannot be mobilized fast enough to prevent the pathogen's activities from interfering with normal body functions. You may have heard of *contagious* diseases. This simply means that the pathogenic agents can be transmitted by direct contact with body fluids secreted or otherwise released (as by explosive wet sneezes) from infected individuals.

During an **epidemic**, a disease spreads rapidly through part of a population for a limited time, then the outbreak subsides. What happens when an epidemic breaks out in several countries around the world at the same time? We call this a **pandemic**. AIDS, an incurable disease, is a prime example. Millions of people around the world are infected by the causative agent, HIV (short for the *Human Immunodeficiency Virus*).

Sporadic diseases such as whooping cough break out irregularly and affect few people. *Endemic* diseases pop up more or less continuously, but they don't spread far in large populations. Tuberculosis is like this. So is impetigo, a highly contagious bacterial infection that often spreads no further than, say, a single day-care center.

AN EVOLUTIONARY VIEW   Now look at disease in terms of the pathogen's prospects for survival. Just like you, *a pathogen stays around only for as long as it has access to outside sources of energy and raw materials.* To a microscopic organism or viral particle, a human host is a veritable treasurehouse of both. With such a bountiful supply of resources, it can multiply or replicate itself to amazing population sizes. And, evolutionarily speaking, the ones that leave the most descendants win.

There are two significant barriers to complete world dominance by pathogens. First, any species with a history of being attacked by a particular pathogen has coevolved with it and has built-in defenses against it. The premier example is the vertebrate immune system, as described in Chapter 40. Second, if a pathogen is too good at rapidly killing an individual host, it may vanish before having time to infect a new one. This is one reason why most pathogens have less-than-fatal effects on a host. After all, an infected individual that lives longer can spread more of the next generation of a pathogen and so contribute to its reproductive or replicative success.

Usually, a host dies only if a pathogen enters its body in overwhelming numbers, if it is a novel host (one with no coevolved defenses against that pathogen)—or if a mutant strain of the pathogen emerges and can surmount the host's current array of defenses.

Being equipped with the evolutionary perspective, you probably can work out the numbers on your own. To wit: *The greater the population density of host individuals, the greater will be the kinds and frequencies of infectious diseases transmitted among them.* And this brings us to the bad news.

EBOLA AND OTHER EMERGING PATHOGENS   Thanks to planes, trains, and automobiles, people now travel often and in droves all around the world. Among their more exotic destinations are virgin tropical forests and other remote environments where the human body was an infrequent (or nonexistent) opportunity for pathogens.

Today, strange and often dangerous pathogens are opportunistic about the novel, two-legged treasurehouses of metabolic machinery and nutrients that are walking smack into their habitats. Human travelers can become infected within a matter of hours and taxi such pathogens far away, and eventually back home.

We know little about the deadly, **emerging pathogens**. Maybe they have been around for a long time and only now are taking great advantage of the increased presence of the novel human hosts. Maybe they are newly mutated strains of existing species.

Consider *Ebola*, one of the deadliest of the viruses that cause hemorrhagic fever. *Ebola*, an RNA virus, appears either filamentous or circular in electron micrographs (Figure 22.18). It may have coevolved with monkeys in tropical forests in Africa. By 1976 and possibly earlier, it was infecting humans. It kills between 70 and 90 percent of its victims. There is no vaccine against it. There is no treatment for the disease, and what eventually happens is not pretty. High fever and flu-like aches mark the onset of the disease. Within a few days, nausea, vomiting, and diarrhea begin. Cells making up the lining of blood vessels are destroyed. Blood is free to seep out from the circulatory system, into the surrounding tissues and out through all of the body's orifices. The liver and the kidneys may rapidly turn to mush. Patients often become deranged, and soon after that they die of circulatory shock. There have been four *Ebola* epidemics since 1976. Each time, government agencies around the world were mobilized; quarantine procedures were implemented to limit the spread of the disease.

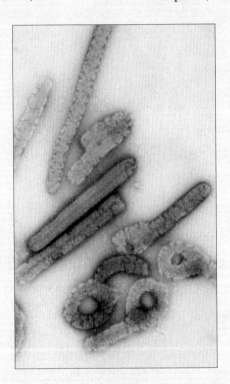

Figure 22.18   Transmission electron micrograph of particles of *Ebola* virus.

ON THE DRUG-RESISTANT STRAINS There is an old saying that, when you attack nature, it comes back at you with a pitchfork. We have already considered antibiotic resistance in several contexts, in Sections 1.4, 1.6, and 18.4. Here we reinforce the point that infectious diseases are on the upsurge.

A mere twenty-five years ago, antibiotics and vaccines were considered invincible, and the Surgeon General of the United States announced we could "close the book on infectious diseases." There are now more than 5.8 billion people. Most live in crowded cities, and at any time, as many as 50 million are on the move within and between countries in search of a better life (Section 46.6). Even putting aside the emerging pathogens, is it any wonder that pathogens responsible for cholera, tuberculosis, and other familiar diseases are striking with a vengeance?

Consider how one of these pathogens is having a field day. Because of economic pressures over the past several decades, the number of working mothers has skyrocketed in the United States. So has the number of preschoolers enrolled in day-care centers. In effect, each of these centers is a population of host individuals whose immune systems are still developing and vulnerable to contagious diseases. Now think of *Streptococcus pneumoniae*. This bacterium causes pneumonia, meningitis, and middle-ear infections in people of all ages; about 40,000 to 50,000 cases end in death every year. The risk of infection is 36 times greater in large day-care centers than it is for children cared for at home. As recently as 1988, drug-resistant strains of *S. pneumoniae* were practically unheard of in the United States. They are becoming the rule, not the exception.

PATHOGENS LURKING IN THE KITCHEN Each year, food poisoning hits more than 80 million people in the United States alone. It kills about 9,000 of them, mainly the very young, the very old, or those with compromised immune systems. Yet the cases reported to health agencies number only in the thousands; people often dismiss their misery as "the 24-hour flu." *Salmonella enteriditis* as well as certain strains of *E. coli* are prominent among the culprits. They typically are ingested with undercooked beef or poultry, contaminated water, unpasteurized milk or cider, and vegetables grown in fields fertilized with manure.

A few examples: In 1966, *S. enteriditis* slipped into ingredients used to make a popular ice cream. In a single outbreak, possibly as many as 224,000 people suffered from stomach cramps, diarrhea, and other symptoms of *Salmonella*. In 1993 in Washington State, more than 500 people became sick and three children died after eating restaurant hamburgers tainted with a pathogenic strain of *E. coli*. A recent outbreak of food poisoning was traced to unpasteurized apple juice.

Outbreaks such as these certainly grab our attention, yet pathogens lurking in the kitchen probably account for at least half of all cases of food poisoning. Examine wood or plastic cutting boards or sleek, stainless steel

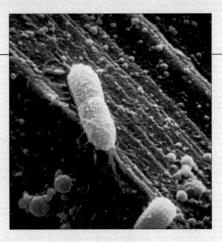

Figure 22.19 Among microscopic bits of food, *Pseudomonas* cells that are attaching to stainless steel by means of pili at their surface—and think about *that* the next time you use an unwashed kitchen knife previously used on raw poultry, beef, or seafood.

knives with electron microscopes and you see nooks and crannies where microbes can lurk (Figure 22.19). Carlos Enriquez and his colleagues at the University of Arizona sampled 75 dishrags and 325 sponges in several homes. Most harbored colonies of *Salmonella* and *E. coli*, as well as *Pseudomonas* and *Staphylococcus*. Bacteria can live two weeks in a wet sponge. Use the sponge to wipe down the kitchen, and you spread them about. Antibacterial soaps, detergents, and weak solutions of household bleach get rid of them. You can sanitize sponges simply by putting them through a dishwasher cycle.

Do you think food poisoning is of small concern? The annual cost of treating the infections caused by food-borne pathogens ranges between 5 and 22 *billion* dollars.

PRIONS LURKING IN SHEEP AND COWS Also weighing in as pathogenic heavyweights are the prions (pronounced PREE-ons), those proteinaceous infectious particles you read about in Section 22.7. In 1996, 150,000 cattle in Great Britain were staggering about, drooling, and showing other symptoms of *BSE* (bovine spongiform encephalopathy), or *mad cow disease*. Prions had triggered conversion of normal proteins in their neurons, and spongy holes and amyloid deposits had formed in the cattle brains. BSE is fatal.

How did it happen? Ground-up tissues of sheep that had died of scrapie—another prion-caused disease—had been used as a supplement in cattle feed. The practice was banned in 1988, but prions were already in the food chain.

Medical researchers already knew that, each year, 1 in 1 million people around the world develops Creutzfeldt-Jakob disease (CJD). Prions are linked to this fatal disease, as well. The symptoms, which include loss of vision and speech, rapid mental deterioration, and spastic paralysis, may take decades to develop. However, at the same time as the outbreak of BSE in Great Britain, ten people were diagnosed with a variant of CJD; four had already died. Genetic analysis and interviews linked all of the cases to exposure to BSE. Coincidentally, a variant of CJD may have caused the death of a woman in Stockton, California, in 1996, also.

Ten other countries had reported cases of BSE, possibly as a result of importing cattle feed from Great Britain. The incidence of BSE in Great Britain is now declining since strict precautionary measures have been instituted. In the United States, importation of live cattle or meat products from BSE-affected countries has been banned since 1989.

SUMMARY

1. A major divergence occurred soon after the origin of life. One lineage gave rise to the eubacteria, and the other to the common ancestors of archaebacteria and eukaryotic cells.

a. By far, the most common existing prokaryotic cells are eubacteria.

b. Archaebacteria (the methanogens, halophiles, and extreme thermophiles) live in extreme environments, like those in which life probably originated. They differ from eubacteria in wall structure and other features, and they share some features with eukaryotic cells.

2. All bacteria are prokaryotic cells; they have no nucleus. Some have membrane infoldings and other structures in the cytoplasm, but none has the profusion of organelles that is typical of eukaryotic cells.

a. Three basic bacterial shapes are common: cocci (spheres), bacilli (rods), and spirilla (spirals).

b. Nearly all eubacteria have a cell wall composed of peptidoglycan. It protects the plasma membrane and helps it resist rupturing. The composition and structure of the cell wall help identify particular bacterial species.

c. A sticky mesh of polysaccharides (glycocalyx) may surround the wall as a capsule or slime layer. It helps a bacterium attach to substrates and sometimes to resist a host organism's infection-fighting mechanisms.

d. Some species have bacterial flagella, structures that rotate a bit like a propeller and function in motility. Many species have pili, filamentous proteins that help cells adhere to a surface or facilitate conjugation.

3. Prokaryotic fission is a cell division mechanism used only by bacteria. It involves replication of the single bacterial chromosome and division of a parent cell into two genetically equivalent daughter cells.

4. Many species have plasmids, small circles of DNA that are replicated independently of the single, circular bacterial chromosome. Certain genes on some plasmids confer antibiotic resistance. Certain genes on one type confer the ability to engage in bacterial conjugation. Plasmids are transmitted to daughter cells and may be transferred, by way of bacterial conjugation, to cells of the same species or a different species.

5. Bacteria as a group show great metabolic diversity, as in their modes of acquiring energy and carbon.

a. *Photoautotrophs* use sunlight and carbon dioxide during photosynthesis. They include cyanobacteria and other oxygen-releasing species of eubacteria. They also include green nonsulfur and purple nonsulfur bacteria that do not produce oxygen; these obtain electrons from sulfur and other inorganic compounds, not from water.

b. *Photoheterotrophs* use sunlight energy and organic compounds as sources of carbon. They include certain archaebacteria and eubacteria.

c. *Chemoautotrophs*, including the nitrifying bacteria, use carbon dioxide but not sunlight. Different kinds get energy by stripping electrons from various organic or inorganic substances.

d. The *chemoheterotrophs* are either parasites (which draw carbon and energy from living hosts) or saprobes (which feed on the organic products, wastes, or remains of other organisms). Most bacteria are in this category. They include major decomposers and pathogens.

6. Bacteria can make behavioral responses to stimuli, as when they move toward regions with more nutrients.

7. Viruses are nonliving, noncellular agents that infect particular species of nearly all organisms.

a. Each virus particle consists of a core of DNA or RNA and a protein coat that sometimes is enclosed in a lipid envelope. Many glycoprotein spikes project from the envelopes. Coats of complex viruses have sheaths, tail fibers, and other accessory structures.

b. A virus particle cannot reproduce on its own. Its genetic material must enter a host cell and direct the cellular machinery to synthesize the materials necessary to produce new virus particles.

8. Nearly all viral multiplication cycles include five steps: attachment to a suitable host cell, penetration of it, viral DNA or RNA replication and protein synthesis, assembly of new viral particles, and release.

9. Two pathways are common in the multiplication cycle of bacteriophages (bacteria-infecting viruses). In a lytic pathway, multiplication is rapid and new viral particles are released by lysis. In a lysogenic pathway, the infection enters a latent period. The host cell is not killed outright, and the viral nucleic acid may undergo genetic recombination with a host cell chromosome.

10. Multiplication cycles of viruses are diverse. They proceed rapidly or enter a latent phase. Penetration and release of most enveloped types occur by endocytosis and exocytosis. The DNA viruses spend part of the cycle in the nucleus of a host cell. The RNA viruses complete the cycle in the cytoplasm. The viral RNA is the template for mRNA synthesis and for protein synthesis.

Review Questions

1. Label the structures on this generalized bacterial cell: *22.1*

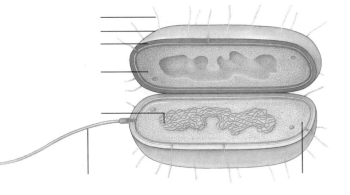

2. Describe the key metabolic and structural features of bacteria. Make sketches of the three basic shapes of bacterial cells. *22.1*

3. Compared to your own bodily growth, how is bacterial growth measured? *22.2*

4. With respect to bacterial classification, what are some of the pitfalls of numerical taxonomy? How are comparisons of rRNA sequences from different bacterial groups assisting classification efforts? *22.3, 22.6*

5. Name a few of the photoautotrophic, chemoautotrophic, and chemoheterotrophic eubacteria. Describe some that are likely to give humans the most trouble, medically speaking. *22.4*

6. What is a virus? Why is a virus considered to be no more alive than a chromosome? *22.7*

7. Distinguish between:
   a. microorganism and pathogen *CI*
   b. infection and disease *22.9*
   c. epidemic and pandemic *22.9*

Self-Quiz *(Answers in Appendix IV)*

1. Nondividing bacteria have _____ chromosome(s) and may have extra circles of _____ called plasmids.
   a. one; RNA   c. one; DNA
   b. two; RNA   d. two; DNA

2. _____ live in habitats much like those of the early Earth.
   a. Cyanobacteria   c. Archaebacteria
   b. Eubacteria   d. Protozoans

3. Which of the following is not grouped with archaebacteria?
   a. halophiles   c. thermophiles
   b. cyanobacteria   d. methanogens

4. Bacteria reproduce by _____ .
   a. mitosis   c. prokaryotic fission
   b. meiosis   d. longitudinal fission

5. Eubacterial cell walls are composed of _____ ; and a sticky mesh of polysaccharides, a _____ , often surrounds the wall.
   a. peptidoglycan; plasma membrane
   b. cellulose; glycocalyx
   c. cellulose; plasma membrane
   d. peptidoglycan; glycocalyx

6. Most bacteria are _____ and include major decomposers and pathogens.
   a. photoautotrophs   c. chemoautotrophs
   b. photoheterotrophs   d. chemoheterotrophs

7. Viruses are _____ .
   a. the simplest living organisms   d. both a and b
   b. infectious particles   e. both b and c
   c. nonliving

8. Viruses have a _____ and a _____ .
   a. DNA core; carbohydrate coat
   b. DNA or RNA core; plasma membrane
   c. DNA-containing nucleus; lipid envelope
   d. DNA or RNA core; protein coat

9. Match the organisms with the most suitable descriptions.
   ____ archaebacteria   a. bacterial cell division mechanism
   ____ eubacteria   b. nonliving infectious particle with
   ____ virus   nucleic acid core and protein coat
   ____ plasmid   c. small circle of bacterial DNA
   ____ prokaryotic   d. methanogens, halophiles,
          fission   extreme thermophiles
                          e. most common prokaryotic cells

Critical Thinking

1. *Salmonella* bacteria, which cause a form of food poisoning, often live in poultry and eggs, but not in newly hatched chicks until chicks eat bacteria-laden feces of healthy adult chickens. Harmless bacteria ingested this way colonize the surface of intestinal cells, leaving no place for *Salmonella* to take hold. Some farmers raise thousands of chicks in confined quarters with no adult chickens. Should they consider feeding the chicks a known mixture of bacteria from a lab or a mixture of unknown bacteria from healthy adult chickens? Devise an experiment to test which approach may be more effective.

2. In 1852 in the Coorong district of Australia, a rubbery substance was discovered near a lake that had formed after a period of heavy rainfall. The substance, later named coorongite, consisted of the same kinds of hydrocarbons as crude oil. The discovery sparked sixty years of feverish oil drilling, even though geologists knew the place wasn't likely to yield oil. Later, biologists wondered if microorganisms had formed the substance. Drilling continued. Then someone found a green, scummy mat of bacteria (*Botyrococcus braunii*) in a natural lagoon. After water around the mat evaporated, the mat congealed into coorongite. Chemical analysis suggests it may be an early stage in the formation of shale oil. Formulate a hypothesis to explain how the Earth's great oil reserves formed.

3. Reflect on the description of mad cow disease in Section 22.9. The FDA is moving to prohibit all protein supplements for sheep, cattle, and other ruminants. Investigate the extent of this practice in the United States and how it has been monitored.

Selected Key Terms

archaebacterium *22.3*   infection *22.9*
bacillus (bacilli) *22.1*   lysis *22.8*
bacterial chromosome *22.2*   lysogenic pathway *22.8*
bacterial conjugation *22.2*   lytic pathway *22.8*
bacterial flagellum *22.1*   methanogen *22.5*
bacteriophage *22.7*   microorganism *CI*
cell wall *22.1*   numerical taxonomy *22.3*
coccus (cocci) *22.1*   pandemic *22.9*
disease *22.9*   pathogen *CI*
emerging pathogen *22.9*   pilus (pili) *22.1*
endospore *22.4*   plasmid *22.2*
epidemic *22.9*   prion *22.7*
eubacterium *22.3*   prokaryotic cell *22.1*
extreme thermophile *22.5*   prokaryotic fission *22.2*
fruiting body *22.4*   spirillum (spirilla) *22.1*
glycocalyx *22.1*   strain (bacterial) *22.6*
Gram stain *22.1*   viroid *22.7*
halophile *22.5*   virus *22.7*
heterocyst *22.4*

Readings

Brock, T., et al. 1994. *Biology of Microorganisms*. Seventh edition. Englewood Cliffs, New Jersey: Prentice-Hall.

Geunno, B. October 1995. "Emerging Viruses." *Scientific American* 273(4): 56–62.

Hively, W. May 1997. "Looking for Life in All the Wrong Places." *Discover*: 76–85. Account of microbes that live in the most extreme environments on and in the Earth.

Web Site See *http://www.wadsworth.com/biology* for practice quiz questions, hypercontents, BioUpdates, and critical thinking. The Wadsworth Biology Resource Center provides a wealth of information fully organized and integrated by chapter.

23 PROTISTANS

Kingdom at the Crossroads

More than 2 billion years ago, in tidal flats and soils, in estuaries, streams, lakes, and lagoons, prokaryotic cells were inconspicuously changing the world. Ever since the time of life's origin, the Earth's atmosphere had been free of oxygen, and anaerobic bacteria had reigned supreme. Their realm was about to shrink, drastically. An oxygen-producing pathway of photosynthesis had been operating in vast populations of cells. Gaseous oxygen had been dissolving in aquatic habitats and escaping into the air, and its atmospheric concentration was beginning to approach its current level.

The oxygen-enriched atmosphere was a selection pressure of global dimensions. Thereafter, anaerobic species that could not neutralize the highly reactive, potentially lethal gas were restricted to black muds, stagnant waters, and other anaerobic habitats. Others developed oxygen tolerance. It must have been a short evolutionary step from having a capacity to neutralize oxygen to using it in metabolic reactions; some aerobic species emerged in nearly all bacterial groups.

This metabolic innovation was the start of rampant competition for resources. In the presence of so much free oxygen, energy-rich organic compounds could no longer accumulate by geochemical processes. Organic compounds formed by living cells became the premier source of carbon and energy. Through evolutionary experimentation, novel ways of acquiring and using the compounds developed. Many diverse bacteria engaged in new kinds of partnerships, predations, and parasitic interactions, as outlined in Section 21.4. In less than 500 million years, some of the experiments led to the first eukaryotic cells. And in short order, eukaryotic lineages were branching in different directions (Figure 23.1).

Of all existing organisms, **protistans** are most like the earliest, structurally simple eukaryotes. Although comparative biochemistry is now yielding insights into their phylogenies, it is still said that protistans are most easily classified by what they are *not*. In later chapters, you will read about structural and functional traits that distinguish protistans from plants, fungi, and animals. By a process of elimination, "protistans" are organisms that do not show these traits and that are not bacteria.

Like all other eukaryotes, protistans differ from the bacteria in several respects. At the least, they have a nucleus, large ribosomes, mitochondria, endoplasmic reticulum, and Golgi bodies. Their chromosomes are DNA molecules with a great many histones and other proteins. Protistans assemble microtubules for use in a cytoskeleton, in spindles that move chromosomes, and in the 9 + 2 core of flagella or cilia. Many species have chloroplasts. And depending on the species, protistans reproduce by way of mitosis, meiosis, or both.

Figure 23.1 A sampling of the confounding diversity among protistans. (**a**) Thriving in the surf pounding against a rocky shoreline, a stand of *Postelsia palmaeformis*, one of the multicelled brown algae. (**b**) *Micrasterias*, a single-celled freshwater green alga, here dividing in two. (**c**) *Physarum*, a plasmodial slime mold. This aggregation of yellow-gold cells is migrating along a rotting log. (**d**) Mealtime for *Didinium*, a ciliated protozoan with a big mouth. *Paramecium*, a different ciliated protozoan, is poised at the mouth (*left*) and swallowed (*right*).

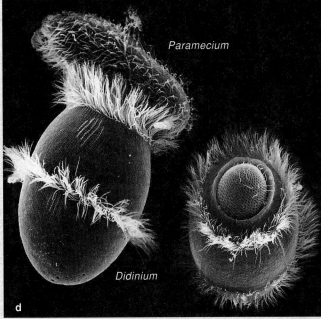

Paramecium

Didinium

50 µm

The photosynthetic types range from microscopic single cells to giant seaweeds. The saprobes resemble some bacteria and fungi. Some of the predators and parasites resemble animals. The vast majority are single-celled, yet nearly every lineage has multicelled forms, and some have very close evolutionary ties to other kingdoms. This chapter introduces the major lineages. As you will see, the chytrids, water molds, slime molds, protozoans, and sporozoans are heterotrophs. Among the euglenoids, chrysophytes, and dinoflagellates are species that are photosynthetic, heterotrophic, or both. Most red, brown, and green algae are evolutionarily committed to photosynthesis.

Opinions differ on how to classify many of these confoundingly diverse organisms. For example, many biologists view the multicelled algae as protistans, and about as many view them as plants. Don't worry about memorizing where the lines of classification schemes are being drawn. Instead, simply become familiar with the groups themselves. In time, biochemical studies will help bring the evolutionary picture into sharper focus.

KEY CONCEPTS

1. The kingdom Protista includes all of the free-living eukaryotic cells and a number of the structurally simple, multicelled species. Some protistan groups have close evolutionary ties to other kingdoms.

2. Protistans are easily distinguishable from prokaryotes (the bacteria) but are difficult to classify with respect to other eukaryotes. They also differ enormously from one another in their morphology and life-styles.

3. Among the fungus-like protistans are parasitic and predatory molds that produce spores. The majority of the chytrids and water molds are single-celled decomposers in aquatic habitats. The phagocytic slime molds live as single amoeboid cells and as aggregations of cells that migrate together and form spore-producing structures.

4. The animal-like protistans include nonphotosynthetic flagellated protozoans, sporozoans, and ciliates. Among their ranks are single-celled predators, grazers, and parasites. In any given year, hundreds of millions of people are affected by infections caused by a few dozen species of flagellated protozoans and sporozoans.

5. Among the plant-like protistans are single-celled, photosynthetic flagellates. Astounding numbers of them live freely or as colonies in the open ocean and other bodies of water. These protistans are important members of phytoplankton, the food producers that are the start of nearly all food webs of aquatic habitats.

6. The protistans most like plants are the red, brown, and green algae; indeed, many botanists prefer placing them in the plant kingdom. Most species are photosynthetic. They range from single-celled to multicelled forms that show great diversity in size, morphology, life-styles, reproductive modes, and habitats.

PARASITIC AND PREDATORY MOLDS

Three groups of protistans have members that resemble fungi in some respects and were once classified with them. These groups are the **chytrids**, **water molds**, and **slime molds**. Like the fungi, all produce spore-bearing structures, and all are heterotrophs. Like fungi, many of their species are saprobic decomposers or parasites. **Saprobes** secrete digestive enzymes that break down organic compounds produced by other organisms, then they absorb the breakdown products. Unlike fungi, the slime molds are phagocytic predators. Also, members of all three groups differ from the fungi in producing motile cells during the life cycle.

Chytrids

The chytrids (Chytridiomycota) are common in marine and freshwater habitats, where they absorb nutrients from plant debris or from necrotic plant tissues. Most of the 575 species are saprobic decomposers, but some are parasites.

When chytrids grow in damp environments, their populations become fuzzy, pale masses that resemble fungal molds. Chytrids are actually similar to fungi in some biochemical respects. Also, chitin reinforces the cell wall of some species, as it does in fungi.

Single-celled species of chytrids produce flagellated asexual spores. When their spores are released into the surroundings, they settle on a host cell, germinate, then develop into globular cells having rootlike absorptive

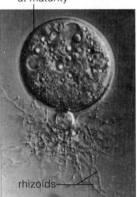

globe-shaped cell that will be a spore-producing structure at maturity

rhizoids

25 µm

Figure 23.2 One of the common chytrids, *Chytridium confervaie.*

structures called rhizoids (Figure 23.2). At maturity, the globe-shaped cells become spore-producing structures.

As in many fungi, complex multicelled species have a mesh of fine absorptive filaments called a **mycelium** (plural, mycelia). Walls of individual cells usually do not cut across the filaments, and cytoplasmic streaming throughout the mycelium distributes enzymes and the nutrients from digested food (compare Section 4.9).

Water Molds

Water molds (Oomycota) may be distantly related to the yellow-green and brown algae. Most of the 580 known types are key saprobic decomposers in aquatic habitats. Of these, many are free-living species that get nutrients from plant debris in ponds, lakes, and streams. Others live inside necrotic tissues of living plants. Some water molds parasitize aquatic organisms. The pale, cottony growths you may have seen on some aquarium fish are mycelia of a parasitic type, *Saprolegnia* (Figure 23.3a).

Most species of water molds produce an extensive mycelium, and some of its filaments differentiate into gamete-producing structures. At fertilization, a male and female gamete fuse and thus form a diploid zygote, which develops into a thick-walled resting spore. After germination, a new mycelium develops from the spore. Water molds also produce flagellated, asexual spores.

Some water molds are major plant pathogens. For example, grapes with *downy mildew* have been attacked by *Plasmopara viticola* (Figure 23.3b). Another pathogen seriously influenced human affairs. Over a century ago, Irish peasants grew potatoes as their main food crop. Between 1845 and 1860, the growing seasons were cool and damp, year after year, and conditions encouraged the rapid spread of *Phytophthora infestans*. This water mold causes *late blight*, a rotting of potato (and tomato) plants. Its abundant spores were dispersed unimpeded through the watery film on the plants. Destruction was rampant. During a fifteen-year period, one-third of the population in Ireland starved to death, died during the outbreak of typhoid fever that followed as a secondary effect, or fled to the United States and other countries.

Slime Molds

All slime molds produce free-living, amoebalike cells during part of the life cycle. The group includes *cellular* slime molds (Acrasiomycota) and *plasmodial* slime molds (Myxomycota). Their cells crawl on rotting plant parts, such as decaying leaves and bark. Like true amoebas, they are phagocytic predators. They engulf bacteria and yeasts, spores, and various organic compounds. When nutrients are scarce and cells are starving, many of the cells aggregate and form a slimy mass that may migrate

Figure 23.3 Effects of two water molds. (**a**) The parasitic water mold *Saprolegnia* has destroyed tissues of this aquarium fish tail. (**b**) *Plasmopara viticola* causes downy mildew in grapes. At times it has threatened huge vineyards in Europe and North America.

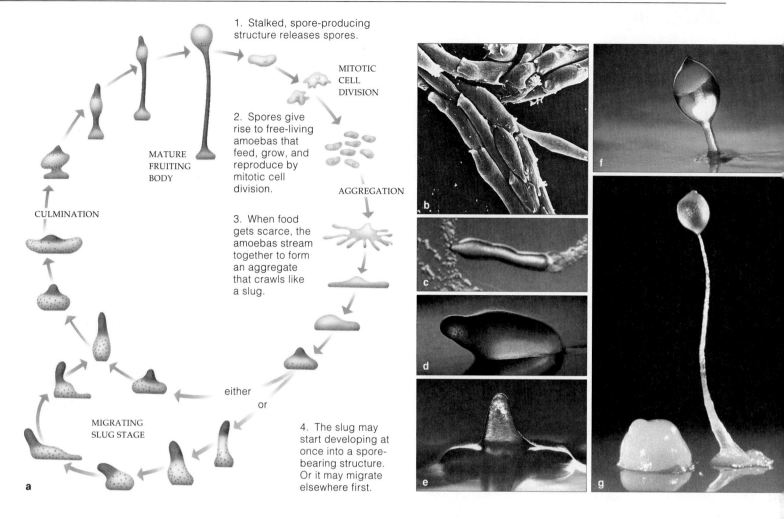

1. Stalked, spore-producing structure releases spores.

MITOTIC CELL DIVISION

2. Spores give rise to free-living amoebas that feed, grow, and reproduce by mitotic cell division.

AGGREGATION

3. When food gets scarce, the amoebas stream together to form an aggregate that crawls like a slug.

MATURE FRUITING BODY

CULMINATION

either

or

4. The slug may start developing at once into a spore-bearing structure. Or it may migrate elsewhere first.

MIGRATING SLUG STAGE

a

Figure 23.4 *Dictyostelium discoideum*, a cellular slime mold. (**a**) The life cycle includes a spore-producing stage. The spores give rise to free-living amoebas, which grow and divide until food (soil bacteria) dwindles. In response to cyclic AMP, a chemical signal that they secrete, amoebas stream toward one another. As the amoebas aggregate, their plasma membranes become sticky and they adhere to one another. A cellulose sheath forms around the aggregation, which starts crawling like a slug (**b–d**). Some slugs may incorporate 100,000 or more amoebas.

As a slug migrates, the amoebas develop into different types of cells: prestalk (*red*), prespore (*white*), and anteriorlike cells (*brown dots*). Prestalk cells secrete ammonia in amounts that vary with temperature and light intensity. The slug moves most rapidly in response to intermediate amounts of ammonia (not too little, not too much), which correspond to warm, moist conditions (not too cold or hot, not soppy or dry). Where such conditions exist at the surface of a substrate, prestalk cells and prespore cells differentiate and form a stalked, spore-bearing structure (**e–g**). Anteriorlike cells, which sort into two groups, may function in elevating the nonmotile spores for dispersal from the top of the spore-bearing structure.

to a new location where conditions are more favorable for growth. Collectively, the contractions of single cells move the mass. Later in the life cycle, the amoebalike cells develop into a few cell types that differentiate and

form a spore-bearing structure. After the spores are released, they germinate on warm, damp surfaces. Each gives rise to one amoebalike cell. Sexual reproduction by way of gametes also is common among slime molds.

Dictyostelium discoideum, one of 70 species of cellular slime molds, is commonly used for laboratory studies of development. Figure 23.4 shows its life cycle. Figure 23.1*c* shows one of the 500 species of plasmodial slime molds. When amoebalike cells of the members of that phylum aggregate, each cell's plasma membrane breaks down. Cytoplasm flows unimpeded and so distributes nutrients and oxygen through the mass. The streaming mass (the plasmodium) often will occupy several square meters and migrate if food runs out. You may have seen one crossing a lawn or a road, or even climbing a tree.

Most chytrids and water molds are saprobic decomposers of aquatic habitats. Some are single cells. Like slime molds, many are free-living predators some of the time and simple experiments in multicellularity at other times.

During a slime mold life cycle, amoeboid cells aggregate to form a migrating mass. Cells in the mass differentiate, then they form reproductive structures and spores or gametes.

23.2 THE ANIMAL-LIKE PROTISTANS

About 65,000 named species of single-celled protistans are informally called the **protozoans** ("first animals"), because they are thought to resemble the single-celled, heterotrophic protistans that gave rise to animals. In the next four sections we limit our sampling to amoebas, animal-like flagellates, sporozoans, and ciliates. Among their ranks are predators, parasites, and grazers that actively move about to secure a meal (Table 23.1).

Table 23.1 Major Groups of Animal-Like Protistans

SARCODINA

Amoeboid protozoans. Soft or shelled bodies; locomotion by pseudopods. Free-living or endosymbiotic heterotrophs.

> RHIZOPODS. Naked amoebas, foraminiferans.

> ACTINOPODS. Radiolarians, heliozoans

MASTIGOPHORA

Animal-like flagellates. Includes free-living and parasitic heterotrophs. Locomotion by one or more flagella.

APICOMPLEXA

Parasitic heterotrophs with a complex of structures, such as rings, tubules, and cones, at the head end. The most familiar members of a group commonly called sporozoans.

CILIOPHORA

Diverse free-living, sessile, and motile heterotrophs with numerous cilia. Predators or symbionts, some parasitic.

Free-living species live in damp soil, in ponds, lakes, and other freshwater habitats, and in marine habitats. The symbionts and parasites live in or on moist tissues of other organisms. Some species are major pathogens.

Protozoans reproduce either asexually or sexually, although many species alternate between reproductive modes in response to environmental conditions. Most commonly, asexual reproduction is by **binary fission**, a process by which the individual's body divides in two. The division plane is random for amoebas, longitudinal for flagellates, and transverse for ciliates. Budding from a parent organism also occurs. Some species undergo **multiple fission**. More than two nuclei form, then each nucleus and a bit of cytoplasm separate as a daughter cell. Many parasitic types go through an encysted stage. The **cyst**, a resistant body covering made of their own secretions, helps them wait out stressful conditions.

Protozoans are single-celled protistans that are animal-like in some respects, as in the way they actively move about to secure their food. They include predators, parasites, and grazers. Some species are pathogens that cause serious diseases in humans and other animals.

23.3 AMOEBOID PROTOZOANS

Let's first look at **amoeboid protozoans** (Sarcodina), which are naked amoebas, foraminiferans, heliozoans, and radiolarians. None of them has permanent motile structures. Each moves by a combination of cytoplasmic streaming and formation of **pseudopods** ("false feet"). As described in Section 4.9, pseudopods are temporary cytoplasmic extensions of the cell body. Figure 23.5 shows the rather thick pseudopods of a naked amoeba.

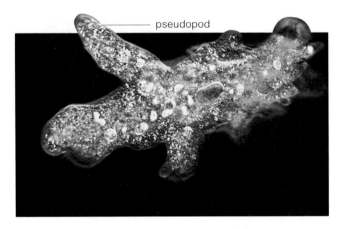
— pseudopod

Figure 23.5 *Amoeba proteus*, one of the naked amoebas. Compared with most other kinds of amoeboid protozoans, the pseudopods projecting from its body are stubbier. This species is a favorite for laboratory experiments in biology classes.

Rhizopods

The naked amoebas and foraminiferans are members of a group known as **rhizopods**. The constantly changing cytoskeleton alone lends structural support to a naked amoeba's soft body, which never shows any symmetry. You find these protistans in damp soil, freshwater, and saltwater. Most species are free-living phagocytes that engulf other protozoans and bacteria. Some live in the gut of certain invertebrates or vertebrates and normally do no harm. But a few are opportunistic parasites, and so are certain free-living species that happen to enter the gut with food or water. When conditions in the gut fan their population growth, these types cause intestinal problems.

Entamoeba histolytica is a pathogenic type that must complete its life cycle in human intestines. The infective stage becomes encysted and leaves the body in feces. The encysted stage is inactive in water or soil. After it enters a new host, the trophozoite—the active, motile feeding stage—emerges from it. Trophozoites multiply rapidly by binary fission and release lysozymes, which destroy the cells making up the lining of the colon and rectum. Painful abdominal cramps and diarrhea are symptoms of this infectious disease, called *amoebic dysentery*.

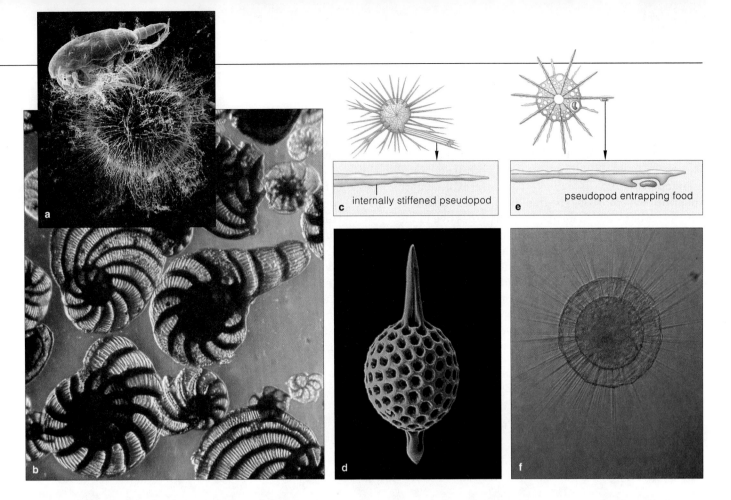

Figure 23.6 (**a**) Foraminiferan with prey. (**b,c**) Foraminiferan skeletons and body plan. (**d**) Radiolarian skeleton. (**e,f**) Body plan and micrograph of a living heliozoan. As in radiolarians, a core of microtubules reinforces the many fine pseudopods.

Amoebic dysentery is most prevalent in subtropical and tropical regions with contaminated water supplies and poor sanitation. Where cyst-harboring water, fruits, or vegetables abound, between 10 and 50 percent of the human population may be infected. The severity of an infection depends on the virulence of the *E. histolytica* strain and the state of an individual's immune system. People who recover show full or partial resistance to subsequent infection. Worldwide, amoebic dysentery is a leading cause of death among infants and toddlers, whose immune systems are not yet well developed.

Unlike naked amoebas, the foraminiferans have a highly perforated external skeleton of secreted organic substances that are hardened with calcium carbonate (Figure 23.6 *a–c*). Thin pseudopods extend through the perforations, and prey becomes trapped in mucus that covers them. Most foraminiferans live on the seafloor. All but 1 percent of the named species are extinct. The fossilized, compacted shells of countless foraminiferans make up distinctive sedimentary rock layers, including the famed white cliffs of Dover, England. We now use their calcified remains in cement and blackboard chalk.

Actinopods

Actinopods ("ray feet") include the radiolarians and heliozoans. Their name refers to the numerous slender, reinforced pseudopods that radiate from the body. Like foraminiferans, the radiolarians are well represented in the fossil record, for they have ornate, silica-hardened parts that resist degradation (Figure 23.6 *d*). Many of these skeletal parts are thin and project outward with the stiffened pseudopods. Nearly all radiolarians drift with ocean currents, as part of **plankton**. The word refers to aquatic communities of passively drifting or motile organisms, mostly microscopic in size. A few species of radiolarians form colonies in which numerous cells are cemented together.

Heliozoans ("sun animals") have many pseudopods radiating like the sun's rays (Figure 23.6 *e,f*). Most are free-floating or bottom-dwellers in freshwater habitats. As in radiolarians, a membrane or capsule divides the single-celled body into two biochemically distinct zones. The inner zone contains the nucleus. The outer zone has so many vacuoles in which digestion occurs that it appears almost frothy. The vacuoles impart buoyancy to the cell, which can remain suspended in the water.

Amoeboid protozoans are soft-bodied single cells, some with hardened skeletal elements. All form pseudopods for use in motility and prey capture. Free-living and gut-dwelling species, some parasitic, belong to this group.

Animal-Like Flagellates

Animal-like flagellates (Mastigophora) are free-living predators and parasites bearing one to several flagella. Members of one group are equipped with flagella *and* pseudopods, which is evidence of close evolutionary links with the amoebas. The free-living types abound in freshwater or marine habitats. Parasitic types live in the moist tissues of plants and animals, including humans.

Giardia lamblia (Figure 23.7*a*) is a notorious internal parasite. In any given year, it may infect as many as 10 percent of the human population in the United States alone. It usually causes mild intestinal upsets, such as diarrhea, but it has severe effects on a few susceptible people. Wild animals and foraging cattle often contain the parasite. It leaves the body encysted in feces, then

Representative Sporozoans

Sporozoan is an informal name for parasitic protistans that complete part of the life cycle *inside specific cells of host organisms*. These parasites form sporozoites, a type of motile infective stage; some have encysted stages. At one end of the body is a complex of structures that function in penetrating host cells. The sporozoans are often grouped as phylum Apicomplexa.

Many sporozoans cause human diseases. *Plasmodium* causes malaria (Section 23.5). *Cryptosporidium* strains that cause serious intestinal disorders now contaminate most water supplies. One outbreak in 1993 hospitalized thousands of people in Milwaukee. Symptoms normally subside after ten days. However, people with weakened immune systems can suffer for months or years.

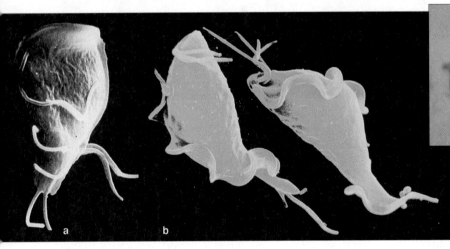

trypanosome   red blood cell

Figure 23.7 Three notorious animal-like flagellates. (**a**) *Giardia lamblia* causes severe intestinal ailments. (**b**) *Trichomonas vaginalis* causes trichomoniasis, one of the sexually transmitted diseases. (**c**) *Trypanosoma brucei* causes African sleeping sickness.

infects people who drink water or ingest food tainted with the feces. Because even remote streams may harbor cysts, streamwater should be boiled before being drunk.

Many trichomonads are among the parasitic members of the group. *Trichomonas vaginalis*, a worldwide nuisance, can infect new human hosts during sexual intercourse (Figure 23.7*b*). Unless trichomonad infections are treated, they can damage the urinary and reproductive tracts.

Trypanosomes also are parasites. One, *Trypanosoma brucei*, causes *African sleeping sickness* (Figure 23.7*c*). Bites of the tsetse fly transmit it to hosts, where it may damage the nervous system. *T. cruzi*, prevalent in South America and Mexico, causes *Chagas disease*. Bugs pick up the parasite when they feed on infected humans and other animals. The parasite multiplies in the insect gut, then it may be excreted onto a host. Scratches in skin invite infection. The liver and spleen enlarge, eyelids and the face swell up, then the brain and heart become severely damaged. There is no treatment or cure.

Toxoplasma completes the sexual phase of its life cycle in cats and asexual phases in humans, cattle, and other animals. Flies and cockroaches can spread cysts from cat feces to the table. Humans also can get infected if they eat undercooked or raw meat that contains cysts. The resulting disease, *toxoplasmosis*, is not prevalent, but an infection can cause birth defects. Pregnant women who become infected may transmit the parasite to the fetus, which may suffer brain damage or die. They should never empty litterboxes or clean up after cats.

Fewer than two dozen species of the animal-like flagellates and sporozoans cause serious diseases in humans. But they infect hundreds of millions each year. There are no effective vaccines against them.

Animal-like flagellates are motile heterotrophs, free-living or parasitic. Sporozoans are internal parasites that produce infective, motile stages. Members of both phyla often form encysted stages.

23.5 MALARIA AND THE NIGHT-FEEDING MOSQUITOES

Each year in Africa alone, about a million people die of *malaria*, a long-lasting disease that four different parasitic species of *Plasmodium* can cause. This type of sporozoan currently has infected more than 100 million people.

Only female mosquitoes of the genus *Anopheles* can transmit the parasites to human hosts. Their bite delivers sporozoites (the infective, motile stage) to the host's blood. The bloodstream transports sporozoites to the liver, where they undergo multiple fission. Some of the resulting cells, called merozoites, penetrate and asexually reproduce in red blood cells, which they destroy. Others develop into male and female gametocytes, which will mature into gametes (Figure 23.8).

Symptoms of malaria begin after infected cells abruptly rupture and release merozoites, metabolic wastes, and cellular debris into the individual's bloodstream. Shaking, chills, a burning fever, and drenching sweats are classic symptoms. After one episode, symptoms subside for a few weeks or months. Infected individuals might even feel normal, but relapses inevitably recur. In time, anemia and gross enlargement of the liver and the spleen may result.

Amazingly, the *Plasmodium* life cycle is sensitive to the body temperature and oxygen levels inside humans and mosquitoes. The gametocytes cannot mature in humans, who are warm-bodied and have little free oxygen in their bloodstream (most is bound to hemoglobin). They are induced to mature inside the mosquito, which has a lower body temperature and which incidentally slurps in oxygen from the air along with the gametocyte-containing blood.

Mature gametes fuse to form zygotes. These repeatedly divide and form many sporozoites, which migrate to the female mosquito's salivary glands and await the next bite.

Malaria has been around for a long time. People were describing its symptoms more than 2,000 years ago. It received its name in the seventeenth century, when Italians made a connection between the disease and noxious gases from swamps near Rome, where mosquitoes flourished (*mal*, bad; *aria*, air). By severely incapacitating so many people, malaria contributed to the decline of the ancient empires of Greece and Rome. Much later in history, it also incapacitated soldiers during the United States Civil War, World War II, and the Korean and Vietnam conflicts.

Throughout human history, malaria has been most prevalent in tropical and subtropical parts of Africa. Today, however, the numbers of cases reported in North America and elsewhere are increasing dramatically, owing to the hordes of globe-hopping travelers and unprecedented levels of human immigration.

Travelers who intend to visit countries with high rates of malaria are advised to use antimalarial drugs such as chloroquine. However, certain strains of *Plasmodium* are now resistant to the drugs, and a vaccine has been difficult to develop. *Vaccines* are preparations that induce the body to build up resistance to a specific pathogen. Experimental vaccines for malaria are not equally effective against all the different stages that develop during sporozoan life cycles. This is generally the case for vaccines that researchers hope to develop against most parasites with complex life cycles.

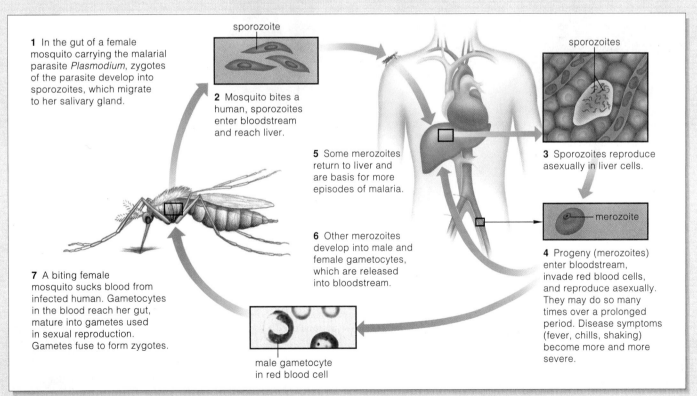

1 In the gut of a female mosquito carrying the malarial parasite *Plasmodium*, zygotes of the parasite develop into sporozoites, which migrate to her salivary gland.

sporozoite

2 Mosquito bites a human, sporozoites enter bloodstream and reach liver.

5 Some merozoites return to liver and are basis for more episodes of malaria.

6 Other merozoites develop into male and female gametocytes, which are released into bloodstream.

7 A biting female mosquito sucks blood from infected human. Gametocytes in the blood reach her gut, mature into gametes used in sexual reproduction. Gametes fuse to form zygotes.

male gametocyte in red blood cell

sporozoites

3 Sporozoites reproduce asexually in liver cells.

merozoite

4 Progeny (merozoites) enter bloodstream, invade red blood cells, and reproduce asexually. They may do so many times over a prolonged period. Disease symptoms (fever, chills, shaking) become more and more severe.

Figure 23.8 Life cycle of one of the sporozoans (*Plasmodium*) that causes malaria.

A SAMPLING OF CILIATED PROTOZOANS

Ciliated protozoans (Ciliophora) typically have profuse arrays of cilia at their surface. You read about these fine motile structures in Section 4.9. Most ciliates use them for swimming through freshwater and marine habitats, where they prey on bacteria, tiny algae, and one another, as Figure 23.1*d* so aptly suggested. The cilia beat in such a synchronized pattern over the body surface, they call to mind a soft wind through a field of tall grasses.

Paramecium is typical of the group (Figure 23.9). A fully grown cell is about 150 to 200 micrometers long. Outer membranes form a **pellicle**, a body covering that may be rigid or quite flexible, depending on how those membranes are organized. Inside the gullet, which starts at an oral depression at the body surface, some cilia sweep food-laden water into the cell body. There, the food becomes enclosed in enzyme-filled vesicles and is digested. Like the amoeboid protozoans, *Paramecium*'s internal solute concentrations are greater than those of its surroundings. Therefore, it must continually counter

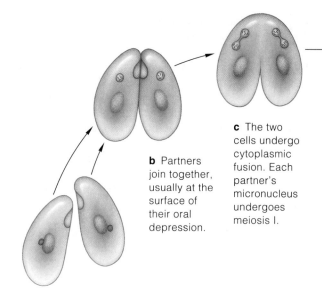

c The two cells undergo cytoplasmic fusion. Each partner's micronucleus undergoes meiosis I.

b Partners join together, usually at the surface of their oral depression.

a Prospective partners meet up.

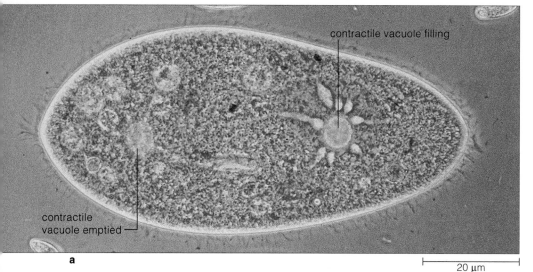

contractile vacuole filling

contractile vacuole emptied

a

20 μm

Figure 23.9 Representatives of two groups of ciliated protozoans.

Left and below: Paramecia. (**a**) Light micrograph of a living paramecium. (**b**) Generalized diagram of the body plan for the genus *Paramecium*. (**c**) Components of the pellicle. Trichocysts, which span the pellicle, are organelles that discharge threads when the cell is irritated. They may be a defense against predators.

Facing page: (**d**) Scanning electron micrograph of the arrays of cilia at the surface of these single-celled predators. Hypotrichs. (**e**) This free-living species lives in the Bahamas. The hypotrichs are the most animal-like of the ciliates. They run around on leglike tufts of cilia. Some have a "head" end with modified sensory cilia.

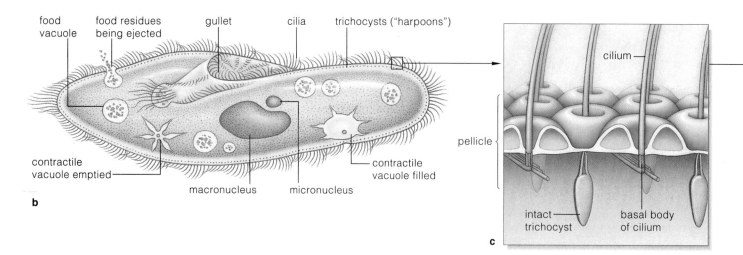

food vacuole | food residues being ejected | gullet | cilia | trichocysts ("harpoons")

contractile vacuole emptied

macronucleus | micronucleus

contractile vacuole filled

b

pellicle

cilium

intact trichocyst | basal body of cilium

c

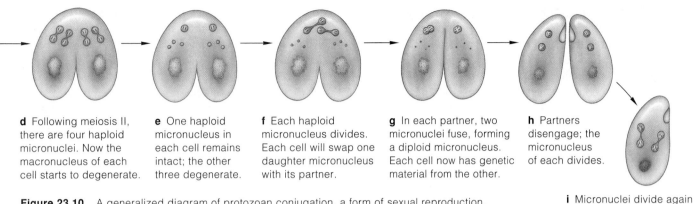

d Following meiosis II, there are four haploid micronuclei. Now the macronucleus of each cell starts to degenerate.

e One haploid micronucleus in each cell remains intact; the other three degenerate.

f Each haploid micronucleus divides. Each cell will swap one daughter micronucleus with its partner.

g In each partner, two micronuclei fuse, forming a diploid micronucleus. Each cell now has genetic material from the other.

h Partners disengage; the micronucleus of each divides.

i Micronuclei divide again in each cell; the original macronucleus degenerates.

Figure 23.10 A generalized diagram of protozoan conjugation, a form of sexual reproduction, as demonstrated by the two ciliates that first encounter each other in (**a**).

water's tendency to diffuse inward by osmosis. (Here you may wish to refer to Section 5.4.) Like the amoebas, it depends on **contractile vacuoles**. Tiny tubes extend from the center of these organelles and collect excess water that moves osmotically into the cell body. A filled vacuole contracts and forces water through a small pore to the outside.

As is true of other protistan groups, the ciliates show diversity in life-styles. Like *Paramecium*, about 65 percent of the known species are free-living and motile. Others attach permanently or temporarily to substrates, often by stalks. Some form colonies. About 30 percent live in or on other organisms, as symbionts. *Balantidium coli* is the largest protozoan parasite of humans, with effects like those of *Entamoeba histolytica* infections.

Ciliates can reproduce sexually and asexually, and things get interesting because each cell commonly has

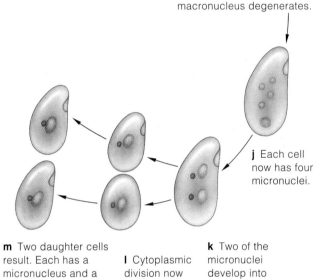

j Each cell now has four micronuclei.

k Two of the micronuclei develop into macronuclei.

l Cytoplasmic division now begins.

m Two daughter cells result. Each has a micronucleus and a macronucleus.

two types of nuclei. When a cell reproduces asexually by binary fission, a small, diploid *micro*nucleus divides by mitosis. And the large *macro*nucleus lengthens and splits in two a bit sloppily; some of the DNA may spill out. Most often, sexual reproduction is by a unique form of **conjugation**, as shown in Figure 23.10. In brief, partner ciliates repeatedly divide their micronuclei, swap two daughter micronuclei, and allow two others to fuse and form a diploid macronucleus to replace the one that disappears. And you probably thought sex among the single-celled critters was simple!

Ciliated protozoans bear a profusion of cilia. These motile structures are used for swimming and for beating food into an oral cavity. Each ciliate usually has two types of nuclei, which are distributed to daughter cells in unique ways during asexual and sexual reproduction.

Like other protistans, the ciliates show great diversity in life-styles and diverse variations on the basic body plan.

e

d

The rest of this chapter surveys the protistans, largely photosynthetic, that are informally called "the algae." Let's first consider the euglenoids, chrysophytes, and dinoflagellates. Single-celled species dominate all three groups, which is not the case for the red, brown, and green algae (Table 23.2). The majority are members of **phytoplankton**: communities of aquatic photosynthetic species, mainly microscopic, that drift or swim weakly through water. As food producers, they are the basis for nearly all food webs in aquatic habitats. Diverse species occur in great numbers in inland seas, rivers, streams, lakes and ponds, and the open ocean (Section 7.8).

Euglenoids

The **euglenoids** (Euglenophyta) are a classic example of evolutionary experimentation. Euglenoids are free-living, flagellated cells that abound in freshwater or stagnant ponds and lakes. Most of the 1,000 known species are *photosynthetic*. The rest are *heterotrophs*; they subsist on organic compounds dissolved in the water.

Euglena has a profusion of organelles (Figure 23.11). Among these are chloroplasts with chlorophylls *a* and *b* and carotenoids, just like plant chloroplasts. *Euglena* has flagella (one long, one short) and a contractile vacuole, as animal-like protozoans do. Like some protozoans, it has a pellicle, this one being a flexible cover with spiral strips of a translucent, protein-rich material. Pigments form an "eyespot" that partly shields a light-sensitive receptor. By moving its long flagellum, the single cell keeps the receptor exposed to light and so stays where the light is most suitable for its activities.

Generally, "self-feeders" make their own vitamins, which are required for growth. But all photosynthetic euglenoids must get vitamin B_{12} from their surroundings (most cannot make vitamin B_1, either). Probably their chloroplasts originated from endosymbiotic green algae, in the manner described in Section 21.4. Given their array of traits, could euglenoids be very close relatives of some existing protozoans? Maybe, but probably not. Researchers believe their similarities are more likely an outcome of convergent evolution.

Chrysophytes

Chrysophytes (Chrysophyta) are mainly photosynthetic, free-living cells with chlorophylls *a*, c_1, and c_2. They include diatoms, golden algae, and yellow-green algae.

Most **diatoms** are photosynthetic. The silica shells of the 5,600 existing species are two perforated structures that overlap like a pillbox (Figure 23.12a). Substances move to and from the cell's plasma membrane through numerous perforations in the shell. For 100 million years, the shells of at least 35,000 species of now-extinct diatoms accumulated on the bottom of lakes and seas, and many sediments now contain their finely crumbled shells. We use the deposits in insulation, abrasives, and filters. More than 270,000 metric tons are quarried annually near Lompoc, California.

Among the 500 species of **golden algae** are many species with no obvious cell wall; they have silica scales

Table 23.2 The Mostly Photosynthetic Protistans	
EUGLENOPHYTA	Euglenoids. Free-living cells. Autotrophic *and* heterotrophic.
CHRYSOPHYTA	Free-living cells, mostly autotrophic.
	CHRYSOPHYTES (golden algae, yellow-green algae). Mostly photosynthetic; some are heterotrophs.
	DIATOMS. Mostly photosynthetic; some symbionts with foraminiferans.
PYRRHOPHYTA	Dinoflagellates. Free-living cells. Autotrophic *and* heterotrophic.
RHODOPHYTA	Red algae. Mostly multicelled and photosynthetic.
PHAEOPHYTA	Brown algae. Multicelled; all photosynthetic.
CHLOROPHYTA	Green algae. Mostly multicelled and photosynthetic.

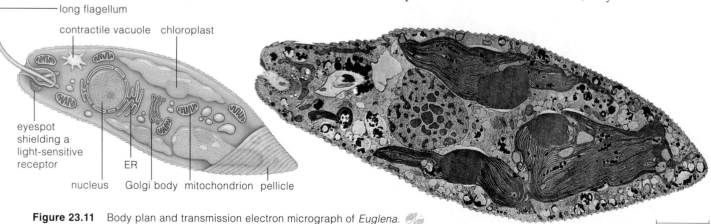

long flagellum

contractile vacuole   chloroplast

eyespot shielding a light-sensitive receptor

nucleus   Golgi body   mitochondrion   pellicle

ER

Figure 23.11   Body plan and transmission electron micrograph of *Euglena*.

5μm

Figure 23.12 Some of the chrysophytes. (**a**) Shells of diatoms. (**b**) *Synura*, a golden alga. Its colonies in phytoplankton have a fishy odor. (**c**) A colonial type of yellow-green alga, *Mischococcus*, which occurs in phytoplankton. (**d**) *Vaucheria*, a yellow-green alga, growing on red sandstone in Arizona.

Figure 23.13 (**a**) Red tide in a central California bay. (**b**) Part of a fish kill that resulted from a dinoflagellate "bloom." (**c**) Scanning electron micrograph of one dinoflagellate, *Gymnodinium breve*, that causes red tides along Florida's coast.

or skeletal elements. As Figure 23.12*b* suggests, their chlorophylls are masked by a golden-brown carotenoid, fucoxanthin. Except for their chloroplasts, some of the amoeboid species closely resemble the amoebas.

Fucoxanthin is absent from the 600 known species of **yellow-green algae** (Figure 23.12*c,d*). Most species are nonmotile, but they have flagellated gametes. Yellow-green algae are common in many aquatic habitats.

Dinoflagellates

More than 1,200 species of **dinoflagellates** live in marine phytoplankton. Most are photosynthetic but many are heterotrophs. Some have flagella and cellulose plates. Dinoflagellates are yellow-green, green, brown, blue, or red, depending on their pigments and endosymbiotic

history. Some undergo population explosions and color the seas red or brown. A few produce a toxin, so the resulting **red tides** can be devastating (Figure 23.13). The toxin accumulates in tissues of mollusks and fishes, and it can poison humans who eat them. Such "algal blooms" may be causing massive die-offs of fish-eating migratory birds in California's Salton Sea, including 150,000 eared grebes in 1992–1993 and nearly 5,000 brown pelicans, an endangered species, in 1996. Raw sewage and fertilizer-enriched runoff from croplands fan such algal blooms, which are worldwide occurrences.

Nearly all single-celled photosynthetic protistans, including most euglenoids, chrysophytes, and dinoflagellates, belong to phytoplankton—the food-producing base of aquatic habitats.

Of 4,100 known species of **red algae** (Rhodophyta), nearly all are marine; only 200 live in freshwater. Red algae are most abundant in warm currents and tropical seas, often at surprising depths (265 meters below the surface) in clear water. A few occur in phytoplankton. Some encrusting types contribute to the formation of coral reefs and banks (Section 49.12).

Cells of some red algae have walls hardened with calcium carbonate. Different species appear red, green, purple, or nearly black, depending on which accessory pigments (phycobilins, mostly) mask their chlorophyll *a*. Phycobilins are good at absorbing the green and blue-green wavelengths, which can penetrate deep waters. (Chlorophylls are more efficient at absorbing red and blue wavelengths, which may not reach as far below the water's surface.) The chloroplasts of red algae resemble cyanobacteria, which suggests endosymbiotic origins.

The life cycle of most species includes multicelled stages, but these lack tissues or organs (Figure 23.14). Reproductive modes are diverse, with complex asexual and sexual phases. Figure 23.15 shows one example.

Red algae have a flexible, slippery texture owing to mucous material in their cell walls. Agar is made from extracts of wall material of several species. This inert, gelatinous substance is used as a moisture-preserving agent in baked goods and cosmetics, as a setting agent

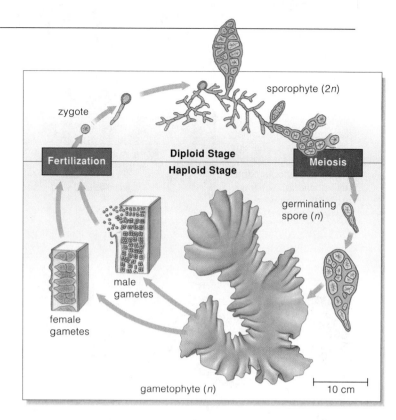

Figure 23.15 Life cycle of *Porphyra*. From 1623 to the 1950s, Japanese fishermen cultivated and harvested a species of this red alga only in the early fall. The rest of the year, it seemed to disappear. Then Kathleen Drew-Baker studied the sheetlike form of *P. umbilicus* in the laboratory. She observed gametes forming in packets interspersed between vegetative cells near the sheet margins; the sheet was a gametophyte (gamete-producing body). She also observed the gametes in a petri dish. The zygotes that formed after fertilization grew on pieces of shell in the dish—and each developed into a tiny branching, filamentous form. This was how *Porphyra* spent most of the year! People already knew about the filamentous form as common pinkish growths on shells. But it was so different from the sheetlike form that it had been classified as a separate species. It is the alga's diploid sporophyte (spore-producing body). Drew-Baker's discovery of these alternating stages in the alga's life history, and the realization that the diploid stage could be grown on shells or other calcium-rich surfaces, revolutionized the nori industry. Within a few years, researchers had worked out the life cycle of *P. tenera*, the harvested species. By 1960, *nori cultivation* was a billion-dollar industry.

for jellies, and as culture gels. We also shape it into soft capsules for delivery of drugs and food supplements. Carrageenan, extracted from *Eucheuma*, is a stabilizer in paints, dairy products, and many other emulsions.

Humans find different species of *Porphyra* tasty as well as nutritious. You may know it from sushi bars as *nori*, a wrapping for rice and fish (Figure 23.15).

Red algae are photosynthetic protistans with phycobilins and other accessory pigments that mask their chlorophyll *a*. Usually, multicelled stages develop during the life cycles. Most species are aquatic. They show great diversity in size, morphology, life-styles, reproductive modes, and habitats.

Figure 23.14 (a) A red alga (*Bonnemaisonia hamifera*). Such a growth pattern (branching, filamentous) is the most common. (b) From a tropical reef, a red alga showing sheetlike growth.

BROWN ALGAE

Walk along a rocky coast at low tide and you may come across brownish seaweeds (Figure 23.16*a*). They are among the 1,500 species of **brown algae** (Phaeophyta). Nearly all thrive in cool or temperate marine waters, from the intertidal zone to the open ocean. Extensive masses of a floating brown alga, *Sargassum,* are major ecosystems in the Sargasso Sea, which lies between the Azores and the Bahamas.

Different species appear olive-green, golden, or dark brown, depending on their

a

array of pigments. Like the chrysophytes, to which they may be related, brown algae contain fucoxanthin and other carotenoids as well as chlorophylls a, c_1, and c_2. They range from microscopic, filamentous *Ectocarpus* to the giant kelps, twenty to thirty meters long. Like the red algae, they have diverse and complex life cycles, with sexual and asexual phases. In their life cycles, too, gametophytes alternate with sporophytes.

Macrocystis, *Laminaria*, and the other giant kelps are the largest, most complex protistans. Their multicelled sporophytes have stipes (stemlike parts), blades (leaflike parts), and holdfasts (anchoring structures). Hollow, gas-filled bladders impart buoyancy to the stipes and blades and keep them upright in the water. Within the stipes are tubelike arrangements of elongated cells. The tubes rapidly carry dissolved sugars and other products of photosynthesis to living cells throughout the body. Flowering plants also transport sugars through similar kinds of tubes. This is a case of convergent evolution in two unrelated groups of large, multicelled organisms.

Giant kelp beds function as productive ecosystems. Think of them as underwater forests within which great numbers of diverse bacteria and protistans, as well as fishes and other animals, carry out their lives. The sizes of the kelp forests are variable, depending on changes in ocean currents and in the abundances of sea urchins (which feed on kelp debris) and sea otters (which feed on sea urchins). For example, in the late 1950s, a warm current displaced the cooler currents off the California coast. *Macrocystis* did not fare well with the temperature shift, and extensive beds off La Jolla and Palos Verdes almost disappeared. Without organic debris for the sea urchins to feed upon, these invertebrates fed directly on the kelps instead. Quantities of quicklime were dumped in the water to reduce the number of sea urchins, and the kelps made a comeback. So did fishes, lobster, abalone,

— bladder

— blade

— stipe

— holdfast

b

Figure 23.16 (**a**) Closer view of *Postelsia*, the brown alga shown in Figure 23.1*a*. You will find it thriving along coasts exposed to heavy surf from Vancouver Island on down to central California. More than a hundred deeply grooved blades top a highly resilient stipe, which is fastened to rocky substrates by a mound of anchoring structures. Its spores never disperse far from the parent sporophyte. At low tide, they simply drip onto rocks from the grooved blades. (**b**) Diagram of *Macrocystis* and an underwater view of a kelp forest. A few species live in the coastal waters of North and South America, New Zealand, Tasmania, and most of the islands in subantarctic waters.

and other species that make the kelp beds their home.

Macrocystis and certain other brown algae are commercially harvested. Extracts from them are used in ice cream, pudding, jellybeans, salad dressings, beer, canned and frozen foods, cough syrup, toothpaste, cosmetics, floor polish, and paper. Alginic acid in the cell walls of some species is used to make algins, which are added to various products as thickening, emulsifying, and suspension agents. In the Far East especially, people harvest kelps as sources of food and mineral salts, and as a fertilizer for crops.

The brown algae range in size from the microscopic to the largest of all of the multicelled protistans. Nearly all species live in marine habitats, ranging from the intertidal zone to the surface waters of the open ocean.

Of all protistans, the **green algae** (Chlorophyta) are structurally and biochemically most like the plants and may be their closest relatives. All are photosynthetic. Like plants, their chlorophylls are the molecular types designated *a* and *b*, and they, too, store carbohydrates as starch grains inside their chloroplasts. The cell walls of some species are composed of cellulose, pectins, and other polysaccharides typical of plants.

Figure 23.17 is a sampling of the more than 7,000 known species of green algae. They include single-celled, colonial, filamentous, sheetlike, and tubular forms. You won't be able to see many species without the aid of a microscope. Most, including the *Micrasterias* cell shown in Figure 23.1*b*, live in freshwater. *Micrasterias* is one of thousands of species of desmids, which are important food producers in nutrient-poor ponds and peat bogs. Green algae also grow at the ocean surface, in marine sediments, just below the surface of soil, and on rocks, tree bark, other organisms, and snow. Some types are symbionts with fungi, protozoans, and a few marine animals. A colonial form (*Volvox*) is a hollow whirling sphere of 500 to 60,000 flagellated cells. Those beautiful white, powdery beaches in tropical regions are largely the work of uncountable numbers of green algal cells

Figure 23.17 Representative green algae. (**a**) One of the marine species (*Codium*) with a pronounced branching form. Although many green algae are microscopic, one member of this genus (*C. magnum*) is taller than you are. In 1956 still another species of *Codium* from Puget Sound, Washington, was accidentally introduced into the Connecticut River where it empties into the Atlantic. In Puget Sound, populations of this alga were kept in check by herbivores that evolved with them. In its new habitat, there were no native *Codium* species or herbivores that were adapted to grazing on it. The introduced alga spread along the coast as far north as Maine and as far south as North Carolina in a little more than two decades.

(**b**) Also from a marine habitat, *Acetabularia*, fancifully called the mermaid's wineglass. Each individual in the cluster is a multinucleate cell mass with a rootlike structure, stalk, and cap in which gametes form.

(**c**) Sea lettuce (*Ulva*) grows in estuaries and attaches to kelps in the seas. Reproductive cells form within and are released from the margins of the sporophyte and gametophyte, both of which have the same form shown in the diagram.

(**d**) *Volvox*, a colony of interdependent green algal cells that bear resemblances to free-living, flagellated cells of the genus *Chlamydomonas*.

(**e**) Above the summer timberline in Utah, phycologist Ron Hoham's footprint in "red snow" reveals the presence of dormant cells of *Chlamydomonas nivalis*, one of the snow algae. Many red accessory pigments protect the chlorophylls of these cells.

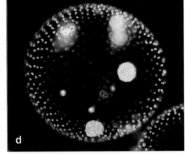

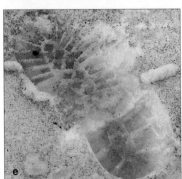

dormant cell

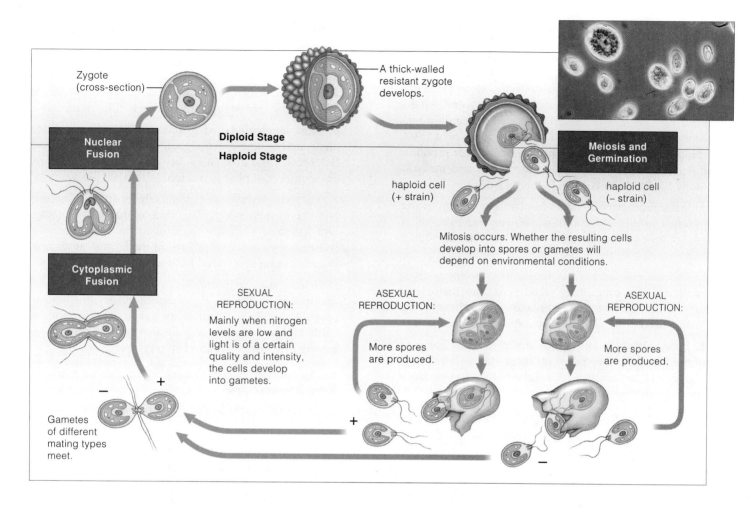

Figure 23.18 Life cycle of a species of *Chlamydomonas*, one of the most common green algae of freshwater habitats. This single-celled species reproduces asexually most of the time and sexually under certain environmental conditions.

Labels within figure:

Zygote (cross-section)

A thick-walled resistant zygote develops.

Diploid Stage

Haploid Stage

Nuclear Fusion

Meiosis and Germination

haploid cell (+ strain)

haploid cell (− strain)

Cytoplasmic Fusion

Mitosis occurs. Whether the resulting cells develop into spores or gametes will depend on environmental conditions.

SEXUAL REPRODUCTION:

Mainly when nitrogen levels are low and light is of a certain quality and intensity, the cells develop into gametes.

ASEXUAL REPRODUCTION:

More spores are produced.

ASEXUAL REPRODUCTION:

More spores are produced.

Gametes of different mating types meet.

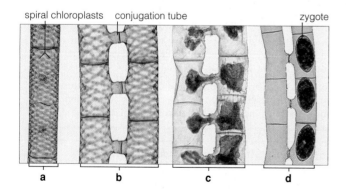

Figure 23.19 One mode of sexual reproduction in *Spirogyra*, or watersilk. (**a**) This green alga has spiral, ribbonlike chloroplasts. (**b**) A conjugation tube forms between cells of adjacent haploid filaments of different mating strains. (**c**,**d**) The cellular contents of one strain pass through the tubes into cells of the other strain, where zygotes form. Zygotes develop thick walls. They undergo meiosis when they germinate and give rise to haploid filaments.

Labels: spiral chloroplasts, conjugation tube, zygote, a, b, c, d

(*Halimeda*) that formed calcified cell walls, then died and disintegrated. Someday, green algae might even accompany astronauts in space. They can grow in very small spaces, give off vital oxygen, and take up carbon dioxide exhaled by the aerobically respiring crew.

Green algae employ diverse modes of reproduction. *Chlamydomonas* provides a classic example. This single-celled alga is twenty-five or so micrometers wide. It can reproduce sexually, but most of the time it engages in asexual reproduction, with up to sixteen daughter cells forming by mitotic cell division within the confines of the parent cell wall. Daughter cells may live at home for a while, but sooner or later they leave by secreting enzymes that digest what's left of their parent. Figure 23.18 shows the life cycle of one species. To give a final example, Figure 23.19 shows a filamentous green alga, *Spirogyra*, reproducing sexually.

The green algae show great diversity in size, morphology, life-styles, and habitats. The structure and biochemistry of some groups indicate they have close evolutionary ties to the plant kingdom.

SUMMARY

1. Recall that protistans and other eukaryotes differ from bacteria in key respects. At the least, eukaryotic cells have a double-membraned nucleus, mitochondria, endoplasmic reticulum, and large ribosomes. They have microtubules, which have uses as cytoskeletal elements, as spindles for chromosome movements, and in the 9 + 2 core of flagella and cilia. These cells contain two or more chromosomes, each of which consists of DNA complexed with numerous histones and other proteins. Eukaryotic cells alone engage in mitosis and meiosis. Many have chloroplasts and other plastids.

2. Many protistans are heterotrophs. They include water molds, chytrids, slime molds (cellular and plasmodial types), protozoans (the amoebas, animal-like flagellates, and ciliated species), and sporozoans. See Table 23.3.

3. The chytrids and water molds are decomposers or parasites. Like fungi, they secrete enzymes that digest organic matter, then absorb breakdown products. Some form a mycelium (a mesh of absorptive filaments).

4. Like animals, slime molds are predatory. For part of the life cycle, they crawl on substrates as free-living, phagocytic, amoebalike cells. Like fungi, they produce spore-producing structures.

5. Like animals, protozoans are predators, grazers, or parasites. The amoebas, foraminiferans, heliozoans, and radiolarians are amoeboid protozoans with or without skeletal elements. Animal-like protozoans are internal parasites or live freely in aquatic habitats. Many cause human diseases. Sporozoans are internal parasites that may form cysts during the life cycle. An example is *Plasmodium*, which causes malaria. Ciliated protozoans such as *Paramecium* typically use their cilia for motility.

6. Most euglenoids, chrysophytes (diatoms, golden algae, and yellow-green algae), dinoflagellates, and the red, brown, and green algae are photoautotrophs. Many are members of phytoplankton.

7. The red, brown, and green algae differ from one another in their body plan, size, reproductive modes, and habitats. Most are multicelled photosynthesizers. Some species of brown algae are the largest protistans. Plants may be descended from early green algae.

8. It has taken us more than one chapter to consider features of protistans, the simplest eukaryotes, and to speculate on their evolutionary ties to other kingdoms. Table 23.4 summarizes key similarities and differences among the prokaryotes and eukaryotes.

Table 23.3	Summary of Major Groups of Protistans
Group	Common Name

HETEROTROPHIC (decomposers, predators, grazers, parasites):

Group	Common Name
Chytridiomycota	Chytrids
Oomycota	Water molds
Acrasiomycota	Cellular slime molds
Myxomycota	Plasmodial slime molds
Sarcodina	Amoeboid protozoans:
	Rhizopods (naked amoebas, foraminiferans)
	Actinopods (heliozoans, radiolarians)
Mastigophora	Animal-like flagellated protozoans
Apicomplexa	Major group of sporozoans
Ciliophora	Ciliated protozoans

AUTOTROPHIC AND HETEROTROPHIC:

Group	Common Name
Euglenophyta	Euglenoids
Pyrrhophyta	Dinoflagellates

MOSTLY AUTOTROPHIC (photosynthesizers; some parasites):

Group	Common Name
Chrysophyta	Chrysophytes (golden algae, yellow-green algae, diatoms)
Rhodophyta	Red algae
Phaeophyta	Brown algae
Chlorophyta	Green algae

Review Questions

1. Outline some of the general characteristics of protistans. Then explain what is meant by this statement: Protistans are often classified by what they are *not*. CI

2. Review Table 23.2, then cover it with a sheet of paper. Now list the common names for the major categories of protistans. CI, *Table 23.3*

3. Correlate some structural features of a protistan from each of the groups listed with conditions in their environments:
   a. chytrids and water molds *23.1*
   b. slime molds *23.1*
   c. amoeboid protozoans *23.3*
   d. animal-like flagellates and sporozoans *23.4, 23.5*
   e. ciliated protozoans *23.6*
   f. euglenoids, chrysophytes, and dinoflagellates *23.7*
   g. red, brown, and green algae *23.8–23.10*

4. Select three different protistan species, then briefly explain how they adversely affect our affairs, such as crop yields and human health. *23.1, 23.3, 23.4–23.7*

5. Select and briefly explain how the activities of photosynthetic protistan species have positive benefits for some communities of organisms, including human communities. *23.7–23.10*

Self-Quiz (*Answers in Appendix IV*)

1. Most chytrids and water molds are _____ decomposers in aquatic habitats.
   a. parasitic   c. autotrophic
   b. saprobic   d. chemosynthetic

2. Free-living, amoebalike cells crawl on rotting plant parts as they engulf bacteria, spores, and organic compounds. This description best fits the _____ .
   a. water molds   c. sporozoans
   b. amoeboid protozoans   d. slime molds

Table 23.4 Comparison of Prokaryotes With Eukaryotes

	Prokaryotes	Eukaryotes
Organisms represented:	Bacteria only	Protistans, fungi, plants, and animals
Ancestry:	Two major lineages (archaebacteria and eubacteria) that evolved more than 3.5 billion years ago	Equally ancient prokaryotic ancestors gave rise to forerunners of eukaryotes, which evolved more than 1.2 billion years ago.
Level of organization:	Single-celled	Protistans, single-celled or multicelled. Nearly all others are multicelled, with a division of labor among differentiated cells, tissues, and often organs.
Typical cell size:	Small (1–10 micrometers)	Large (10–100 micrometers)
Cell wall:	Most have distinctive walls.	Cellulose or chitin; none in animal cells
Membrane-bound organelles:	Very rarely	Typically profuse
Modes of metabolism:	Both anaerobic and aerobic	Aerobic modes predominate.
Genetic material:	Bacterial chromosome (and sometimes plasmids)	Complex chromosomes (DNA, many associated proteins) within a nucleus
Mode of cell division:	Prokaryotic fission, mostly; also budding	Nuclear division (mitosis, meiosis, or both), associated with one of various modes of cytoplasmic division

3. During a _____ life cycle, amoeboid cells aggregate and form a migrating mass. Cells in the mass then differentiate, forming reproductive structures and spores or gametes.
   a. slime mold   c. protozoan
   b. water mold   d. chytrid

4. Amoebas, foraminiferans, radiolarians, and heliozoans are all classified as _____ .
   a. ciliated protozoans   c. amoeboid protozoans
   b. animal-like protozoans   d. sporozoans

5. Parasitic, flagellated protozoans of tropical regions known as trypanosomes are associated with which disease(s)?
   a. African sleeping sickness and Chagas disease
   b. toxoplasmosis
   c. malaria
   d. amoebic dysentery

6. Euglenoids and chrysophytes are mostly _____ .
   a. photoautotrophic   c. heterotrophic
   b. chemoautotrophic   d. omnivorous

7. Single-celled photosynthetic protistans, including most of the euglenoids, chrysophytes, and dinoflagellates, are members of the _____ , the "pastures" of most aquatic habitats.
   a. zooplankton   c. brown algae
   b. red algae   d. phytoplankton

8. Algin is used in ice cream, pudding, salad dressing, jelly beans, beer, cough syrup, toothpaste, cosmetics, and other products. Certain _____ are sources of algin.
   a. green algae   c. red algae
   b. brown algae   d. dinoflagellates

Critical Thinking

1. Suppose you decide to vacation in a developing country where sanitation practices and standards of personal hygiene are poor. Having read about some of the parasitic protozoans that lurk in water and damp soil, what would you consider safe to drink, once you arrive there? Which kinds of foods might be best to avoid and what kinds of food preparations might make them safe to eat?

2. As you read in this chapter, red tides are associated with "algal blooms." Such blooms may follow enrichment of aquatic habitats with water that drains into them from heavily fertilized croplands or from concentrated sources of raw sewage. After thinking about it, do you accept the resulting destruction of aquatic species, birds, and other forms of wildlife as an unfortunate but necessary side effect of human activities? If you do not accept it, how would you stop the water pollution? And if you could stop it, what sorts of measures would you take to feed the enormous human population, which is now extremely dependent on high-yield (and very heavily fertilized) crops? How would you propose to dispose of or prevent the accumulation of fecal matter and other wastes of 7.5 billion people?

Selected Key Terms

actinopod 23.3
amoeboid protozoan 23.3
animal-like flagellate 23.4
binary fission 23.2
brown alga 23.9
chrysophyte 23.7
chytrid 23.1
ciliated protozoan 23.6
conjugation (protozoan) 23.6
contractile vacuole 23.6
cyst 23.2
diatom 23.7
dinoflagellate 23.7
euglenoid 23.7
golden alga 23.7
green alga 23.10
multiple fission 23.2
mycelium 23.1
pellicle 23.6
phytoplankton 23.7
plankton 23.3
protistan CI
protozoan 23.2
pseudopod 23.3
red alga 23.8
red tide 23.7
rhizopod 23.3
saprobe 23.1
slime mold 23.1
sporozoan 23.4
water mold 23.1
yellow-green alga 23.7

Readings

Bold, H., and M. Wynne. 1985. *Introduction to the Algae.* Second edition. Englewood Cliffs, New Jersey: Prentice-Hall. Includes descriptions of the economic importance of major algal groups.

Margulis, L. 1993. *Symbiosis in Cell Evolution.* Second edition. New York: Freeman. Paperback.

Margulis, L., and K. Schwartz. 1992. *Five Kingdoms.* Second edition. New York: Freeman. Paperback.

Satchell, M. 28 July 1997. "The Cell From Hell." *U.S. News and World Report.* 26–28. Among other cases, correlates hundreds of millions of gallons of raw sewage from North Carolina's factory-like log farms with a bloom of the dinoflagellate *Pfiesteriia* that killed 14 million fish.

Web Site See *http://www.wadsworth.com/biology* for practice quiz questions, hypercontents, BioUpdates, and critical thinking. The Wadsworth Biology Resource Center provides a wealth of information fully organized and integrated by chapter.

24 FUNGI

Dragon Run

The first winter storm is blasting through southeastern Virginia, and Dragon Run—part swamp, part marsh, part old-growth woodland—pays tribute to the wind. Oaks, maples, gums, and beeches release dead leaves by the millions and these shower to the earth, where they pile up as crisp, ankle-deep mounds. Branches snap off trees; sheaves of dead grasses sink into marsh

b Big laughing mushroom (*Gymnophilus*)

c Purple coral fungus (*Clavaria*)   **d** Rubber cup fungus (*Sarcosoma*)

Figure 24.1 Fungal species from southeastern Virginia. This small sampling merely hints at the rich diversity within the kingdom Fungi.

muds, and other plants buckle into the shallow waters of the bottomlands. The storm kills uncounted numbers of insects, some birds, a few squirrels. By late December, Dragon Run is partially buried in organic debris.

Dig down through the debris and you will discover the accumulated litter of many past seasons—moist cushions of decayed leaves, bits of spiders and insects, mouldering branches, and carcasses of small mammals.

A new growing season begins with the warm rains of February. Buds open on shrubs and trees, including a massive silver beech that has been leafing out every spring for three centuries. Woodland violets are the first to sprout in the thawing soil. By April, the resurgence of growth has imparted such a sharp, green freshness to Dragon Run that you might overlook the organisms on which the resurgence largely depends. On logs, under leaves, and in soil, fungi are commandeering resources and engaging in vegetative growth (Figure 24.1).

Fungi are premier decomposers. Like every other heterotroph, they feast on organic compounds that other organisms produce. But few organisms besides fungi digest their dinner out on the table, so to speak. As fungi grow in or on organic matter, they secrete enzymes that digest it into bits that individual cells can

a Sulfur shelf fungus (*Polyporus*) on a living tree

e Scarlet hood (*Hygrophorus*)　　**f** Frost's bolete (*Boletus*)

g Yellow coral fungus (*Clavaria*)　　**h** Trumpet chanterelle (*Craterellus*)

absorb. Such a mode of nutrition is called **extracellular digestion and absorption**. Some of the carbon and other nutrients from organic compounds also can be absorbed by plants, the primary producers of Dragon Run and nearly all other ecosystems.

Keep the global perspective in mind. Why? As you will see, the metabolic activities of some fungi do cause diseases in humans, pets and farm animals, ornamental plants, and important crop plants. Some species are notorious spoilers of food supplies. Others have uses in the commercial production of substances ranging from antibiotics to excellent cheeses. We tend to assign "value" to fungi and other organisms in terms of their direct effect on our lives. There is nothing wrong with battling dangerous species and admiring beneficial ones—as long as we do not lose sight of the greater roles of fungi or any other kind of organism in nature.

KEY CONCEPTS

1. Fungi are heterotrophs. Together with heterotrophic bacteria, they are decomposers of the biosphere. Saprobic types obtain nutrients from nonliving organic matter. Parasitic types obtain them from tissues of living hosts.

2. Fungi secrete enzymes that digest food outside their body, then fungal cells absorb breakdown products. Their metabolic activities release carbon dioxide to the atmosphere and return many nutrients to the soil, where they become available to producer organisms.

3. Most fungi are multicelled. A mycelium, which is the food-absorbing portion of a fungal body, develops during the fungal life cycle. Each mycelium is a mesh of hyphae, which are elongated filaments that grow and develop by repeated mitotic cell divisions.

4. Commonly, a number of fungal hyphae become modified and weave together to form a reproductive structure in or upon which fungal spores develop. A "mushroom" is such a structure. Germinating spores grow and develop into a new mycelium.

5. Many fungi are symbionts with other organisms. Some are partners with algae and other organisms, in lichens. Many are part of mycorrhizae; they are locked in mutually beneficial relationships with young roots of land plants. Metabolic activities of cells making up the fungal hyphae provide the plants with nutrients, and the plants provide the fungi with carbohydrates.

6. We tend to assign value to plants and fungi in terms of their direct effect on our lives. Our battles with "bad" ones and reliance on the "good" ones should start from a solid understanding of their long-established roles in nature.

Major Groups of Fungi

For many of us, "fungi" are drab mushrooms sold in grocery stores. These are simply fungal body parts, and they are produced by a fungus with stunningly diverse relatives. The few species in Figure 24.1 don't do justice to the 56,000 fungal species we know about—and there may be at least a million more we don't know about!

We know, from the fossil record, that fungi evolved before 900 million years ago. Some accompanied plants onto the land 430 million years ago. About 100 million years later, three major lineages were well established. We call them the **zygomycetes** (Zygomycota), **sac fungi** (Ascomycota), and **club fungi** (Basidiomycota). Other, puzzling kinds known as "imperfect fungi" are lumped together but are not a formal taxonomic group. The vast majority of species in all these groups are multicelled.

Nutritional Modes

Fungi are heterotrophs, meaning they require organic compounds that other organisms synthesize. Most are **saprobes**; they obtain nutrients from nonliving organic matter and so cause its decay. Others are **parasites**; they extract nutrients from tissues of a living host. When the cells of any species grow in or on organic matter, they secrete digestive enzymes and then absorb breakdown products. Again, their extracellular digestion benefits plants, which absorb some of the released nutrients as well as the carbon dioxide by-products. Without the fungi and heterotrophic bacteria, communities would slowly become buried in their own garbage, nutrients would not be cycled, and life could not go on.

Key Features of Fungal Life Cycles

Fungi reproduce asexually most often, but given the opportunity, they also reproduce sexually. They form great numbers of nonmotile spores. As in plants, their **spores** are reproductive cells or multicelled structures, often walled, that germinate after dispersal from the parent. In multicelled species, spores give rise to a mesh of branched filaments. The mesh, a **mycelium** (plural, mycelia), rapidly grows over or into organic matter and has a good surface-to-volume ratio for food absorption. Each filament in a mycelium is called a **hypha** (plural, hyphae). Hyphal cells commonly have chitin-reinforced walls. Their cytoplasm interconnects, so that nutrients flow unimpeded throughout the mycelium.

Fungi are major decomposers that engage in extracellular digestion and absorption of organic matter. Multicelled types form absorptive mycelia and spore-producing structures.

A Sampling of Spectacular Diversity

Fungal life cycles and life-styles show dizzying variety. The most we can do here is to sample a few species, starting with the club fungi. The 25,000 or so club fungi include mushrooms, shelf fungi, coral fungi, puffballs, and stinkhorns. Figures 24.1, 24.2, and 1.6d show fine examples. Some of the saprobic species are important decomposers of plant debris. As you will see, other species are symbionts that live in close association with the young roots of forest trees. The fungal rusts and smuts can destroy fields of wheat, corn, and other crop plants. Cultivation of the common mushroom (*Agaricus brunnescens*) is a multimillion-dollar business. It is the mushroom of grocery-store and pizza-topping fame. Yet some of its relatives produce toxins that can kill you or any other organism that nibbles on them.

Have you ever wondered which organisms are the oldest and the largest? *Armillaria bulbosa* is one of them. The mycelium of one individual, discovered in a forest in northern Michigan, extends through fifteen hectares of soil. (Each hectare is the equivalent of 10,000 square meters.) By some estimates, this fungus weighs more than 10,000 kilograms and has been spreading beneath the forest floor for more than 1,500 years!

Figure 24.2 Two club fungi. (**a**) The light-red coral fungus *Ramaria*. (**b**) Closer view of the shelf fungus *Polyporus* shown in Figure 24.1a. With the exception of the rubber cup fungus, all of the species shown in Figure 24.1 are club fungi.

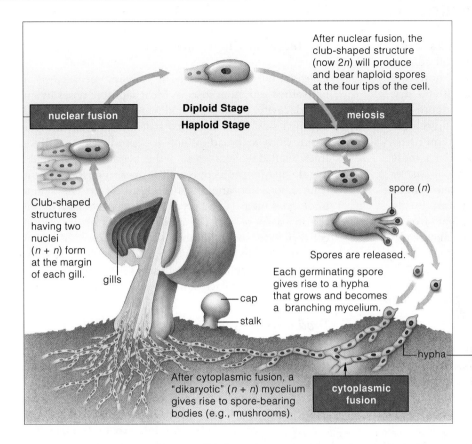

After nuclear fusion, the club-shaped structure (now 2n) will produce and bear haploid spores at the four tips of the cell.

nuclear fusion

Diploid Stage

Haploid Stage

meiosis

Club-shaped structures having two nuclei (n + n) form at the margin of each gill.

gills

spore (n)

Spores are released.

cap

stalk

Each germinating spore gives rise to a hypha that grows and becomes a branching mycelium.

hypha

After cytoplasmic fusion, a "dikaryotic" (n + n) mycelium gives rise to spore-bearing bodies (e.g., mushrooms).

cytoplasmic fusion

Figure 24.3 Generalized life cycle for many club fungi. When two compatible mating strains grow together, the cytoplasm (but not nuclei) of two hyphal cells may fuse. Cell divisions result in a dikaryotic mycelium (its cells each have two nuclei). Mushrooms form, with club-shaped spore-bearing structures on the surface of the gills. The two nuclei in each structure fuse to form a diploid zygote, which starts the cycle anew.

When you see mushrooms or any other fungus growing outdoors, think twice before devouring them. Unlike cultivated species, such as the common mushroom, many are toxic, including the species shown in Figure 24.4b. *No one should eat any fungi that were gathered in the wild unless they have been accurately identified as edible.* As the saying goes, there are old mushroom hunters and bold mushroom hunters, but no old bold mushroom hunters.

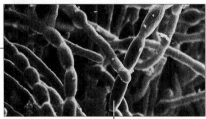

hypha in mycelium

a   b

Figure 24.4 (**a**) Fly agaric mushroom (*Amanita muscaria*) causes hallucinations when it is eaten. It was used to induce trances in ancient rituals in Central America, Russia, and India. (**b**) From California, *A. ocreata*. It can be fatal if ingested. A close relative, the death cap mushroom (*A. phalloides*), lives up to its name. Nibble even as little as five milligrams of its toxin and vomiting and diarrhea will begin eight to twenty-four hours later. Your liver and kidneys will degenerate; death can follow within a few days.

Example of a Fungal Life Cycle

Probably you already know what a common mushroom (*A. brunnescens*) looks like, so let's use it as our example of a fungal life cycle. Like most of the other club fungi, this species produces short-lived reproductive bodies—**mushrooms**—that are merely its aboveground parts; its living mycelium is buried in soil or decaying wood. A mushroom has a stalk and a cap. Fine tissue sheets, or gills, line the cap's inner surface. On these gills, club-shaped, spore-bearing structures form. The spores that form here are **basidiospores**. When a spore dispersed from a particular strain of mushroom lands on a suitable site, it germinates and gives rise to a haploid mycelium.

Suppose hyphae of two compatible mating strains make contact. They may undergo cytoplasmic fusion, but their nuclei do not fuse at once. The fused part may be the start of a *dikaryotic* mycelium, in which hyphal cells have one nucleus of each mating type (Figure 24.3). An extensive mycelium forms. And when conditions are favorable, mushrooms will form. Each spore-producing structure of the mushroom is initially dikaryotic, but then its two nuclei fuse to form a short-lived zygote. The zygote undergoes meiosis, haploid spores develop, and then air currents disperse them.

Club fungi, the fungal group with the greatest diversity, produce spores inside distinctive club-shaped structures.

SPORES AND MORE SPORES

A fungus has a thing about spores. It produces sexual spores, asexual spores, or both, depending on contact with a suitable hypha, food availability, and how cool or damp conditions are. Its spores are small and dry, and air currents easily disperse them. Each spore that germinates can be the start of a hypha and a mycelium. Stalked reproductive structures may develop on many of the hyphae and produce asexual spores. After these spores germinate, each may be the start of still *another* extensive mycelium. In no time at all, that one fungus and staggering numbers of its descendants are busily decomposing organic stuff or pirating nutrients from a host! Look at what can happen to a slice of stale bread:

Each fungal class produces unique sexual spores. Club fungi form basidiospores, zygomycetes form spores by way of zygosporangia, and sac fungi form ascospores.

Figure 24.5 Life cycle of the black bread mold *Rhizopus stolonifer.* Asexual phases are common. Different mating strains (+ and −) also reproduce sexually. Either way, haploid spores form and give rise to mycelia. Chemical attraction between a + hypha and a − hypha causes them to fuse. Two gametangia form, each with several haploid nuclei. Later their nuclei fuse to form a zygote. The zygote develops a thick wall, so becoming a zygosporangium, and may remain dormant for several months. Meiosis occurs as the zygosporangium germinates, and sexual spores form.

Producers of Zygosporangia

Consider the zygomycetes. The parasitic species feed on houseflies and other insects. Most saprobic types live in soil, decaying plant or animal material, and stored food. You just saw what *Rhizopus stolonifer*, the black bread mold, does to bread. When it reproduces sexually, a thick wall develops around the zygote. Together, the zygote *and* its protective wall form a **zygosporangium** (plural, zygosporangia), as in Figure 24.5a. The zygote undergoes meiosis and gives rise to a specialized hypha bearing a spore sac. The entire contents of the sac are converted to some number of spores, each of which may give rise to a new mycelium. Stalked hyphae grow out of such mycelia. Asexual spores form in a spore sac perched on top of each stalk (Figure 24.5b).

Producers of Ascospores

We know of more than 30,000 kinds of sac fungi. Most form sexual spores called **ascospores** inside sac-shaped cells. They alone form these cells, called asci (singular, ascus). The asci of multicelled species are enclosed in reproductive structures, which consist of tightly interwoven hyphae. The structures resemble diverse flasks, globes, and shallow cups, as in Figure 24.6.

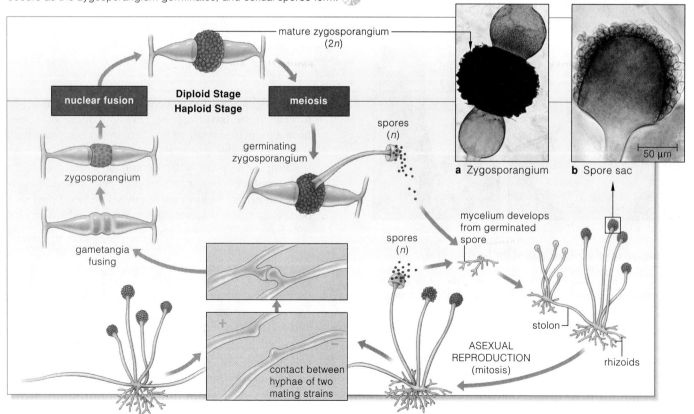

a Zygosporangium   **b** Spore sac

ascospore (sexual spore)

spore sac

b ascocarp   **c** ascocarp   **d** conidia (chains of asexual spores)   **e** budding yeast cell

spore-bearing hypha of this ascocarp

a

Figure 24.6 Sac fungi. (**a**) Diagram and (**b**) photograph of *Sarcoscypha coccinia,* the scarlet cup fungus. Saclike structures on the cup's inner surface produce sexual spores (ascospores) by meiosis. (**c**) One of the morels (*Morchella esculenta*). This edible species has a poisonous relative. (**d**) From *Eupenicillium,* chains of asexual spores of a type called conidiospores. These drift away from the chains, like dust, even after the slightest jiggling. "Conidia" means dust. (**e**) Cells of *Candida albicans,* agent of "yeast" infections of the vagina, mouth, intestines, and skin.

The vast majority of sac fungi are multicelled. They include highly prized truffles and morels (Figure 24.6*c*). The truffles are underground symbionts with oak and hazelnut tree roots. Trained pigs and dogs snuffle out truffles in the wild. In France, truffles are now being cultivated on the roots of inoculated seedlings.

Other multicelled sac fungi include certain species of *Penicillium* that "flavor" Camembert and Roquefort cheeses and species that make penicillins, widely used as antibiotics. We use *Aspergillus* to make citric acid for candies and soft drinks, and to ferment soybeans for soy sauce. Most of the red, bluish-green, and brown fungal molds that spoil stored food also are multicelled. One species, the salmon-colored *Neurospora sitophila,* wreaks havoc in bakeries and in research laboratories. It produces so many spores that it is extremely difficult to eradicate. One of its relatives, *N. crassa,* is an important organism in genetic research.

Sac fungi also include about 500 species of single-celled yeasts (although still other yeasts are classified as club fungi). Yeasts reproduce sexually when two cells fuse and become a spore-producing sac. Some types live in the nectar of flowers and on fruits and leaves. Bakers and vintners put fermenting by-products of vast populations of yeasts to use. For example, the carbon dioxide by-products of *Saccharomyces cerevisiae* leaven bread. The commercial production of wine and beer depends on its ethanol end product. Many yeast strains with desirable properties have been developed through artificial selection and genetic engineering. Then again, *Candida albicans,* a notorious relative of "good" yeasts, causes vexing infections in humans (Figure 24.6*e*).

roundworm   noose formed by hypha

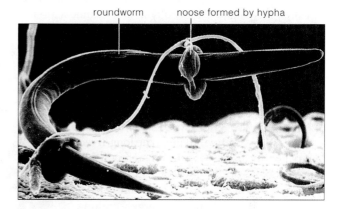

Figure 24.7 *Arthrobtrys dactyloides,* an imperfect fungus. This predatory species forms a nooselike ring that swells rapidly with turgor pressure when stimulated. The "hole" in the noose shrinks and captures a worm, into which hyphae will grow.

Elusive Spores of the Imperfect Fungi

Imperfect fungi are set aside in a taxonomic holding station not because they are somehow defective but mainly because no one has yet discovered what kind of sexual spores they produce (if any). Figure 24.7 shows one of the species, a puzzling predatory fungus, that is awaiting formal classification. Investigators recently reunited the previously orphaned *Aspergillus, Candida,* and *Penicillium* with their kin—other sac fungi.

Through their exuberant and rapid production of asexual and sexual spores, fungi take quick advantage of available organic matter, whether it has been discarded or is part of a living or dead organism. Their penchant for spore production is central to their success as decomposers and parasites.

Recall, from earlier chapters, that **symbiosis** refers to species that live in close association. The word literally means "living together." In many cases, one species is a victim, not a partner, of a parasite. In other cases, called **mutualism**, the interaction benefits both partners or at least does one of them no harm. We find classic examples of mutualists among fungi that enter into partnerships with algae and cyanobacteria, as well as with tree roots.

Lichens

A **lichen** is a single vegetative body in which a fungus has become intimately intertwined with one or more photosynthetic organisms. The fungal part of a lichen is the *myco*biont (after the Greek *myes*, meaning fungus). The photosynthetic component is the *photo*biont. About 13,500 types of lichens have been identified. Nearly half of the mycobionts are sac fungi. However, only 100 or so species function as photobionts, and most commonly they are green algae (Chlorophyta) and cyanobacteria.

Lichens typically colonize sites that most organisms would find hostile. We find one type or another growing slowly in habitats ranging from the Antarctic to the Arctic. They can colonize almost any substrate, including sunbaked or frozen rocks, recently hardened lava, fence posts, gravestones, and the bark or leaves of a variety of plants. The long-lived Galápagos tortoises have lichens clinging to their shells; so do the tops of giant Douglas firs in the Pacific Northwest. Certain lichens even withstand periodic submergence along seashores and riverbanks.

In almost every instance, the fungus is the largest part of the lichen. Cyanobacterial cells typically reside in a separate structure inside or outside the main body. The fungus benefits by having a long-term source of nutrients that it absorbs from cells of the photobiont. By giving up some nutrients, the photobiont suffers a bit in terms of its own growth, although it might benefit a bit from the lichen's sheltering effect. In cases where more than one fungus is present, the added species might be a mycobiont. Then again, it might be parasitizing the lichen or merely using the lichen as a substrate.

A lichen forms after the tip of a fungal hypha binds with a suitable host cell. Either both lose their wall and undergo cytoplasmic fusion or the hypha induces its host cell to cup around it. Now the mycobiont and the photobiont grow and multiply together. The structure of the lichen depends on how cells of the photobiont become distributed among fungal cells. Sometimes they are distributed more or less uniformly throughout the lichen. In most cases, however, the lichen has distinct layers, as in Figure 24.8a. The overall growth pattern

may be leaflike, flattened, pendulous, or erect. Figure 24.8b–e provides examples of these growth patterns.

Lichens absorb mineral ions from their substrates and nitrogen from the air. Among the products of their metabolic activities are antibiotics against bacteria that might decompose the lichen body and toxins that inhibit the larval development of invertebrates that might graze on them. Coincidentally, the products also contribute to

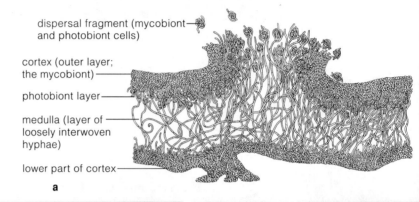

dispersal fragment (mycobiont and photobiont cells)

cortex (outer layer; the mycobiont)

photobiont layer

medulla (layer of loosely interwoven hyphae)

lower part of cortex

a

b

c

Figure 24.8   Lichens. (**a**) Diagram of a stratified lichen, cross-section. Notice the distinct layers. (**b**) Leaflike lichen on birch tree bark. (**c**) Encrusting lichens. (**d**) *Usnea* (old man's beard), a pendant lichen. (**e**) An erect, branching lichen, *Cladonia rangiferina*, sometimes called reindeer moss. Reindeer and caribou feed mainly on this species during winter.

the formation or enrichment of soils. When lichens are pioneers in new habitats, such as soil newly exposed by a retreating glacier, their metabolic activities can set the stage for colonization by different species. Actually, lichens may have been among the first invaders of the land. Also, some ecosystems depend on cyanobacteria-containing lichens as nitrogen sources. For example, up to 20 percent of the nitrogen entering old-growth forests of the Pacific Northwest starts with the lichen *Lobaria*.

Lichens also serve as early warnings of deteriorating environmental conditions. They absorb but cannot rid themselves of toxins. From extensive studies in New York City and in England, we know that when lichens die around cities, air pollution is getting bad.

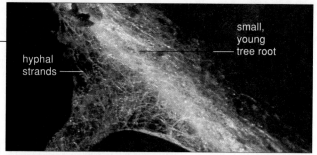

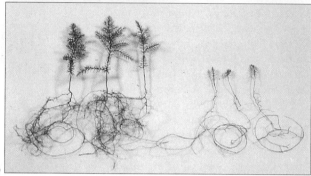

small,
young
tree root

hyphal
strands

Figure 24.9 (**a**) Mycorrhiza from a hemlock tree. (**b**) Effects of the presence or absence of mycorrhizae on plant growth. The juniper seedlings at the left are six months old. They were grown in sterilized, phosphorus-poor soil with a mycorrhizal fungus. The seedlings at the right were grown without the fungus.

About 5,000 fungal species enter the associations. Most are club fungi, including those truffles described earlier.

The more common *endo*mycorrhizae form in about 80 percent of all vascular plants. These fungal hyphae penetrate plant cells, as they do in lichens. Fewer than 200 species of zygomycetes serve as the fungal partner. Their hyphae branch extensively, forming tree-shaped absorptive structures within cells. They also extend for several centimeters into the surrounding soil. Section 30.2 provides a closer look at these beneficial species.

Mycorrhizae

Fungi also enter into symbiotic interactions with young tree roots in which benefits flow both ways. This form of mutualism is a **mycorrhiza** (plural, mycorrhizae), which means "fungus-root." The fungus benefits by absorbing carbohydrates from the plant, which benefits by absorbing minerals from the fungus. Collectively, the fungal hyphae have a huge surface area for absorbing water and dissolved mineral ions. The fungus can take up ions of phosphorus and other minerals when they are abundant in soil and release them to the plant when ions are scarce. Many plants do not grow as efficiently in the absence of mycorrhizae.

The example in Figure 24.9*a* is an *exo*mycorrhiza, in which hyphae form a dense net around living cells in roots but don't penetrate them. Other hyphae form a velvety wrapping around the root, and the mycelium radiates outward from it. Exomycorrhizae are common in temperate forests, and they help the trees withstand adverse seasonal changes in temperature and rainfall.

As Fungi Go, So Go the Forests

Since the early 1900s, collectors have been recording data on the wild mushroom populations in the forests of Europe. A look at the records tells us that the number and diversity of fungi are now declining at alarming rates. Certainly mushroom gatherers aren't to blame; toxic as well as edible species are vanishing. But the decline does correlate with rising air pollution. Vehicle exhaust, smoke from coal burning, and emissions from nitrogen fertilizers pump ozone, nitrogen oxides, and sulfur oxides into the air. Normally, as a tree ages, one species of mycorrhizal fungus gives way to another in predictable patterns. When fungi die, the tree loses its support system and becomes vulnerable to severe frost and drought. Are the North American forests at risk? Conditions there are deteriorating in comparable ways.

Lichens and mycorrhizae are both symbiotic associations between fungi and other organisms, with mutual benefits.

A LOOK AT THE UNLOVED FEW

You know you are a serious student of biology when you view organisms objectively in terms of their place in nature, not in terms of their impact on humans generally and you in particular. As a student you salute saprobic fungi as vital decomposers and praise parasitic fungi that help keep populations of harmful insects and weeds in check. The true test is when you open the fridge to get a bowl of high-priced raspberries and discover that a fungus beat you to them. The true test is when a fungus starts feeding on warm, damp tissues between your toes and turns skin scaly, reddened, and cracked (Figure 24.10*a*).

Which home gardeners wax poetic about black spot or powdery mildew on their roses? Which farmers happily give up millions of dollars each year to sac fungi that attack corn, wheat, peaches, and apples (Figure 24.10*b*)? Who cares that the sac fungus *Cryphonectria parasitica* blitzed most of the chestnut trees in eastern North America, leaving them to sprout as stubby versions of their former selves?

And who willingly would inhale airborne spores of *Ajellomyces capsulatus*? These are dimorphic beasties. When they alight on soil, they form mycelia. When they alight in moist lung tissues, however,

they form populations of yeastlike cells that cause the respiratory disease *histoplasmosis*. Often, aggregations of macrophages that converge on and defend the infected tissues become packed with the invading cells. Usually the defenders eliminate the invaders, and their aggregations

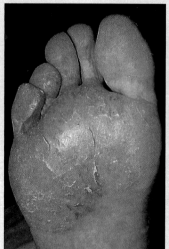 

a   b

Figure 24.10 Love those fungi! (**a**) Athlete's foot, courtesy of *Epidermophyton floccosum*. (**b**) Apple scab, trademark of *Venturia inaequalis*.

become calcified as the tissue heals. With heavy exposure to the spores, however, pneumonia develops. In rare cases, the calcified masses form in the lymph nodes, liver, and other organs. Progressive histoplasmosis usually ends in death. Where nitrogen-rich droppings of many chickens, starlings, pigeons, and other birds accumulate, the fungus thrives. For example, kick up dust from spore-containing regions drained by the Ohio and Mississippi rivers, and you put yourself at risk. Go spelunking in caves where roosting bats busily pile up droppings on the cave floor, and you similarly invite the spores into your lungs.

Certain fungi even have had impact on human history. Consider one notorious species, *Claviceps purpurea*, which parasitizes rye and other cereal grains. Give it credit; we use some of its by-products (alkaloids) to treat migraine headaches and, following childbirth, to shrink the uterus to prevent hemorrhaging. However, the alkaloids are toxic when ingested in large amounts. Eat a lot of bread made with tainted rye flour and you end up with *ergotism*. The symptoms include vomiting, diarrhea, hallucinations, hysteria, and convulsions. Untreated, the disease can turn limbs gangrenous, and it can end in death.

Ergotism epidemics were common in Europe during the Middle Ages, when rye was a major crop. Ergotism also thwarted Peter the Great, the Russian czar who was obsessed with conquering ports along the Black Sea for his nearly landlocked empire. Soldiers laying siege to the ports ate mostly rye bread and fed rye to their horses. The former went into convulsions and the latter into "blind staggers." Possibly, outbreaks of ergotism were used as an excuse to launch witch-hunts in colonial Massachusetts and elsewhere.

Table 24.1 Some Pathogenic and Toxic Fungi*

ZYGOMYCETES

Rhizopus	Food spoilage

ASCOMYCETES

Aspergillus (certain species)	Aspergilloses (allergic reactions, sinus, ear, and lung infections; *A. flavus* toxin linked to cancers)
Ajellomyces capsulatus	Histoplasmosis
Candida albicans	Infection of mucous membranes
Claviceps purpurea	Ergot of rye, ergotism
Coccidioides immitis	Valley fever
Cryphonectria parasitica	Chestnut blight
Ophiostoma ulmi	Dutch elm disease
Microsporum, Trichophyton, Epidermophyton	Various species cause ringworms, of scalp, body, nails, beard, athlete's foot (Figure 24.10*a*)
Monilinia fructicola	Brown rot of peaches and other stone fruits
Venturia inaequalis	Apple scab (Figure 24.10*b*)
Verticillium	Plant wilt

BASIDIOMYCETES

Amanita (some species)	Severe mushroom poisoning
Pneumocystis carinii	A virulent form of pneumonia in individuals with weakened immune systems, as with AIDS
Puccinia graminis	Black stem wheat rust
Tilletia indica	Smut of cereal grains
Ustilago maydis	Smut of corn

* After C. Alexopoulos, C. Mims, and M. Blackwell. 1996. *Introductory Mycology*, fourth edition. Wiley.

SUMMARY

1. Fungi are heterotrophs and major decomposers. The saprobes feed on nonliving organic matter; the parasites feed on the tissues of living organisms. Some species of fungi are symbiotic partners with other organisms. The cells of all species secrete digestive enzymes that break down food into small molecules, which the cells absorb.

2. Nearly all fungi are multicelled. The food-absorbing portion, the mycelium, consists of a mesh of filaments (hyphae). Aboveground reproductive structures, such as mushrooms, form from tightly interwoven hyphae.

3. The major groups of fungi are the zygomycetes, the ascomycetes (sac fungi), and the basidiomycetes (club fungi). Each is characterized by distinctive sexual and asexual spores. When a sexual phase cannot be detected or is absent from the life cycle, a fungus is assigned to an informal category called the imperfect fungi.

4. Lichens are mutualistic associations of fungi with photosynthetic partners (green algae and cyanobacteria, mostly). Mycorrhizae are mutualistic associations of a fungus and the young roots of plants. Fungal hyphae provide nutrients for their symbiont, which provides the fungus with carbohydrates.

Review Questions

1. Describe the fungal mode of nutrition, and explain how the structure of mycelia facilitates this mode. *24.1*

2. How does a lichen differ from a mycorrhiza? *24.4*

Self-Quiz (*Answers in Appendix IV*)

1. New mycelia form after _____ germinate.
   a. hyphae   c. mycelia
   b. spores   d. mushrooms

2. A "mushroom" is _____ .
   a. the food-absorbing part of a fungal body
   b. the part of the fungal body not constructed of hyphae
   c. a reproductive structure
   d. a nonessential part of the fungus

3. A mycorrhiza is a _____ .
   a. fungal disease of the foot   c. parasitic water mold
   b. fungus-plant relationship   d. fungus endemic to barnyards

4. Parasitic fungi obtain nutrients from _____ .
   a. tissues of living hosts   c. only living animals
   b. nonliving organic matter   d. none of the above

5. Saprobic fungi derive nutrients from _____ .
   a. nonliving organic matter   c. root hairs
   b. living organisms   d. both b and c

6. Match the terms appropriately.
   _____ zygomycetes   a. mushrooms, shelf fungi
   _____ conidia   b. some yeasts and morels
   _____ hypha   c. chains of asexual spores
   _____ basidiomycetes   d. each filament in a mycelium
   _____ asci   e. black bread mold
   _____ ascomycetes   f. sac-shaped cells in which
                              asexual spores are produced

Figure 24.11 Reproductive structures of *Pilobolus*, a Greek word meaning "hat-thrower." The "hats" are spore sacs.

Critical Thinking

1. *Pilobolus* is a type of fungus that commonly dines on horse dung. Each morning, stalked reproductive hyphae emerge from irregularly spaced piles of dung. By early afternoon, they have completed the task of dispersing spores to sunlit grasses where horses feed. The spores pass through the horse gut unharmed and exit with their very own pile of dung. At the tip of each stalked hypha is a dark-walled, spore-containing sac (Figure 24.11). Just below the sac, the stalk is differentiated into a vesicle, swollen with a fluid-filled central vacuole. At the base of the vesicle is a ring of light-sensitive, pigmented cytoplasm. The stalk bends as it grows until its wall is parallel with the sun's rays and light strikes all of the ring. When that happens, turgor pressure builds up inside the central vacuole until the vesicle ruptures. The forceful blast can propel spore sacs two meters away—which is amazing, considering that the stalk is less than ten millimeters tall. Reflect on the examples of fungi discussed in this chapter. Would you say that *Pilobolus* is a zygomycete, club fungus, or sac fungus?

2. Diana discovers in the laboratory that a fungus (*Trichoderma*) can grow well in distilled water. It continues to do so even after she rigorously treats the water and glassware to remove all traces of organic carbon. This fungus is not a photoautotroph. Suggest a metabolic life-style that would allow it to grow under these conditions.

3. Some strains of *Trichoderma* are being tested as a natural pest control agent. Laboratory experiments demonstrated that the strains of this sac fungus combat other fungi that cause plant diseases. Some even promote seed germination and plant growth. In one set of twenty trials, workers increased lettuce yields by 54 percent. What concerns must be addressed before *Trichoderma* can be released into the environment for commercial applications?

Selected Key Terms

ascospore *24.3*	hypha *24.1*	sac fungus
basidiospore *24.2*	lichen *24.4*	(ascomycetes) *24.1*
club fungus	mushroom *24.2*	saprobe *24.1*
(basidiomycetes) *24.1*	mutualism *24.4*	spore (fungal) *24.1*
extracellular digestion	mycelium *24.1*	symbiosis *24.4*
and absorption *C1*	mycorrhiza *24.4*	zygomycetes *24.1*
fungus *CI*	parasite *24.1*	zygosporangium *24.3*

Reading

Moore-Landecker, E. 1990. *Fundamentals of the Fungi*. Third edition. Englewood Cliffs, New Jersey: Prentice-Hall.

Web Site See *http://www.wadsworth.com/biology* for practice quiz questions, hypercontents, BioUpdates, and critical thinking. The Wadsworth Biology Resource Center provides a wealth of information fully organized and integrated by chapter.

25

PLANTS

Pioneers In a New World

Seven hundred million years ago, no shorebirds stirred and noisily announced the dawn of a new day. There were no crabs to clack their claws together and skitter off to burrows. The only sounds were the rhythmic muffled thuds of waves in the distance, at the outer limits of another low tide. More than 3 billion years before, life had its beginning somewhere in the waters of the Earth—and now, quietly, the invasion of the land was under way.

Astronomical numbers of photosynthetic cells had come and gone, and the oxygen-producing types had slowly changed the atmosphere. High above the Earth, the sun's energy had converted much of the oxygen into a dense ozone layer. That layer became a shield against lethal doses of ultraviolet radiation, which had kept early organisms beneath the water's surface.

Were cyanobacteria the first to adapt to intertidal zones, where mud dried out with each retreating tide? Were they the first to spread into shallow, freshwater streams flowing down to the coasts? Probably so. From fossil evidence, we know that later in time, green algae and fungi made the same journey together.

Every plant around you is a descendant of ancient species of green algae that lived near the water's edge or made it onto the land. Diverse fungi still associate with nearly all of them. Together, the plants and fungi became the basis of communities in coastal lowlands, near the snow line of high mountains, and just about all places in between (Figure 25.1).

We have a few tantalizing fossils of the first pioneers. We also are learning about them through comparative biochemistry and studies of existing species. Today, as in the late Precambrian, cyanobacteria and green algae grow in mats in nearshore waters and on the banks of freshwater streams (Figure 25.1a). After a volcano erupts or a glacier retreats, cyanobacteria are the first to colonize the barren rocks. Then, symbiotic associations between green algae and fungi follow. Gradually their organic products and remains accumulate and create pockets of soil. Mosses and other plant species can take hold in the newly forming soil and further enrich it.

With this chapter we turn to the plant kingdom. With only a few exceptions, plants are multicelled photoautotrophs. They absorb energy from the sun, carbon dioxide from the air, and some minerals dissolved in water to synthesize organic compounds. These metabolic wizards also can split water molecules. In doing so, they obtain the stupendous numbers of the electrons and hydrogen atoms required for growth into multicellular

a

Figure 25.1 (a) Filaments of a green alga, massed in a shallow stream. More than 400 million years ago, green algal species that may have been ancestral to all plants, past and present, lived in similar streams that meandered down to the shores of early continents. (b) One land-dwelling descendant of those ancestral forms—a Ponderosa pine high above the floor of Yosemite Valley in the Sierra Nevada of California. (c) Flowers of one of the most highly prized flowering plants—orchids—growing on a branch of a living tree in a tropical rain forest. With this chapter, we turn to the beginning—and ends of the line—of some ancient lineages.

b

c

KEY CONCEPTS

1. With very few exceptions, the plant kingdom consists of multicelled photoautotrophs. From earlier chapters, you know that plants use chlorophylls *a* and *b* as their main photosynthetic pigments. In this respect they are like green algae, which are their closest relatives.

2. Unlike their algal ancestors, which were adapted to aquatic habitats, nearly all existing plants live on land.

3. In general, plants are structurally adapted to intercept sunlight, absorb water and mineral ions, and conserve water. Their lignin-reinforced tissues permit upright growth. Root systems mine the soil for water and ions, and internal tissues conduct water and solutes to all living cells in belowground and aboveground parts.

4. Land plants are reproductively adapted to withstand dry periods. During the life cycle, a sporophyte develops roots, stems, and leaves. It holds onto its developing gametes and supplies them with water and food resources. And it disperses the new generation in ways that are responsive to the conditions in specific habitats.

5. Early divergences gave rise to the bryophytes, then seedless vascular plants, and then seed-bearing vascular plants. Of these categories, the seed producers were the most successful in radiating into drier environments.

6. The seed-bearing vascular plants called gymnosperms include the cycads, ginkgo, gnetophytes, and conifers. The angiosperms, another group of vascular plants, bear flowers as well as seeds. There are two classes of flowering plants, informally called the dicots and monocots.

forms as tall as the giant redwoods, as extensive as an aspen forest that is one continuous clone.

We know of at least 295,000 kinds of existing plants. Be glad their ancient ancestors left the water. Without them, we humans and other land-dwelling animals never would have made it onto the evolutionary stage.

EVOLUTIONARY TRENDS AMONG PLANTS

Overview of the Plant Kingdom

The plant kingdom includes at least 295,000 species of photoautotrophs and a few heterotrophs. Most kinds are **vascular plants**, with internal tissues that conduct water and solutes through the plant body. Such plants have roots, stems, and leaves, defined in part by the presence of vascular tissues. Fewer than 19,000 species are *non*vascular plants called **bryophytes**. Plants, like photoautotrophic bacteria and protistans, are producers of organic compounds for entire communities.

Liverworts, hornworts, and mosses are bryophytes. The whisk ferns, lycophytes, horsetails, and ferns are *seedless* vascular plants. Cycads, ginkgo, gnetophytes, and conifers belong to a group of *seed-bearing* vascular plants called **gymnosperms**. The **angiosperms**, another group of vascular plants, bear flowers as well as seeds. Dicots and monocots are two classes of flowering plants.

The ancestors of plants evolved in the seas by 700 million years ago. About 265 million more years passed before simple stalked plants appeared along coasts and streams. Evolutionarily speaking, the pace picked up after that. Within a mere 60 million years, plants had radiated through much of the land. Some long-term changes in their structure and reproductive events help explain how the diversity came about.

Evolution of Roots, Stems, and Leaves

Simple underground structures started to evolve when plants first colonized the land. In lineages that led to vascular plants, they developed into **root systems**. Most root systems have a number of underground absorptive structures that collectively afford a large surface area. These rapidly take up soil water and dissolved mineral ions; often they anchor the plant. Aboveground, **shoot systems** evolved. Shoot systems have stems and leaves, which efficiently absorb energy from the sun's rays and carbon dioxide from the air. Stems grew and branched extensively only after plants developed the biochemical capacity to synthesize and deposit **lignin**, an organic compound, in cell walls. Cells with lignified walls can structurally support stems, which grow in patterns that increase the total light-intercepting surface of leaves.

Many plants became equipped with cellular pipelines for water and solutes. The pipelines were key factors in the evolution of roots, stems, and leaves. They evolved as components of two vascular tissues called xylem and phloem. **Xylem** distributes water and dissolved ions to all living cells throughout the plant. **Phloem** distributes dissolved sugars and other photosynthetic products.

Life on land also depended on water conservation, which wasn't a problem in most aquatic habitats. Shoots became protected by a **cuticle**, a waxy coat that helps conserve water on hot, dry days. **Stomata** (singular, stoma), tiny openings across the surfaces of leaves and some stems, became strategic in controlling absorption of carbon dioxide and restricting evaporative water loss. Later chapters describe these tissue specializations.

From Haploid to Diploid Dominance

As early plants radiated into higher, drier places, their life cycles changed. Think about the gametes of algae, which cannot get together without the presence of liquid water. As shown in earlier chapters, the *haploid* (n) phase, in the form of **gametophytes** (gamete-producing bodies), dominates their life cycle. The diploid (2n) phase is the zygote, which forms when gametes fuse at fertilization.

Now look at Figure 25.2. *In most plant life cycles, the diploid phase dominates.* After a diploid zygote forms at fertilization, mitotic cell divisions and cell enlargements transform it into a multicelled diploid body, of a type called a **sporophyte**. Pine trees are an example. In time, some cells of the sporophyte undergo meiosis and give rise to haploid cells of a type called **spores** (sporophyte means spore-producing body). Later, the spores divide by way of mitosis and give rise to the gametophytes.

The shift to diploid dominance was an adaptation to habitats on land, most of which show seasonal changes in the availability of free water and dissolved nutrients. Long ago in those challenging habitats, natural selection must have favored sporophytes with well-developed root systems. Young roots of such systems interact with fungal symbionts (Section 24.4). The association, called a mycorrhiza, enhances the plant's uptake of water and scarce minerals, even during dry seasons.

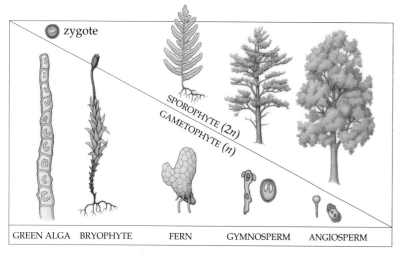

zygote

SPOROPHYTE (2n)

GAMETOPHYTE (n)

GREEN ALGA BRYOPHYTE   FERN   GYMNOSPERM   ANGIOSPERM

Figure 25.2 Evolutionary trend from gametophyte (haploid) dominance to sporophyte (diploid) dominance, represented here by existing species ranging from a green alga (*Ulothrix*) to a flowering plant. This trend occurred when early plants were colonizing habitats on land. See also Section 10.5.

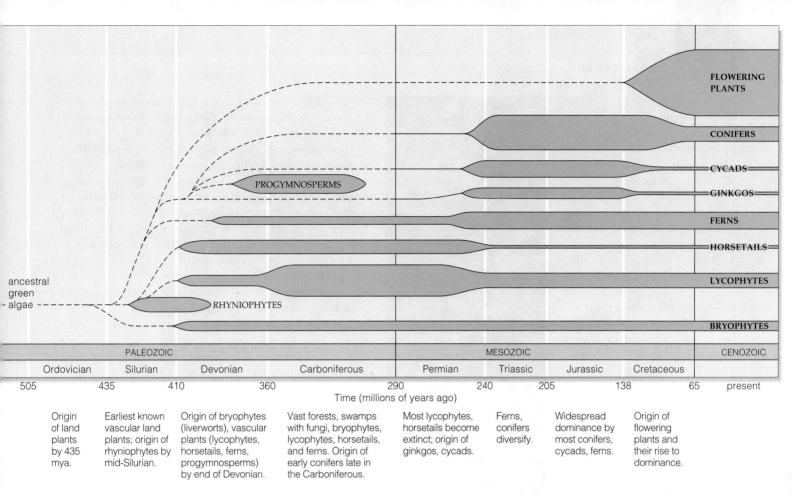

			PALEOZOIC				MESOZOIC			CENOZOIC

Ordovician | Silurian | Devonian | Carboniferous | Permian | Triassic | Jurassic | Cretaceous |

Time (millions of years ago)
505   435   410   360   290   240   205   138   65   present

| Origin of land plants by 435 mya. | Earliest known vascular land plants; origin of rhyniophytes by mid-Silurian. | Origin of bryophytes (liverworts), vascular plants (lycophytes, horsetails, ferns, progymnosperms) by end of Devonian. | Vast forests, swamps with fungi, bryophytes, lycophytes, horsetails, and ferns. Origin of early conifers late in the Carboniferous. | Most lycophytes, horsetails become extinct; origin of ginkgos, cycads. | Ferns, conifers diversify. | Widespread dominance by most conifers, cycads, ferns. | Origin of flowering plants and their rise to dominance. |

Figure 25.3 Milestones in plant evolution.

Unlike algae and bryophytes, vascular plants have a sporophyte that is larger and structurally more complex than the gametophyte. As you will see, the gametophytes of seedless vascular plants develop independently of the sporophyte that produces them. But they protect their gametes, and, after fertilization, they nourish the embryo sporophytes. The sporophyte became most dominant among the gymnosperms and, later, the angiosperms. It retains, nourishes, and protects developing gametophytes as well as the young sporophytes. *And it does so until environmental conditions are suitable for fertilization and for dispersal of the new generation.*

Evolution of Pollen and Seeds

Like some seedless species, seed-bearing plants produce not one but two kinds of spores. This condition is called *hetero*spory, as opposed to *homo*spory. In gymnosperms and angiosperms, one type of spore gives rise to **pollen grains**, which will develop into mature, sperm-bearing male gametophytes. The other type of spore develops into female gametophytes, where eggs form and later become fertilized. The pollen grains hitch rides on air currents, insects, birds, and so on; they do not require free-standing water to reach the eggs. In this respect, they differ enormously from the algae. The evolution of pollen grains is one reason for the successful radiation of seed-bearing plants into high and dry habitats.

Seed production also was adaptive in drier habitats. Female gametophytes (and eggs) of seed-bearing plants form inside nutritive tissues and a jacket of cell layers. Each **seed** consists of an embryo sporophyte, nutritive tissues, and a protective coat (which develops from the jacket of cell layers). Seeds withstand hostile conditions. Not coincidentally, the seed plants rose to dominance in Permian times, when shifts in climate were extreme.

Before turning to the spectrum of diversity among plants, take a look at Figure 25.3. You can use it as a map for tracking the branching evolutionary roads.

The plant kingdom includes multicelled, photosynthetic species called bryophytes, seedless vascular plants, and seed-bearing vascular plants. Most of these live on land.

In most lineages, structural adaptations to life on land included root and shoot systems, waxy cuticles, stomata, vascular tissues, and lignin-reinforced tissues.

Sporophytes with well-developed roots, stems, and leaves came to dominate the life cycle of most land plants. Parts of these complex sporophytes nourish and protect fertilized eggs and embryos through unfavorable conditions.

Some plants started producing two types of spores, not one. This led to the evolution of male gametes adapted for dispersal without liquid water and to the evolution of seeds.

The bryophyte lineage includes about 18,600 existing species known as **mosses**, **liverworts**, and **hornworts**. Most of these nonvascular plants are adapted to grow in fully or seasonally moist habitats, although you will find some mosses growing in deserts and on the bitterly cold, windswept plateaus of Antarctica. Mosses especially are sensitive to air pollution. Where the air quality is poor, mosses are often few or absent.

Bryophytes are generally small plants, less than twenty centimeters (eight inches) tall. They do have leaflike, stemlike, and rootlike parts, but these don't contain xylem or phloem. Bryophytes, like lichens and some algae, can dry out, then revive after they absorb some

moisture. Most have rhizoids, which are elongated cells or threadlike structures that attach the gametophytes to the soil and serve as absorptive structures.

Bryophytes are the simplest plants to display three features that emerged early in plant evolution. *First*, a cuticle prevents water loss from aboveground parts. *Second*, a cellular jacket around the parts that produce sperm and eggs holds in moisture. *Third*, of all plants, bryophytes alone have large gametophytes that hold onto sporophytes and do not depend on them for their nutrition. Embryo sporophytes start to develop inside the gametophyte tissues—and even at maturity, they remain *attached to* the gamete-producing body and still gain some nutritional support from it.

With 10,000 species, mosses are the most common bryophytes. Gametophytes of some species grow in clusters and form low, cushiony mounds (Figure 25.4a). Those of others commonly grow in branched, feathery

Figure 25.4 (**a**) Moss-covered rocks near a small stream. (**b**) Photograph and life cycle of a representative bryophyte, a moss (*Polytrichum*). The moss sporophyte remains attached to the gametophyte and depends upon it for water and nutrients.

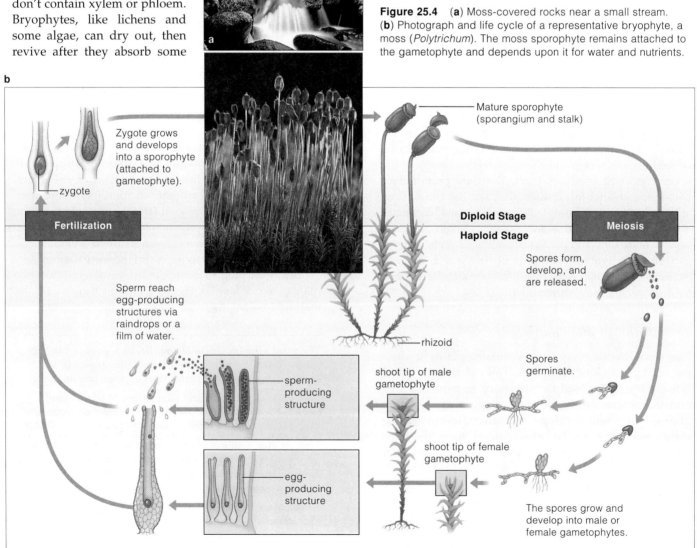

Zygote grows and develops into a sporophyte (attached to gametophyte).

zygote

Fertilization

Sperm reach egg-producing structures via raindrops or a film of water.

sperm-producing structure

egg-producing structure

Mature sporophyte (sporangium and stalk)

Diploid Stage

Haploid Stage

Meiosis

Spores form, develop, and are released.

rhizoid

shoot tip of male gametophyte

shoot tip of female gametophyte

Spores germinate.

The spores grow and develop into male or female gametophytes.

a

b

Figure 25.5 (**a**) A peat bog in Ireland. A family is cutting blocks of peat, and stacking them to dry, as a fuel source for their home. Peat also is harvested on a scale large enough to generate electricity in peat-burning power plants. (**b**) The gametophyte of a peat moss (*Sphagnum*) with a few sporophytes attached. The sporophytes are the brown, jacketed structures on the white stalks. Because of its good antiseptic properties and high absorbency, peat moss was used as an emergency poultice on the wounds of soldiers during World War I.

male gametophyte　　female gametophyte　　gemmae

a

b

c

Figure 25.6 *Marchantia*, one of the liverworts. Like other liverworts, this nonvascular plant can reproduce sexually. Unlike other types, it forms (**a**) male reproductive structures and (**b**) female reproductive structures on different plants. (**c**) *Marchantia* also reproduces asexually by way of gemmae, multicelled vegetative bodies that develop in tiny cups on the plant body. Gemmae grow and develop into individual plants after splashing raindrops transport them to suitable sites.

patterns on tree trunks and branches in humid climates. Eggs and sperm develop in tiny, jacketed vessels at the shoot tips of gametophytes. Sperm reach the eggs by swimming through a film of water on plant parts. After fertilization, the zygotes give rise to sporophytes, each consisting of a stalk and a jacketed structure in which spores will develop.

Figure 25.5 shows one of 350 kinds of peat mosses (*Sphagnum*). Whereas most bryophytes grow slowly, the peat mosses can grow fast enough to yield twelve metric tons of organic matter per hectare, which is about twice as much as corn plants yield. They soak up five times as much water as cotton does, owing to large, dead cells in their leaflike parts. They also produce acids that inhibit the growth of bacterial and fungal decomposers. Their remains accumulate into compressed, excessively moist mats called **peat bogs**. In cold and temperate regions, peat bogs cover an area equal to one-half of the United States. Only the most acid-tolerant plants, such as larch,

cranberries, blueberries, and Venus flytraps, can grow in the bogs, which can be as acidic as vinegar.

Nearly all of the peat harvested and dried in Ireland and elsewhere is burned to generate electricity in power plants. Compared with coal burning, peat fires generate fewer pollutants. Every so often, peat harvesters come across the exceedingly well-preserved bodies of humans that lived 2,000 to 3,000 years ago. Some ancient bogs apparently were sites of ceremonial human sacrifices.

So as not to dwell on the macabre, let us leave this section with the liverworts and their interesting ways of reproducing, as described in Figure 25.6.

Bryophytes are nonvascular plants with flagellated sperm that require liquid water to reach and fertilize the eggs.

A sporophyte of these plants develops within gametophyte tissues. It remains attached to the gametophyte and receives some nutritional support from it.

Figure 25.7a shows one of the early seedless vascular plants. Descendants of certain lineages are still with us; we call them **whisk ferns**, **lycophytes**, **horsetails**, and **ferns**. Like their ancestors, they differ from bryophytes in three key respects. The sporophyte does not remain attached to a gametophyte; it has true vascular tissues; and it is the larger, longer lived phase of the life cycle.

Most seedless vascular plants live in wet, humid places, and their gametophytes lack vascular tissues. Water droplets clinging to the plants are the only means by which flagellated sperm can reach the eggs. The few species living in dry habitats reproduce sexually during brief, seasonal pulses of heavy rains. Thus, whisk ferns, lycophytes, horsetails, and ferns are the "amphibians" of the plant kingdom. They have not fully escaped the aquatic habitats of their ancestors.

Whisk Ferns

Whisk ferns (Psilophyta), which are not ferns, resemble whisk brooms. Florist suppliers commonly cultivate them in Hawaii, Texas, Louisiana, Florida, Puerto Rico, and other tropical or subtropical regions. One genus,

Psilotum, is a unique vascular plant, for its sporophytes have no roots or leaves. The photosynthetic, branched stems have xylem, phloem, and scalelike projections (Figure 25.7b). Belowground are **rhizomes**, branching, short, mostly horizontal stems that serve in absorption.

Lycophytes

About 350 million years ago, lycophytes (Lycophyta) included tree-sized members of swamp forests. About 1,100 far tinier species exist today. The most familiar are club mosses—members of communities in the Arctic, the tropics, and regions in between. Many form mats on forest floors. One type, called the resurrection plant, is common in Texas, New Mexico, and Mexico.

Sporophytes of most club mosses have leaves and a branching rhizome that gives rise to vascularized roots and stems. Some have nonphotosynthetic, cone-shaped clusters of leaves that bear spore sacs (Figure 25.7c). Each cluster is a **strobilus** (plural, strobili). After spores disperse, they germinate and then develop into small, free-living gametophytes. In one genus (*Selaginella*), two kinds of spores develop in the same strobilus.

Horsetails

Tree-sized sphenophytes (Sphenophyta) flourished in ancient swamp forests. Twenty-five or so smaller species of one genus (*Equisetum*) made it to the present. These are the horsetails. Of all existing plants, they may be the oldest, and their body plan changed little over the past 300 million years. They grow in streambank mud and vacant lots, roadsides, and other disrupted habitats. Figure 25.7d–f shows the fertile stems and vegetative, photosynthetic stems of one species. Its spores give rise

Figure 25.7 (a) *Cooksonia*, one of the earliest known vascular plants, no more than a few centimeters tall. It probably grew in mud flats. Its upright, branching stems had a cuticle. Its spores formed in sporangia at stem tips. Compare Figure 21.2c. (b) Sporophytes of a whisk fern (*Psilotum*), a seedless vascular plant. Pumpkin-shaped, spore-producing structures form at the ends of stubby branchlets. (c) Sporophyte of one of the lycophytes (*Lycopodium*). (d) Vegetative stem of *Equisetum*, which resembles a horsetail. (e) Fertile, nonphotosynthetic stems of *Equisetum*. (f) Closer look at the spore-bearing structure of a fertile stem.

a

strobilus, an aggregation of spore-producing structures, at tip of a vegetative shoot of a horsetail sporophyte

f Closer look at a strobilus. Inside each petal-shaped structure, many spores have been forming by way of meiotic cell divisions.

b   c   d   e

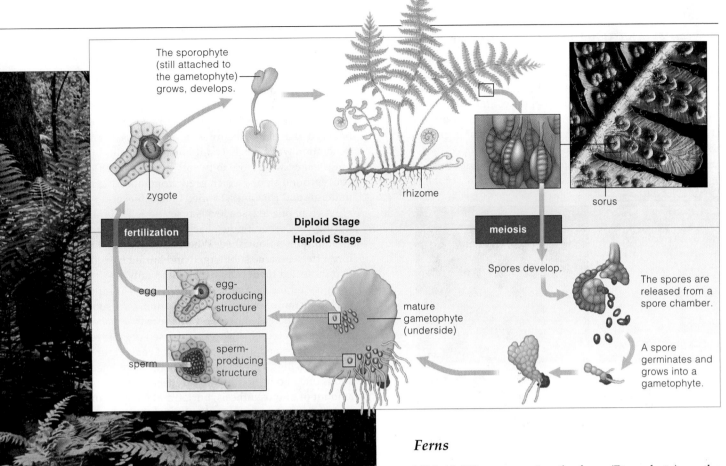

The sporophyte (still attached to the gametophyte) grows, develops.

zygote

rhizome

sorus

fertilization

Diploid Stage

Haploid Stage

meiosis

egg

egg-producing structure

sperm

sperm-producing structure

mature gametophyte (underside)

Spores develop.

The spores are released from a spore chamber.

A spore germinates and grows into a gametophyte.

Figure 25.8 Life cycle of a fern. The photograph shows ferns growing in a moist habitat in Indiana. Ferns with finely divided fronds are in the foreground.

to free-living gametophytes 1 millimeter to 1 centimeter across. Sporophytes of most horsetails have rhizomes, hollow photosynthetic stems, and scale-shaped leaves. Clusters of xylem and phloem are arrayed as a ring in the stems. Silica-reinforced ribs structurally support the stems and give them a texture like sandpaper. Pioneers of the American West, who had to travel light, gathered horsetails along the way to use as pot scrubbers.

Ferns

With 12,000 or so species, the ferns (Pterophyta) are the largest and most diverse group of seedless vascular plants. All but about 380 are native to the tropics, but they are popular houseplants all over the world. Their size range is stunning. Some floating species are less than 1 centimeter across. Some tropical tree ferns are 25 meters (82 feet) tall. One climbing fern has a modified leaf stalk about 30 meters long.

Most ferns have vascularized rhizomes that give rise to roots and leaves. Exceptions include tropical tree ferns and epiphytes. (*Epiphyte* refers to any aerial plant that grows attached to tree trunks or branches.) While they develop, young fern leaves are coiled, rather like a fiddlehead. At maturity, the leaves (fronds) commonly are divided into leaflets.

You may have noticed rust-colored patches on the lower surface of many fern fronds. Each patch, a cluster of spore chambers, is called a sorus (plural, sori). At dispersal time, the chambers snap open and cause the spores to catapult through the air. Each germinating spore develops into a small gametophyte, such as the green, heart-shaped type shown in Figure 25.8.

Seedless vascular plants (whisk ferns, lycophytes, horsetails, and ferns) have sporophytes adapted to conditions on land. Yet they have not entirely escaped their aquatic ancestry. When they reproduce sexually, their flagellated sperm cannot reach the eggs unless liquid water is clinging to the plant.

ANCIENT CARBON TREASURES

Three hundred million years ago, about halfway through the Carboniferous, mild climates prevailed and swamp forests carpeted the wet lowlands of the continents. The absence of pronounced seasonal swings in temperature favored plant growth through much of the year. The plants having lignin-reinforced tissues and well-developed root and shoot systems had the competitive edge under these growth conditions, and some of them

Lepidodendron

evolved into giants. Massive-stemmed lycophyte trees—the giant club mosses—topped out at nearly forty meters (Figure 25.9). Each of their strobili produced as many as 8 billion microspores or several hundred megaspores. And being so high above the forest floor, dispersal of each new generation was a cinch. Giant horsetails, including species of *Calamites*, were close to twenty meters tall. Often their aboveground stems, which grew from rapidly spreading underground rhizomes, formed dense thickets.

As it happened, sea levels rose and fell fifty times during the Carboniferous. Each time the seas receded, the swamp forests flourished. When the seas moved back in, forest trees became submerged and buried in sediments that protected them from decay. Gradually the sediments compressed the saturated, undecayed remains into peat. Each time more sediments accumulated, they subjected the peat to increased heat and pressure that made it even more compact. In this way, compressed organic remains were transformed into great seams of **coal** (Figure 25.9).

With its high percentage of carbon, coal is energy rich and is one of our premier "fossil fuels." It took a fantastic amount of photosynthesis, burial, and compaction to form each major seam of coal in the Earth. It has taken us only a few centuries to deplete much of the world's known coal deposits. Often you will hear about annual "production rates" for coal or some other fossil fuel. But how much do we really produce each year? None. We simply *extract* it from the Earth. Coal is a nonrenewable source of energy.

stem of a giant lycophyte
(*Lepidodendron*)

seed fern (*Medullosa*); probably related to the progymnosperms,
which may have been among the earliest seed-bearing plants

stem of a giant horsetail
(*Calamites*)

Figure 25.9   Reconstruction of a Carboniferous forest. The boxed inset shows part of a seam of coal.

THE RISE OF THE SEED-BEARING PLANTS

About 360 million years ago, as the Devonian gave way to the Carboniferous, the first seed-bearing plants arose. In terms of diversity, numbers, and distribution, they would become the most successful groups of the plant kingdom. Seed ferns, gymnosperms, and (much later) angiosperms were the dominant groups. They differed from seedless vascular plants in three crucial respects.

First, seed-bearing plants produce pollen grains, the sperm-bearing male gametophytes. Remember, these plants produce two types of spores. Their **microspores** give rise to pollen grains. Unlike the spores of seedless vascular plants, they do not have a "tetrad scar," which marks the cleavage planes between four spores that form during meiotic cell division (Figure 25.10).

Like a suitcase, a pollen grain is a means of getting its contents (the sperm) to the eggs even during times of prolonged drought. Seedless vascular plants do not have such an advantage; without predictable rains and moisture, their sperm simply cannot reach the eggs, and this has adverse effects on reproductive success. By contrast, pollen grains of gymnosperms simply drift with air currents. Those of angiosperms also are loaded onto insects, birds, bats, and other animals that truck them to the eggs. **Pollination** is the name for the arrival of pollen grains on the female reproductive structures. By this process, seed-bearing plants escaped dependence on free water for fertilization.

Second, besides microspores, seed-bearing plants also produce **megaspores.** These develop into **ovules,** the female reproductive structures which, at maturity, are seeds (Figure 25.11). Each ovule consists of a female gametophyte (with its egg cell), nutrient-rich tissue, and a jacket of cell layers which, recall, develops into the seed coat. A zygote will form inside the ovule when a sperm reaches and fertilizes the egg. An embryo sporophyte will develop, and when the time comes to say good-bye to the parent plant, the seed coat around it will afford protection for the journey.

Third, compared to seedless vascular plants, these plants had thicker cuticles, stomata recessed below the surface of the leaf, and other traits that gave them competitive advantages in drier and cooler environments. Such environments were ushered in when the Carboniferous gave way to the Permian (Section 21.5). **Seed ferns** of the type shown in Figure 25.9 rose to dominance and then prevailed for about 70 million years. These seed-bearing plants apparently arose from **progymnosperms.** Progymnosperms may have been the first seed-bearing plants, although their seeds may only have been seed-like structures. When the global climate grew cooler and drier, swamplands disappeared, and so did the seed ferns. They were replaced by the cycads, conifers, and other gymnosperms.

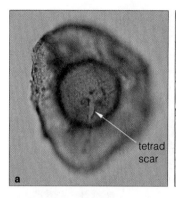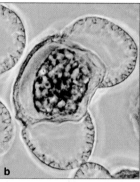

tetrad scar

a

b

Figure 25.10 (**a**) From the Devonian-Carboniferous boundary, a fossilized spore of a lycophyte. Its tetrad scar is typical of the spores of seedless plants. (**b**) A pine pollen grain (*Pinus*). LIke the pollen of other seed-bearing plants, it has no tetrad scar.

ovule, which may mature into a seed

seed (mature ovule); the surrounding ovary matured to form the fleshy fruit

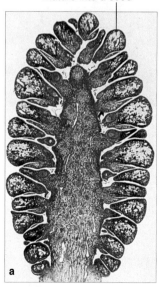

a

b

Figure 25.11
(**a**) Longitudinal section through a seed-bearing cone of a pine. These seeds are not protected by sporophyte tissues; they are exposed on scales of a cone. (**b**) A mature ovule, or seed, of a peach (*Prunus*), one of the angiosperms. It is protected by a seed coat, which is enclosed within the fleshy, edible tissue of the fruit.

In time, gymnosperms were the primary producers that sustained the dinosaurs. What some folks call the Age of Dinosaurs, botanists call the Age of Cycads.

Seed-bearing plants rely on pollen grains, ovules that mature into seeds, and tissue changes adapted to dry conditions.

GYMNOSPERMS—PLANTS WITH "NAKED" SEEDS

With a bit of history behind us, we turn now to a survey of some of the existing gymnosperms. Unlike the seeds of flowering plants, which are enclosed in a chamber called an ovary, gymnosperm seeds are perched, in an exposed way, on a spore-producing structure. (*Gymnos* means naked; *sperma* is taken to mean seed.)

Conifers

Conifers (Coniferophyta) are woody trees and shrubs that have needlelike or scalelike leaves and that bear seeds exposed on cone scales. Conifer **cones** are clusters of modified leaves that surround the spore-producing structures. Figure 25.12 gives examples.

Most conifers shed some leaves all year long yet stay leafy, or *evergreen*. A few are *deciduous*, meaning they shed all of their leaves in the fall. Among the conifers are the most abundant trees of the Northern Hemisphere (pines), the tallest (coast redwoods), as well as the oldest (the bristlecone pine; one 4,725-year-old individual sprouted when Egyptians were building the Great Sphinx). The firs, yews, spruces, junipers, larches, cypresses, bald cypress, dawn redwood, and podocarps also belong to this group.

Lesser Known Gymnosperms

CYCADS About 100 species of **cycads** (Cycadophyta) made it to the present day. Figure 25.13 shows examples of their pollen-bearing and seed-bearing cones, which form on separate plants. Insects or air currents transfer pollen from "male" to "female" plants. At first glance, you might mistake cycad leaves for those of a palm tree; but palms are flowering plants.

These plants mainly inhabit tropical and subtropical areas. Two species (*Zamia*) grow wild in Florida and are planted as ornamentals. Elsewhere, their seeds and a flour made from the trunks are edible after their toxic alkaloids are removed. Many species are vulnerable to extinction.

Figure 25.12 (**a**) Bristlecone pine (*Pinus longaeva*) growing near the timberline in the Sierra Nevada. (**b**) Male pine cones releasing pollen. (**c**) Appearance of a female pine cone at the time of pollination. (**d**) From a juniper (*Juniperus*), cones with a berry-like appearance. These cones are made of fused-together, fleshy scales.

Figure 25.13
(**a**) Pollen-bearing cone of a "male" cycad (*Zamia*).
(**b**) Seed-bearing cone of a "female" cycad. Of all the existing species of gymnosperms, the cycads produce the largest seed-bearing cones. Some of these grow as long as one meter and weigh more than fifteen kilograms.

Figure 25.14 (**a**) Mature ginkgo. (**b**) Fossilized ginkgo leaf compared with a leaf from its existing descendant. The fossil formed at the Cretaceous-Tertiary boundary. Even though 65 million years have passed since then, the leaf structure has not changed much, if at all. (**c**) Pollen-bearing cones and (**d**) fleshy-coated seeds of this type of gymnosperm.

Figure 25.15 (**a**) Sporophyte of *Ephedra* and (**b**) its pollen-bearing cones and (**c**) a seed-bearing cone. (**d**) Sporophyte of *Welwitschia mirabilis* and (**e**) its seed-bearing cones.

GINKGOS The **ginkgos** (Ginkgophyta) were a diverse group in dinosaur times. The only surviving species is the maidenhair tree, *Ginkgo biloba*. Like a few species of larch and some other gymnosperms, these plants are deciduous. Several thousand years ago, ginkgo trees were widely planted around temples in China. Then the natural populations nearly became extinct, even though ginkgos seem hardier than many other trees. Perhaps they became targets for firewood. Today, male ginkgo trees are again widely planted. They have attractive, fan-shaped leaves and are resistant to insects, disease, and air pollutants. Female trees are not favored. Their thick, fleshy seeds, which are the size of small plums, give off an awful stench (Figure 25.14).

GNETOPHYTES At present, there are three genera of woody plants known as the **gnetophytes** (Gnetophyta). Trees and leathery leafed vines of *Gnetum* thrive in the humid tropics. The shrubby *Ephedra* lives in California

deserts and some other arid regions (Figure 25.15*a–c*). Photosynthesis proceeds in its green stems. *Welwitschia mirabilis* grows in hot deserts of south and west Africa. Its sporophyte is mainly a deep-reaching taproot. Its exposed part, a woody disk-shaped stem, has cones and one or two strap-shaped leaves that split lengthwise repeatedly as the plant ages (Figure 25.15*e*).

Conifers, cycads, ginkgos, and gnetophytes are groups of existing gymnosperms. Like their ancestors, they bear their seeds on the exposed surfaces of cones and other spore-producing structures.

A CLOSER LOOK AT THE CONIFERS

Before we leave the gymnosperms, let's use the conifers as an example of reproductive strategies. Depending on the species, a conifer's life cycle lasts a year or more.

Who among us hasn't noticed the woody, shelflike scales of "a pine cone"? The scales are actually parts of a mature female cone in which megaspores formed and developed into female gametophytes. Pine trees also produce male cones, in which microspores form and develop into pollen grains (Figure 25.16). Each spring, millions of pollen grains drift away from the male cones.

Pollination is completed when some land on ovules of female cones. After each germinates, a tubular structure forms from it. This germinating pollen grain, the sperm-bearing male gametophyte, grows toward the egg in the female gametophyte. For species of pines, fertilization occurs months or a year after pollination.

As in other gymnosperms, seed formation begins at the ovule (Figure 25.16). An embryo sporophyte starts developing from the fertilized egg. The outer layers of the jacket around the female gametophyte and embryo mature into a hard coat. The seed coat will protect the embryo sporophyte after it is dispersed from the parent plant. The nutrients will help it through the critical time

Figure 25.16 Life cycle of one conifer, the ponderosa pine.

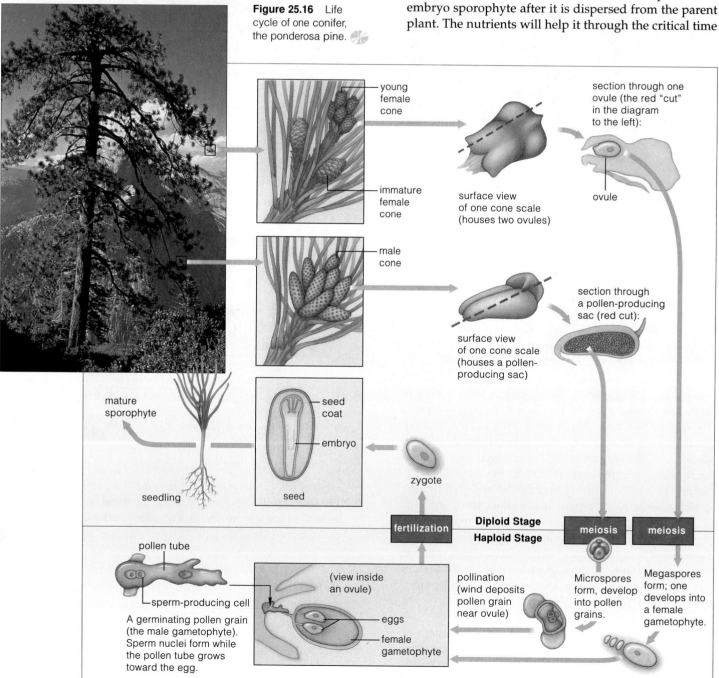

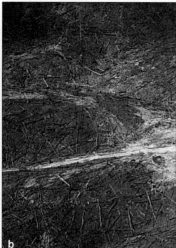

Figure 25.17 A sampling of the rampant deforestation under way around the world. Only deforested tracts in North America are shown; Section 50.4 focuses on the tropical rain forests.

From America's heartland, logged-over acreage in Arkansas (**a**). From the eastern seaboard, a denuded piece of North Carolina (**b**). Clear-cut peaks in Washington (**c**) and Alaska (**d**). These are not isolated examples. In the early 1980s, 400 million board feet of timber were being cut in Washington's Olympic Peninsula every year. In Arkansas, about one-third of the Ouachita National Forest was clear-cut. Its once-diverse forest communities have been replaced by "tree farms" of a single species of pine. Throughout the world, huge tracts of land that were deforested years ago still show no signs of recovery.

of germination, before its roots and shoots become fully functional.

Conifers dominated many land habitats during the Mesozoic, but their slow reproductive pace put them at a competitive disadvantage when the flowering plants began their great adaptive radiation (Section 21.6). Coniferous forests still predominate in the far north, at higher elevations, and in some parts of the Southern Hemisphere. However, existing conifers face more than competition with flowering plants for resources. Now they are vulnerable to **deforestation**: the removal of all trees from large tracts, as by clear-cutting (Figure 25.17). Conifers just have the bad luck to be premier sources of lumber, paper, and other wood products required in human societies. We return to this topic in Chapter 50.

Where flowering plants flourish, conifers are at a competitive disadvantage, partly because they take so long to reproduce. Rampant deforestation isn't helping them one bit, either.

ANGIOSPERMS—THE FLOWERING, SEED-BEARING PLANTS

Only angiosperms produce specialized reproductive structures called **flowers** (Figure 25.18). *Angeion,* which means vessel, refers to the female reproductive parts at the center of a flower. The enlarged base of the "vessel" is the floral ovary, where ovules and seeds develop.

Most flowering plants coevolved with **pollinators**—insects, bats, birds, and other animals that withdraw nectar or pollen from a flower and, in so doing, transfer pollen to its female reproductive parts. The recruitment of animals as assistants in reproduction probably has contributed to the success of flowering plants, which have dominated the land for 100 million years.

At least 260,000 species now live in a great variety of habitats. They range in size from the tiny duckweeds (about a millimeter long) to towering *Eucalyptus* trees, some of which are more than 100 meters tall. A few species, including mistletoes and Indian pipe, are not even photosynthetic; they directly or indirectly feed on other plants.

Figure 25.18 The unique trademark of angiosperms —the flower, a reproductive structure that has roles in pollination and formation of seeds. (**a**) A hummingbird is sipping nectar from this passion flower. (**b**) A few angiosperms, including the water lilies (*Nymphaea*), live in water. (**c**) Dwarf mistletoe (*Arceuthobium*), a parasitic plant, limits the growth of forest trees in the western United States. (**d**) Indian pipe (*Monotropa uniflora*) is one of the few nonphotosynthetic species. It gets nutrients from fungal symbionts that form mycorrhizae with young roots of photosynthetic plants. (**e**) Parts of flowers and seeds.

Dicots and Monocots

There are two great classes of flowering plants, called the **dicots** and **monocots** (formally, the Dicotyledonae and Monocotyledonae). Among the 180,000 dicots are most herbaceous (nonwoody) plants, such as cabbages and daisies; most flowering shrubs and trees, such as oaks and apple trees; water lilies; and cacti. Among the 80,000 or so species of monocots are orchids, palms, lilies, and grasses, such as rye, sugarcane, corn, rice, wheat, and other highly valued crop plants.

In flowers (reproductive structures), seeds develop from ovules and some tissues of the parent sporophyte after fertilization.

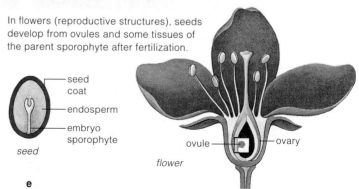

seed coat
endosperm
embryo sporophyte
seed

ovule
ovary
flower

e

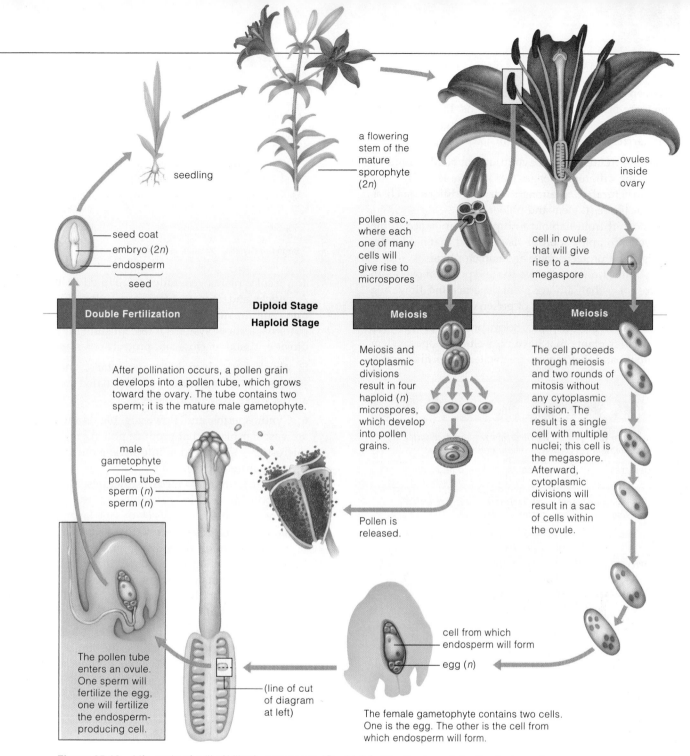

Figure 25.19 Life cycle of a lily (*Lilium*), a monocot. "Double" fertilization is a distinctive feature of flowering plant life cycles. A male gametophyte delivers *two* sperm to an ovule. One sperm fertilizes the egg, and the other fertilizes a cell that gives rise to endosperm, a tissue that will nourish the forthcoming embryo. Figure 32.4 provides a closer look at flowering plant life cycles, using a dicot as the example.

Key Aspects of the Life Cycles

The next unit deals with the structure and function of flowering plants. For now, simply start thinking about the large sporophyte that dominates their life cycles (Figure 25.19). It retains and nourishes gametophytes; its sperm are dispersed within pollen grains. Inside flowering plant seeds, endosperm (a nutritive tissue) surrounds the embryo. Also, as seeds develop, tissues of most ovaries and some other structures mature into **fruits**, which protect and help disperse the embryos.

Angiosperms are the most successful plants in terms of their diversity, numbers, and distribution. They alone produce flowers. Most species coevolved with animal pollinators.

SUMMARY

1. Green algae probably gave rise to plants, nearly all of which are multicelled photoautotrophs. Plants invaded the land more than 430 million years ago. Table 25.1 summarizes and compares the major plant phyla.

2. Several trends in plant evolution can be identified by comparing different lineages (see also Table 25.2):

   a. Structural adaptations to dry conditions, such as vascular tissues (xylem and phloem).

   b. A shift from haploid to diploid dominance in the life cycle. Complex sporophytes evolved that hold onto, nourish, and protect spores and gametophytes.

   c. A shift from one to two spore types (homospory to heterospory) that, among gymnosperms and flowering plants, led to the evolution of pollen grains and seeds.

3. Mosses, liverworts, and hornworts are bryophytes, nonvascular plants that have no well-developed xylem or phloem and that require free water for fertilization.

Table 25.1 Comparison of Major Plant Groups

Nonvascular land plants. Fertilization requires free water. Haploid dominance. Cuticle, stomata present in some.

BRYOPHYTES	18,600 species. Moist, humid habitats.

Seedless vascular plants. Fertilization requires free water. Diploid dominance. Cuticle, stomata present.

WHISK FERNS	7 species, sporophytes with no obvious roots or leaves. *Psilotum.*
LYCOPHYTES	1,100 species with simple leaves. Mostly wet or shady habitats.
HORSETAILS	25 species of single genus. Swamps, disturbed habitats.
FERNS	12,000 species. Wet, humid habitats in mostly tropical, temperate regions.

Gymnosperms—vascular plants with "naked seeds." Diploid dominance. Cuticle, stomata present.

CONIFERS	550 species, mostly evergreen, woody trees and shrubs having pollen- and seed-bearing cones. Widespread distribution.
CYCADS	185 slow-growing species. Tropics, subtropics.
GINKGO	1 species, a tree with fleshy-coated seeds.
GNETOPHYTES	70 species. Limited distribution in deserts and tropics.

Angiosperms—vascular plants with flowers and protected seeds. Diploid dominance. Cuticle, stomata present.

FLOWERING PLANTS

Monocots	80,000 species. Floral parts often arranged in threes or in multiples of three; one seed leaf; parallel leaf veins common.
Dicots	At least 180,000 species. Floral parts often arranged in fours, fives, or multiples of these; two seed leaves; net-veined leaves common.

Table 25.2 Evolutionary Trends Among Plants

Bryophytes	Ferns	Gymnosperms	Angiosperms

Nonvascular ⟶ Vascular ⟶

Haploid dominance ⟶ Diploid dominance ⟶

Spores of one type ⟶ Spores of two types ⟶

Motile gametes ⟶ Nonmotile gametes* ⟶

Seedless ⟶ Seeds ⟶

* Require pollination by wind, insects, etc.

4. Vascular plants generally are adapted to life on land. A cuticle and stomata conserve water. Root systems mine soil for nutrients. Upright and branching growth patterns of shoot systems intercept sunlight and carbon dioxide. Tissues enclose and protect spores and gametes.

5. Whisk ferns, lycophytes, horsetails, and ferns are seedless vascular plants. Their flagellated sperm require ample water to swim to the eggs.

6. Gymnosperms and flowering plants (angiosperms) are vascular plants that produce pollen grains (matured microspores that develop into male gametophytes) and seeds (mature ovules).

   a. Ovules are reproductive structures that contain the egg-producing female gametophytes, the precursor of nutritive tissue, and a jacket of cell layers, the outer portion of which develops into the seed coat.

   b. The evolution of pollen grains freed these plants from dependence on water for fertilization. Their seeds are efficient means of dispersing new generations even during hostile conditions. Pollen grains and seeds were key adaptations in the move to high, dry habitats.

7. Only angiosperms produce flowers. Most coevolved with pollinators, which enhance the transfer of pollen grains to female reproductive parts. Their seeds have a unique nutritive tissue (endosperm) and are usually surrounded by fruits, which aid in dispersal.

Review Questions

1. Identify a few of the structural and reproductive modifications that helped plants invade and diversify in habitats on land. *25.1*

2. Does the haploid phase or diploid phase dominate the life cycles of most plants? *25.1*

3. Name representatives of the following groups of plants and compare their key characteristics: (*also refer to Table 25.1*)
   a. Bryophytes and seedless vascular plants *25.2, 25.3*
   b. Gymnosperms and angiosperms *25.6–25.8*

4. Distinguish between:
   a. Root system and shoot system *25.1*
   b. Xylem and phloem *25.1*
   c. Sporophyte and gametophyte *25.1*
   d. Ovule and seed *25.1, 25.5*
   e. Microspore and megaspore *25.5*

Figure 25.20 Where many conifers end up.

Self-Quiz (Answers in Appendix IV)

1. Which of the following statements is *not* true?
   a. Monocots and dicots are two classes of angiosperms.
   b. Bryophytes are nonvascular plants.
   c. Lycophytes and angiosperms are both vascular plants.
   d. Gymnosperms are the simplest vascular plants.

2. Of all land plants, bryophytes alone have independent _____ and attached, dependent _____ .
   a. sporophytes; gametophytes   c. rhizoids; zygotes
   b. gametophytes; sporophytes   d. rhizoids; stalked sporangia

3. Psilophytes, lycophytes, horsetails, and ferns are _____ plants.
   a. multicelled aquatic   c. seedless vascular
   b. nonvascular seed   d. seed-bearing vascular

4. Which does *not* apply to gymnosperms and angiosperms?
   a. vascular tissues   c. single spore type
   b. diploid dominance   d. all of the above

5. A seed is _____ .
   a. a female gametophyte   c. a mature pollen tube
   b. a mature ovule   d. an immature embryo

6. Match the terms appropriately.
   _____ gymnosperm   a. produces haploid gametes
   _____ sporophyte   b. control water loss
   _____ seedless vascular   c. "naked" seeds
         plant   d. protects, disperses embryo
   _____ ovary   sporophyte
   _____ bryophyte   e. produces haploid spores
   _____ gametophyte   f. nonvascular land plant
   _____ stomata   g. lycophyte
   _____ angiosperm seed   h. usually a fruit at maturity

Critical Thinking

1. Elliot Meyerowitz of the California Institute of Technology has studied the genetic basis of flower formation in *Arabidopsis thaliana*. By inducing mutations in seeds of this small weed, he discovered three genes (call them A, B, and C) that interact in different parts of a developing flower. The gene interactions lead to the formation of different structures—sepals, petals, stamens, and carpels—from the same mass of undifferentiated tissue. From what you know of gene regulation, suggest ways in which the A, B, and C genes might be controlling flower development.

2. Genes nearly identical to the A, B, and C genes of *A. thaliana* also have been isolated from snapdragons and other flowering plants. About 150 million years ago, these plants originated and rose to dominance rather abruptly, in nearly all land habitats.

Speculate on how their rapid, spectacular radiation might have come about.

3. Figure 25.20 shows a forest in the Nahmint Valley of British Columbia, before and after logging. It also shows the wood frames of homes that are in the process of being built. Reflect on these photographs and the ones in Figure 25.17. As a way of stopping the loggers, would you chain yourself to a tree in an old-growth forest scheduled for clear-cutting? If your answer is yes, would you also give up the possibility of owning a wood-frame home (as most homes are in developed countries)? What about forest products, including wood for fireplaces and camp fires, newspapers, and toilet tissue?

4. With respect to question 3, multiply each of your answers by 5.8 billion (there are about that many people in the world) and describe what might happen when, inevitably, we run out of trees. Also describe what you might consider to be some of the pros and cons of tree farms of, say, a single species of pine.

Selected Key Terms

Readings

Gensel, P., and H. Andrews. 1987. "The Evolution of Early Land Plants." *American Scientist* 75: 478–489.

Moore, R., W. D. Clark, and K. Stern. 1995. *Botany.* Dubuque, Iowa: W. C. Brown.

Raven, P., R. Evert, and S. Eichhorn. 1986. *Biology of Plants.* Fourth edition. New York: Worth.

Web Site See *http://www.wadsworth.com/biology* for practice quiz questions, hypercontents, BioUpdates, and critical thinking. The Wadsworth Biology Resource Center provides a wealth of information fully organized and integrated by chapter.

ANIMALS: THE INVERTEBRATES

Madeleine's Limbs

In August of 1994, about 900 million years after the first animals appeared on Earth, Madeleine made *her* entrance. As they are wont to do, grandmothers and aunts made a quick count on the sly—arms, legs, ears, and eyes, two of each; fully formed mouth and nose—just to be sure these were present and accounted for.

One grandmother, having been too long in the company of biologists, experienced an epiphany as she witnessed Madeleine's birth. In that profound instant she sensed ancestral connections, emerging from the distant past and through her, into the future.

Madeleine's body plan did not emerge out of thin air. Thirty-five thousand years ago, people just like us were having children just like Madeleine. And if we are interpreting the fossil record correctly, then five million years ago the offspring of individuals on the road to modern humans resembled her in some respects but not others. Sixty million years ago, the primate ancestors of those individuals were giving birth precariously, up in the trees. Two hundred and fifty million years ago,

mammalian ancestors of those primates were giving birth—and so on back in time to the very first animals, which had no limbs or eyes or noses at all.

We have very few clues to what those first animals looked like, but one thing is clear. By the dawn of the Cambrian, they had given rise to all major groups of invertebrates—animals without backbones—even to Madeleine's backboned but limbless ancestors.

And what stories those Cambrian animals tell! One bunch flourished 530 million years ago, in a submerged basin that had formed between a reef and the coast of an early continent. Protected from ocean currents, the sediments had piled against the steep reef. About 500 feet below the surface, the water was oxygenated and clear. Tiny, well-developed animals lived in, on, and above the dimly lit muddy sediments (Figure 26.1a).

Like castles built from wet sand along a seashore, their living quarters were unstable. Part of the bank above the community slumped abruptly and obliterated it. The sediments from that underwater avalanche kept scavengers from reaching and removing all traces of the dead. Gradually through time, muddy silt rained down on the tomb. The increased pressure and

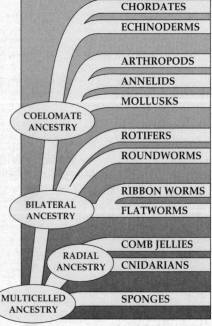

Figure 26.1 (**a**) Reconstruction of a few Cambrian animals, known from the Burgess Shale fossils. (**b**) Presumed evolutionary relationships among major groups of animals. Take a moment to study this evolutionary tree diagram. We will use it repeatedly as our road map through discussions of each group. (**c**) Madeleine.

CHORDATES
ECHINODERMS
ARTHROPODS
ANNELIDS
MOLLUSKS
COELOMATE ANCESTRY
ROTIFERS
ROUNDWORMS
RIBBON WORMS
BILATERAL ANCESTRY
FLATWORMS
COMB JELLIES
RADIAL ANCESTRY
CNIDARIANS
MULTICELLED ANCESTRY
SPONGES

b SINGLE-CELLED, PROTISTAN-LIKE ANCESTORS

chemical changes transformed the sediments into finely stratified shale; and soft parts of the flattened animals became shimmering mineralized films.

Sixty-five million years ago, part of the seafloor was plowing under the North American plate, and western Canada's mountain ranges were slowly rising. By 1909, the fossils were high in the eastern mountains of British Columbia. There, a fossil hunter tripped over a chunk of shale, which split apart into fine layers—and so the Burgess Shale story came to light.

In this chapter and the next, you will be comparing the body plans of different groups of animals. Such comparisons give insight into evolutionary relatedness and help us construct family trees, such as the one in Figure 26.1b. Don't assume that the structurally simple animals of the most ancient lineages are somehow primitive or evolutionarily stunted. As you will see, they, too, are exquisitely adapted to their environment.

As you poke through the branches of the animal family tree, keep the greater evolutionary story in mind. At each branch point, microevolutionary processes gave rise to workable changes in body plans. Madeleine's uniquely human traits, and yours, emerged through modification of certain traits that had evolved earlier in countless generations of vertebrates and, before them, in ancient invertebrate forms.

KEY CONCEPTS

1. All animals are multicelled, aerobic heterotrophs that ingest or parasitize other organisms. Nearly all kinds have tissues, organs, and organ systems, and most are motile during at least part of their life cycle. Animals reproduce sexually and often asexually, and their embryos develop in a series of continuous stages.

2. Animals originated approximately 900 million years ago. More than 2 million existing species have been identified. Of these, more than 1,950,000 are invertebrates (animals with no backbone). Fewer than 50,000 species are vertebrates (animals with a backbone).

3. Comparisons of the body plans of existing animals, in conjunction with the fossil record, reveal that there were several trends in the evolution of certain lineages. The most revealing aspects of an animal's body plan are its type of symmetry, gut, and cavity (if any) between the gut and body wall; whether it has a distinct head end; and whether it is divided into a series of segments.

4. The placozoans and sponges are structurally simple animals with no body symmetry. Both are at the cellular level of body construction. The cnidarians and the comb jellies have radial symmetry. They are at the tissue level of body construction.

5. Flatworms, roundworms, rotifers, and nearly all other animals that are more complex than the cnidarians show bilateral symmetry. They consist of tissues, organs, and organ systems.

6. Not long after the flatworms evolved, divergences gave rise to two major lineages. One evolutionary branching gave rise to the mollusks, annelids, and arthropods. The other gave rise to the echinoderms and chordates.

7. By biological measures, including diversity, sheer numbers, and distribution, the arthropods—especially insects—have been the most successful animal group.

General Characteristics of Animals

What, exactly, are **animals**? We can only define them by a list of general characteristics, not with just a sentence or two. *First*, animals are multicelled, and in most cases their body cells form tissues that are arranged as organs and organ systems. The body cells are diploid in nearly all species. *Second*, animals are heterotrophs that obtain carbon and energy by ingesting other organisms or by absorbing nutrients from them. *Third*, animals require oxygen, for use in aerobic respiration. *Fourth*, animals reproduce sexually and, in many cases, asexually. *Fifth*, most animals are motile during at least part of the life cycle. *Sixth*, their life cycles include a series of stages of embryonic development. In brief, mitotic cell divisions transform the animal zygote into a multicelled embryo. The embryonic cells are the forerunners of **ectoderm**, **endoderm**, and, in most species, **mesoderm**. These are the primary tissue layers which give rise to all tissues and organs of the adult, as described in Section 44.2.

Diversity in Body Plans

Mammals, birds, reptiles, amphibians, and fishes are the most familiar animals. All are **vertebrates**—the only animals with a "backbone." And yet, of probably more than 2 million species of animals, fewer than 50,000 are vertebrates! What we call **invertebrates** are animals with diverse features, but not a backbone.

We group animals into more than thirty phyla. Table 26.1 lists the ones that are described in this book. The characteristics they share with one another arose early in time, before divergences from a common ancestor gave rise to separate lineages. Later, as morphological differences accumulated among them, animal lineages took off in amazingly diverse directions. How might we get a conceptual handle on their modern descendants—on animals as different as flatworms, hummingbirds, platypuses, humans, and giraffes? We can compare their similarities and differences with respect to five basic features. These are body symmetry, cephalization, type of gut, type of body cavity, and segmentation.

BODY SYMMETRY AND CEPHALIZATION With very few exceptions, animals are radial or bilateral. Those with **radial symmetry** have body parts arranged regularly around a central axis, like spokes of a bike wheel. Thus a cut down the center of a hydra (Figure 26.2*a*) divides it into equal halves; another cut at right angles to the first divides it into equal quarters. All radial animals live in water. Their body plan is adapted to intercepting food that is coming toward them from any direction.

Animals with **bilateral symmetry** have right and left halves that are mirror images of each other. Most

Table 26.1   Animal Phyla Described in This Book

Phylum	Some Representatives	Existing Species
PLACOZOA (*Trichoplax*)	Simplest animal; like a tiny plate, but with layers of cells	1
PORIFERA (poriferans)	Sponges	8,000
CNIDARIA (cnidarians)	Hydrozoans, jellyfishes, corals, sea anemones	11,000
CTENOPHORA (comb jellies)	Like jellyfishes with comblike structures of modified cilia	11,000
PLATYHELMINTHES (flatworms)	Turbellarians, flukes, tapeworms	15,000
NEMATODA (roundworms)	Pinworms, hookworms	20,000
NEMERTEA (ribbon worms)	Proboscis-equipped worms closely related to flatworms	800
ROTIFERA (rotifers)	Tiny body with crown of cilia, great internal complexity	2,000
MOLLUSCA (mollusks)	Snails, slugs, clams, squids, octopuses	110,000
ANNELIDA (segmented worms)	Leeches, earthworms, polychaetes	15,000
ARTHROPODA (arthropods)	Crustaceans, spiders, insects	1,000,000+
ECHINODERMATA (echinoderms)	Sea stars, sea urchins	6,000
CHORDATA (chordates)	Invertebrate chordates: Tunicates, lancelets	2,100
	Vertebrates: Fishes Amphibians Reptiles Birds Mammals	21,000 3,900 7,000 8,600 4,500

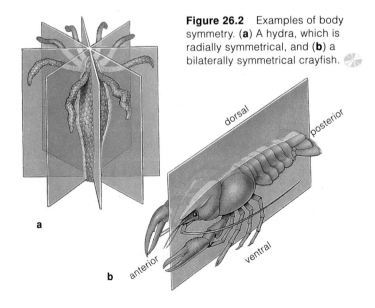

Figure 26.2 Examples of body symmetry. (**a**) A hydra, which is radially symmetrical, and (**b**) a bilaterally symmetrical crayfish.

dorsal   posterior   anterior   ventral

a

b

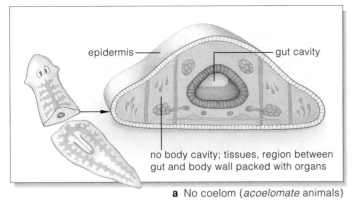

epidermis — — gut cavity

no body cavity; tissues, region between gut and body wall packed with organs

a No coelom (*acoelomate* animals)

epidermis — — gut cavity

unlined body cavity (pseudocoel) around gut

b Pseudocoel (*pseudocoelomate* animals)

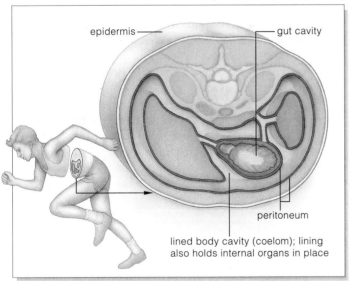

epidermis — — gut cavity

peritoneum

lined body cavity (coelom); lining also holds internal organs in place

c Coelom (*coelomate* animals)

Figure 26.3 Type of body cavity (if any) in animals.

have an *anterior* end (head) and an opposite, *posterior* end. They have a *dorsal* surface (a back) and an opposite, *ventral* surface (Figure 26.2*b*). This body plan evolved among the first forward-creeping species. Mostly, their forward end must have been the first to encounter food and other stimuli. We can imagine there was selection for **cephalization**. By this evolutionary process, sensory

structures and nerve cells became concentrated in the head. The joint evolution of bilateral body plans and cephalization resulted in pairs of muscles and pairs of sensory structures, nerves, and brain regions.

TYPE OF GUT The **gut** is a region inside the body in which food is digested, then absorbed into the internal environment. Saclike guts have one opening (a mouth) for taking in food and expelling residues. Other guts are part of a tubelike system with a mouth and an anus— that is, openings at both ends. Such "complete" digestive systems are very efficient. Different parts of them have specialized functions, such as preparing, digesting, and storing material. The evolution of such systems helped pave the way for increases in body size and activity.

BODY CAVITIES A body cavity separates the gut and the body wall of most bilateral animals (Figure 26.3). One type of cavity, a **coelom**, has a unique tissue lining called a peritoneum. This lining also encloses organs in the coelom and helps hold them in place. For example, your body houses a coelom, which a sheetlike muscle divides into two smaller cavities. Your heart and lungs are positioned inside the upper (thoracic) cavity, and your stomach, intestines, and other organs occupy the lower (abdominal) cavity.

Some invertebrates don't have a body cavity; tissues fill the region between their gut and body wall. Others have a pseudocoel ("false coelom"), a body cavity with no peritoneum. The coelom was a key innovation in the evolution of animals that were larger and more complex than ancestral forms. It favored increases in size and activity by cushioning and protecting internal organs.

SEGMENTATION Segmented animals have a repeating series of body units that may or may not be similar to one another. The many segments of earthworms have a similar outward appearance. Insect segments are fused into three units (head, thorax, and abdomen) and differ greatly from one another. Among insects especially, diverse head parts, legs, wings, and other appendages evolved from less specialized segments.

Animals are multicelled, and most have tissues, organs, and organ systems. The body cells of most species are diploid.

Animals are aerobically respiring heterotrophs that ingest other organisms or absorb nutrients from them.

Animals reproduce sexually and, in many cases, asexually. They go through a period of embryonic development, and most are motile during at least part of the life cycle.

Body plans of animals differ with respect to five features: body symmetry, cephalization, type of gut, type of body cavity, and segmentation.

From telltale tracks and burrows they left behind in marine sediments, we suspect that multicelled animals probably evolved by 900 million years ago. *Where did the first animals come from?* They probably originated with protistan lineages, although we don't know which ones (Sections 21.3 and 21.4).

By one hypothesis, the forerunners of animals were ciliates, much like *Paramecium*, that had multiple nuclei in a single-celled body; over time, each nucleus became compartmentalized in one cell of a multicelled body. Yet no existing animal develops by compartmentalization.

By another hypothesis, multicelled animals arose from spherical colonies of a number of flagellated cells, maybe like the *Volvox* colonies shown in Figure 6.3. In time, as a result of mutations, some cells in the colony became modified in ways that enhanced reproduction and other specialized tasks. So began the division of labor that characterizes multicellularity.

Suppose such colonies became flattened and started creeping about on the seafloor. This may have led to the evolution of layers of cells, such as those of *Trichoplax adhaerens*. This soft-bodied marine animal, shaped a bit like pita bread, has no symmetry and no mouth. It is the only known **placozoan** (Placozoa, after *plax*, meaning plate, and *zoon*, meaning animal). Its several thousand cells are arrayed in two distinctly different layers. When it glides over food, its body briefly humps up (Figure 26.4). Glandular cells present in the lower layer secrete digestive enzymes onto the food, then individual cells absorb the breakdown products. Reproduction might be asexual (by budding or fission) or it might be sexual, by mechanisms not yet understood. In sum, structurally and functionally, *Trichoplax* is as simple as animals get.

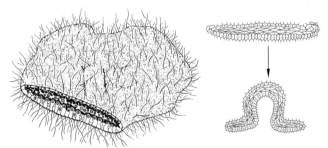

Figure 26.4 Cutaway view of *Trichoplax adhaerens*, an animal with a two-layer body measuring about three millimeters across.

Possibly the question of origins requires more than one answer. It may be that different lineages descended from more than one group of protistan-like ancestors.

Multicelled animals arose from protistan-like ancestors that may have resembled the ciliates, the colonial flagellates, or both.

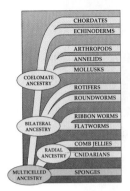

Sponges (Porifera) are animals with no symmetry, tissues, or organs, yet they are one of nature's success stories. They have been abundant in the seas ever since the Precambrian, especially in waters off coasts and along reefs. Of the 8,000 or so known species, only about 100 live in freshwater. Tiny worms, shrimps, and other animals make their home inside or on sponges.

Some sponges are large enough to sit in, and others are as small as a fingernail. Figures 26.5 and 26.6 show a few of the sprawling, flattened, lobed, compact, tubular, cuplike, and vaselike shapes. Regardless of its shape, the body is not symmetrical. Flattened cells line its outer surface and inner cavities, but the linings are not much more than the cell layers of *Trichoplax*, and they differ from the tissues of other animals. Amoeboid cells live in a gelatin-like substance between the two linings (Figure 26.6b). Glasslike spicules (of calcium carbonate or silica), tough fibers (of the protein spongin), or both stiffen the sponge body and impart structure to it.

The skeletal elements may be a reason why sponges as a group have endured so long. Cleveland Hickman put it this way: Most potential predators discover that sampling a sponge is about as pleasant as eating a mouthful of glass splinters embedded in fibrous gelatin. Besides, chemically speaking, many sponges stink.

Water flows into a sponge body through microscopic pores and chambers, and then out through one or more openings known as oscula (singular, osculum). It does so mainly when thousands or millions of **collar cells** are beating their flagella. These cells are components of the sponge body's inner lining. Besides having a flagellum, they have "collars" of food-trapping structures called microvilli (Figure 26.6c). Bacteria and other bits of food dissolved in the water become trapped in the collars, then are engulfed by way of phagocytosis. Some of the

Figure 26.5 A sprawling, red-orange sponge, one of many types that encrust underwater ledges in temperate seas.

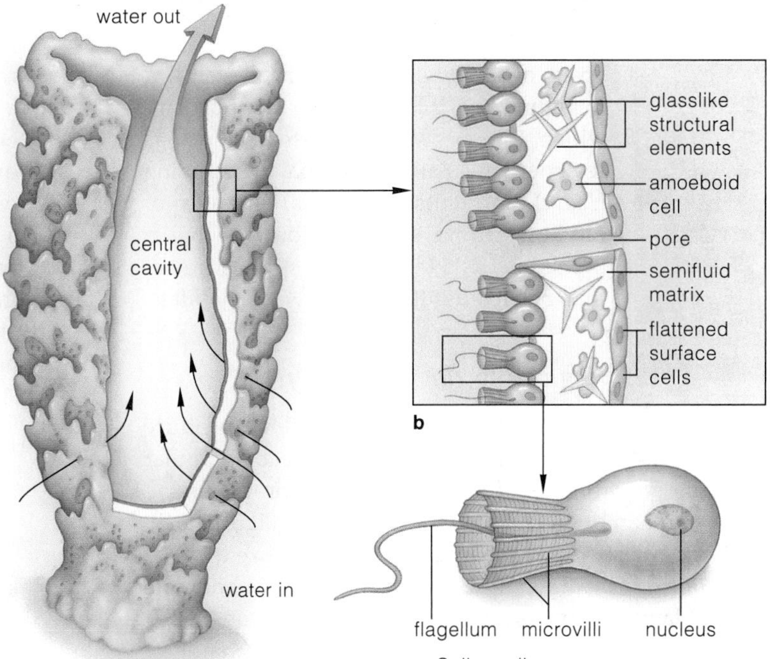

Figure 26.6 (**a**,**b**) Body plan of a simple sponge. The outer lining consists of flattened cells. It also has some contractile cells, arranged as bands around the tiny openings across the body. These cells contract slowly, independently of the other cell types, and so influence water flow through the body. Amoeboid cells inside the gelatin-like matrix between the inner and outer linings secrete materials from which the body's spicules and fibers are constructed. Other amoeboid cells digest and transport food. They do not lose the capacity to divide, and their cellular descendants can differentiate into any other type of sponge cell. They also have roles in asexual processes, such as gemmule formation.

(**c**) A great many flagellated cells line inner canals and chambers of the sponge body. Each has a collar of food-trapping structures called microvilli. Fine filaments connect the microvilli to each other; they form a "sieve" that strains food particles from the water. At the base of the collar, the cell engulfs the trapped food, by phagocytosis.

(**d**) An example of skeletal elements— Venus's flower basket (*Euplectella*). This marine sponge has six-rayed spicules of silica fused in a rigid, elaborate network. A thin layer of cells stretches over all of the interconnecting spicules. At the base of the body is an anchoring tuft of spicules.

(**e**) A basket sponge of the Caribbean, releasing a cloud of sperm into the water.

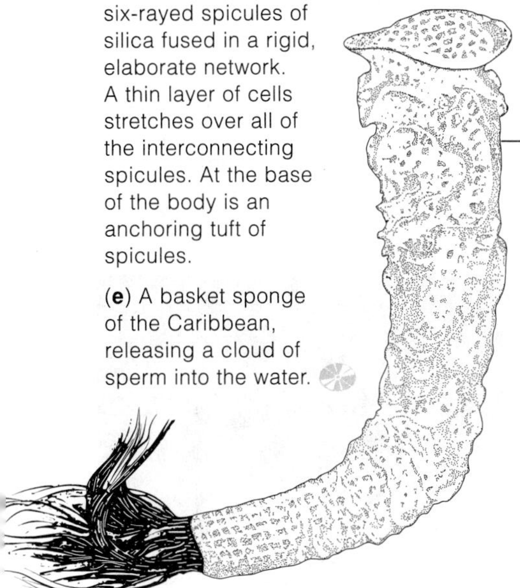

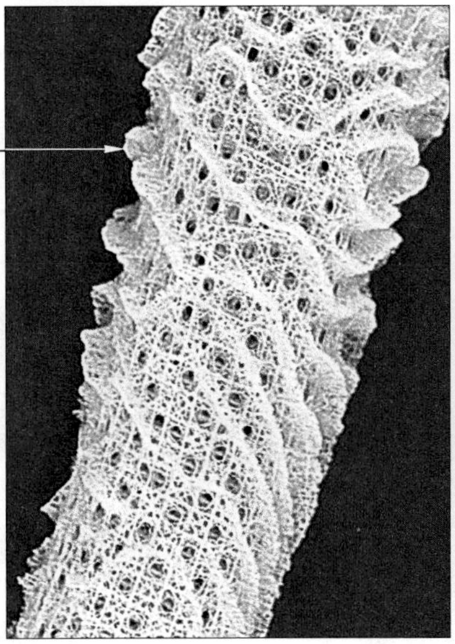

water out

central cavity

water in

a

glasslike structural elements

amoeboid cell

pore

semifluid matrix

flattened surface cells

b

flagellum   microvilli   nucleus

c Collar cell

d

e

engulfed food also gets transferred to the amoebalike cells for further breakdown, storage, and distribution.

Sponges reproduce sexually when some cells give rise to eggs and to sperm. Most release sperm into the water (Figure 26.6*e*), but the eggs typically are retained until after they have been fertilized and embryos start their development. Young sponges proceed through a microscopic, swimming larval stage. A **larva** (plural, larvae) is a sexually immature stage that grows and develops into the sexually mature form of the species, the **adult**. Many animal life cycles include larval stages.

Some kinds of sponges also can reproduce asexually by fragmentation (small fragments break away from the parent and grow into new sponges). Most freshwater species also reproduce asexually by way of gemmules. These are clusters of sponge cells, some of which form a hard covering around others. The clusters inside are protected from extreme cold or drying out. Later, when favorable conditions return, the gemmules germinate and establish a new colony of sponges.

Sponges have no symmetry, tissues, or organs; they are at the cellular level of construction. Yet they have successfully endured through time, possibly because most predators find their spicule-rich and often stinky bodies unappetizing.

CNIDARIANS—TISSUES EMERGE

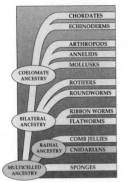

Think of a jellyfish, such as the one shown in Figure 26.7. It is one of the scyphozoans, which, together with anthozoans (such as sea anemones) and hydrozoans (including *Hydra*) are tentacled, radial animals of phylum **Cnidaria**. Most cnidarians live in the seas. Of 11,000 known species, fewer than 50 are adapted to freshwater habitats.

Regarding the Nematocysts

Of all animals, cnidarians alone produce **nematocysts**: capsules that house dischargeable, tubular threads. The threads of some types have prey-piercing barbs and an open tip that delivers toxins (Figure 26.8). Other nematocysts discharge threads that ooze a sticky substance from their tip or long threads that entangle prey. Many swimmers have learned that the toxin-tipped threads of some species can sting. Hence the name of the phylum, Cnidaria, after the Greek word for nettle.

Cnidarian Body Plans

The **medusa** (plural, medusae) and **polyp** are the most common cnidarian body forms. Both have a saclike gut (Figure 26.7a,b). Medusae float. Some look like bells and others like upside-down saucers. The mouth, centered under the bell, may have extensions that assist in prey capture and feeding. Polyps have a tubelike body with a tentacle-fringed mouth at one end. Usually the other end is attached to a substrate. When *Trichoplax* is draped over and digesting food, it is rather like a gut on the run. By contrast, the cnidarian gut is a permanent food-processing chamber. This gut has a gastrodermis, a sheetlike lining with many glandular cells that secrete digestive enzymes. An epidermis lines the rest of the body's surfaces (Figure 26.9).

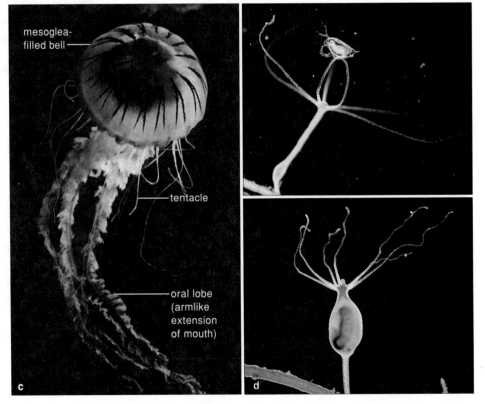

Figure 26.8 One type of nematocyst before and after a prey organism (not shown) touched its trigger. The contact made the capsule more "leaky" to water. Water diffused inward, turgor pressure built up within the capsule, and the thread was forced to turn inside out. The thread's tip pierced the prey's body.

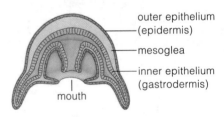

a Medusa, midsection

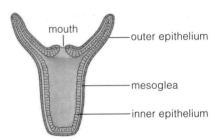

b Polyp, midsection

Figure 26.7 Cnidarian body plans. (**a**,**b**) Diagrams of a medusa and a polyp, sliced through the midsection. (**c**) The medusa of a sea nettle (*Chrysaora*), one of the jellyfishes. (**d**) A hydrozoan polyp (*Hydra*) attached to a substrate, as it captures and digests its prey.

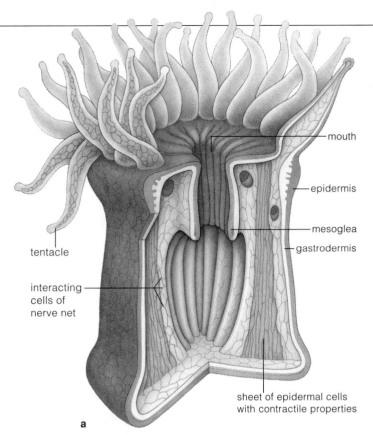

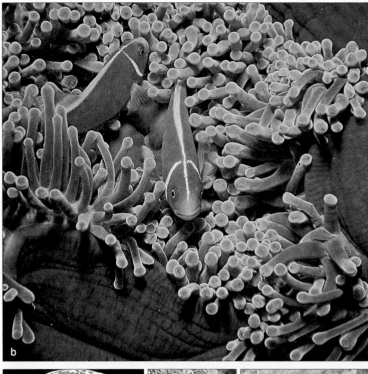

Figure 26.9 (**a**) Tissue organization of a sea anemone. (**b**) Tentacles fringing the mouth of a sea anemone, which eats many fishes but not clownfishes. A clownfish swims out, captures food, and returns to the tentacles, which protect it from predators. The sea anemone eats food scraps falling from the fish mouth. This is a case of mutualism, a two-way flow of benefits between species. (**c**) A sea anemone using its hydrostatic skeleton to escape from a sea star. It closes its mouth, and the force generated by contractile cells in epithelial tissues acts against water in the gut. The body changes shape and the anemone thrashes about, generally in a direction away from the predator.

In the figure (a) labels: mouth, epidermis, mesoglea, gastrodermis, tentacle, interacting cells of nerve net, sheet of epidermal cells with contractile properties

Each lining is an **epithelium** (plural, epithelia), a tissue with a free surface that faces the environment or some type of fluid inside the body. All animals more complex than sponges have epithelia. Cnidarian epithelia house **nerve cells**, which receive signals from receptors that sense changes in the surroundings and send signals to **contractile cells** that can carry out suitable responses. (When stimulated, contractile cells *shorten*, then return to their original length when stimulation stops.) The nerve cells interact as a "nerve net," a simple nervous tissue, to control movement and changes in shape.

Between the epidermis and gastrodermis is a layer of gelatinous secreted material, the mesoglea ("middle jelly"). Jellyfishes contain enough mesoglea to impart buoyancy and serve as a firm yet deformable skeleton against which contractile cells act. Imagine many cells contracting in coordinated ways in a jellyfish bell. Their action narrows the bell and forces water to jet out from underneath it, and the jet propels the jellyfish forward. The bell returns to its original position, cells contract

again, and the animal is propelled forward. Although the swimming movements are not Olympian, they work well enough for an animal that secures food by dragging its tentacles through the water for prey moving past.

Any fluid-filled cavity or cell mass against which contractile cells can act is a **hydrostatic skeleton**. With coordinated contractions, the cavity's volume or mass does not change but rather is shunted about, so that the shape of the body changes. The contractile cells of most polyps, which have little mesoglea, act against water in their gut. As you will see, nearly all animals have some form of skeletal-muscular system of movement.

Cnidarians are radial animals with tentacles, a saclike gut, epithelia, a nerve net, and a hydrostatic skeleton. They alone produce nematocysts.

Cnidarians are at the tissue level of construction; compared with sponges, they have layers of cells interacting in more coordinated fashion in the performance of specific tasks.

Various Stages in Cnidarian Life Cycles

From the preceding section, you might have concluded that the body form of a particular cnidarian is a medusa or a polyp. This is indeed the case for many species. However, the life cycle of *Obelia, Physalia,* and some other cnidarians includes both body forms. By using the life cycle of *Obelia* as an example, you can get a general sense of how these forms grow and develop (Figure 26.10).

The medusa is the sexual stage of the cnidarian life cycle. It has simple **gonads**, which are primary (gamete-producing) reproductive organs. Either the epidermis or gastrodermis houses the gonads, which release gametes by rupturing. Most of the zygotes formed at fertilization develop into **planulas**—a kind of swimming or creeping larva, usually with ciliated epidermal cells. In time, a mouth opens at one end, the larva is transformed into a polyp or medusa, and the cycle begins anew.

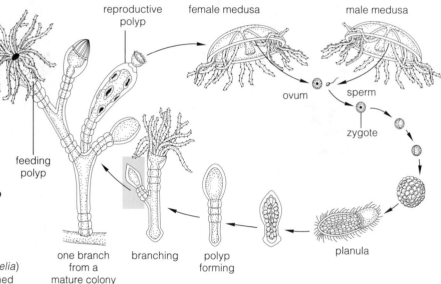

Figure 26.10 (*Right*) Life cycle of a hydrozoan (*Obelia*) that includes medusa and polyp stages. An established colony may contain thousands of feeding polyps.

A Sampling of Colonial Types

The reef-forming corals and other colonial anthozoans are a fine example of variations on the basic cnidarian body plan. The reef formers thrive in clear, warm water that is at least 20°C (68°F). As Figure 26.11 shows, the colonies consist of polyps that have secreted calcium-

Figure 26.11 (**a**) One of the reef-building, colonial corals. External skeletons of their polyps interconnect with one another. Dinoflagellate mutualists live in the polyp tissues. (**b**) Aerial view of a barrier reef, mainly an accumulation of the compacted skeletons of certain coral species. (**c**) Not all corals are reef builders; cup corals such as *Tubastrea* are solitary forms. With tentacles extended, their polyps look like tiny sea anemones.

reinforced external skeletons, which interconnect with one another. Over time, the skeletons accumulate and so become the main building material for reefs. Today the most impressive accumulation, the Great Barrier Reef, parallels the eastern coast of Australia for about 1,600 kilometers.

Reef-building corals receive nutrient inputs from the changing tides and from dinoflagellate mutualists living in their tissues. Masses of these photosynthetic protistans supply corals with oxygen, recycle their mineral wastes, and adjust the pH of the surrounding water in a way that enhances the rate of calcium deposition for their host's skeletons. The host corals reciprocate by giving the dinoflagellates a relatively safe, sunlit habitat that has plenty of dissolved carbon dioxide and mineral ions. In the sunlit parts of a reef, nutrients are cycled quickly, directly, and efficiently.

As a final example of cnidarian diversity, consider *Physalia*, informally called the Portuguese man-of-war. The toxin in the nematocysts of this infamous colonial hydrozoan poses a danger to bathers and fishermen as well as to prey organisms (fish). Although *Physalia* lives mainly in warm waters, currents sometimes move it up to the Atlantic coasts of North America and Europe. A blue, gas-filled float that develops from the planula keeps the colony near the water's surface, where winds move it about (Figure 26.12). Under the float, groups of

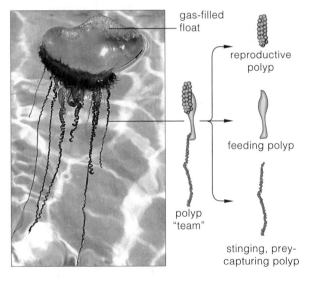

Figure 26.12 Portuguese man-of-war (*Physalia*).

polyps and medusae interact as "teams" in feeding, reproduction, defense, and other specialized tasks.

Cnidarians, including the colonial forms, show notable variation in their body plans and life-styles.

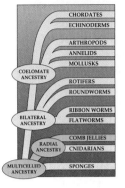

If you have ever been diving in coastal waters, anywhere from the tropics to the poles, you may have observed cnidarians that look a bit like jellyfishes with combs running down the sides. These are **comb jellies**, of phylum Ctenophora (meaning "comb-bearing"). All are weak-swimming predators in planktonic communities, and all show modified radial symmetry. Slice them in two equal halves, then slice them into quarters, and two of the quarters will be mirror images of the other two.

A comb jelly has eight rows of comblike structures made of thick, fused cilia (Figure 26.13). All the combs in a row beat in waves and propel the animal forward, usually (and handily) mouth first. Some species have two long, muscular tentacles with branches that are equipped with sticky cells. Comb jellies do not produce nematocysts, but sometimes they opportunistically save and use the ones from jellyfish they have eaten. Others use their sticky lips to capture prey.

Evolutionarily, comb jellies are interesting because they have cells with multiple cilia. This trait evolved in

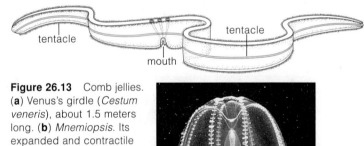

Figure 26.13 Comb jellies. (**a**) Venus's girdle (*Cestum veneris*), about 1.5 meters long. (**b**) *Mnemiopsis*. Its expanded and contractile mouth lobes waft prey into the mouth. It is about 15 centimeters (6 inches) long.

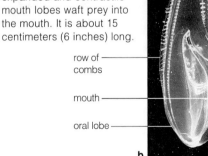

many complex animals. In addition, the comb jellies are the simplest animals with embryonic tissues that are like mesoderm. In complex animals, mesoderm is the embryonic source of muscles as well as organs of the circulatory, excretory, and reproductive systems.

Comb jellies have a modified radial body and comblike structures of modified cilia. They have an embryonic tissue that resembles the mesoderm of more complex animals.

ACOELOMATE ANIMALS—AND THE SIMPLEST ORGAN SYSTEMS

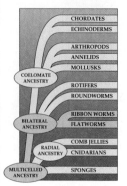

When we move beyond the cnidarians in our survey, we find animals that range from flatworms to humans. All of these animals have simple or complex organs. An **organ** is an association of one or more kinds of tissues, arranged in particular proportions and patterns. Most often, one organ interacts with others to carry out an activity that helps the body function. By definition, two or more organs that are interacting efficiently in the performance of some task represent an **organ system**.

We turn now to the simplest animals at the *organ-system* level of construction. They are not what you would call breathtakingly complex but, as you will see shortly, a few can make our lives breathtakingly miserable.

Flatworms

Among the 15,000 or so known species of **flatworms** (phylum Platyhelminthes) are turbellarians, flukes, and tapeworms. Most of these bilateral, cephalized animals have a flattened body (hence their name) and simple organ systems (Figure 26.14). Their digestive system, for example, has a pharynx (a muscular tube, which is used for feeding) and a saclike and often branching gut. Their reproductive systems vary, but most flatworms

are **hermaphrodites**. That is, an individual has female and male gonads. Two individuals reproduce sexually through the mutual transfer of sperm. Each has a penis (a sperm-delivery structure) and glands that produce a protective capsule around fertilized eggs.

TURBELLARIANS Most turbellarians (class Turbellaria) live in the seas; only planarians and a few others live in freshwater. Some eat tiny animals or suck tissues from dead or wounded ones. Commonly, a planarian will reproduce asexually by transverse fission. It divides in half at its midsection, then each half regenerates the missing parts. Like you, a planarian is able to adjust the composition and volume of its body fluids. Its water-regulating system has one or more tiny, branched tubes called protonephridia (singular, protonephridium). The tubes extend from pores at the body surface to bulb-shaped flame cells in body tissues. When excess water diffuses into the flame cells, a tuft of cilia "flickering" in the bulb drives the water through the tubes and on out to the surroundings (Figure 26.14*b*).

FLUKES Flukes (class Trematoda) are parasitic worms. A parasite, recall, lives in or on a living host and feeds on its tissues. Most do not kill the host, at least not until they have reproduced. Fluke life cycles have sexual and asexual phases and at least two kinds of hosts. As the

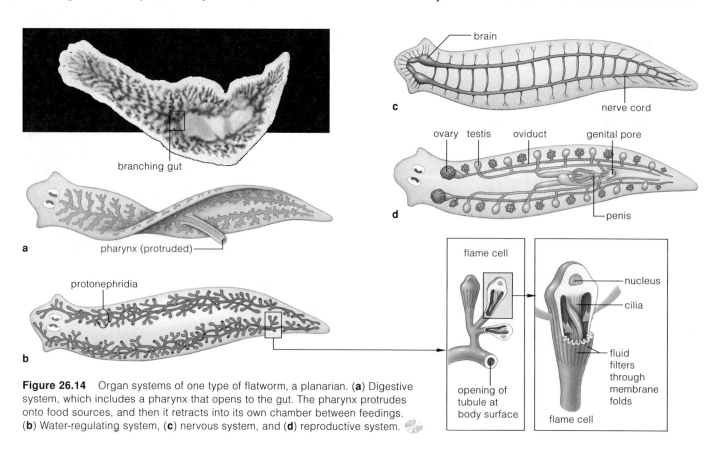

Figure 26.14 Organ systems of one type of flatworm, a planarian. (**a**) Digestive system, which includes a pharynx that opens to the gut. The pharynx protrudes onto food sources, and then it retracts into its own chamber between feedings. (**b**) Water-regulating system, (**c**) nervous system, and (**d**) reproductive system.

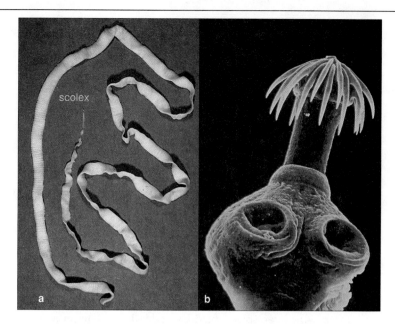

Figure 26.15 (a) Tapeworm scolex. This one attaches to the gut lining of a shorebird, its primary host. (b) Sheep tapeworm.

bilateral animals evolved from planula-like ancestors, through increased cephalization and the development of tissues derived from mesoderm.

Ribbon Worms

Ribbon worms (phylum Nemertea) are bilateral, soft-bodied, elongated predators that swallow or suck tissue fluids from small worms, mollusks, and crustaceans. Most crawl, burrow, or lurk in shallow marine habitats, as in Figure 26.16, but some live in freshwater or humid tropical habitats. Like flatworms, which may be their close relatives, ribbon worms have a ciliated surface, and they secrete mucus and then move by beating their cilia through it. Ribbon worms also resemble flatworms in their tissue organization. However, they differ from flatworms in having a circulatory system, a complete gut, and separation of sexes. Ribbon worms also have a proboscis, which in this case is a tubular, prey-piercing,

examples in Section 26.9 show, the flukes grow and reach sexual maturity in an animal that serves as their *primary* host. Larval stages use an *intermediate* host, in which they either develop or become encysted.

TAPEWORMS Tapeworms (class Cestoda) parasitize intestines of vertebrates. It seems probable that the ancestral tapeworms had a gut but later lost it during their evolution in animal intestines, which happen to be habitats that are rich in predigested food. Their existing descendants attach to the intestinal wall by a scolex, a structure equipped with suckers, hooks, or both (Figure 26.15). **Proglottids** are new units of the tapeworm body that bud just behind the scolex. They are hermaphroditic bodies; they can mate and transfer sperm with one another. The older proglottids (those farthest from the scolex) store fertilized eggs. They break off from younger ones, then leave the body in feces. Later, an intermediate host may meet up with their eggs. Section 26.9 includes a look at proglottid formation in one tapeworm life cycle.

In some respects, the simplest turbellarians, larval flukes, and larval tapeworms resemble the planulas of cnidarians. The resemblance inspires speculation that

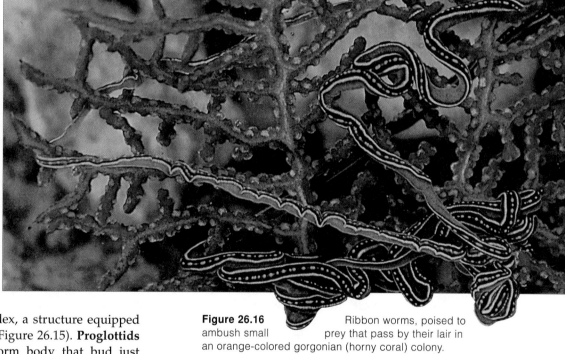

Figure 26.16 Ribbon worms, poised to ambush small prey that pass by their lair in an orange-colored gorgonian (horny coral) colony.

venom-delivering device tucked inside the head end. Muscle contractions force the proboscis inside out and into prey, which the venom paralyzes.

Flatworms are among the simplest bilateral, cephalized animals with organ systems. Ribbon worms may be related to them, but they have a complete gut, a circulatory system, and other traits that are notable departures from flatworms.

26.8 ROUNDWORMS

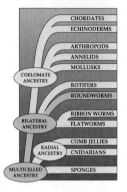

Roundworms (phylum Nematoda) are pseudocoelomate worms that live nearly everywhere and are probably the most abundant of all multicelled animals alive today. A million nematodes in each square meter of sediments is typical in shallow saltwater and freshwater. Thousands of them may occupy a handful of rich soil, in which scavenging types make fast work of dead earthworms or rotting vegetation. We know of 20,000 species, but a hundred times that many may yet be discovered. Each has a bilateral, cylindrical body, usually tapered at both ends and protected by a cuticle. In animals, a **cuticle** is a tough, often flexible body covering (Figure 26.17). The roundworms are the simplest animals outfitted with a complete digestive system. Between the gut and the body wall is a false coelom, typically jam-packed with reproductive organs. The cells in all tissues absorb nutrients from coelomic fluid and give up wastes to it.

Parasitic roundworms can do extensive damage to their hosts, which include humans, cats, dogs, cows, and sheep as well as soybeans, potatoes, and other valued crop plants. They are all thin but can grow long; some in female sperm whales are nine *meters* long. One of the most rapidly spreading diseases in the world, elephantiasis, is the handiwork of a few species of roundworms (Section 26.9).

Unfortunately, the parasitic types have given roundworms everywhere a bad name. Yet most are harmless, free-living types that do beneficial work, as when they help cycle nutrients in a variety of communities. One roundworm, *Caenorhabditis elegans*, is a key experimental organism for laboratory studies into inheritance, development, and aging.

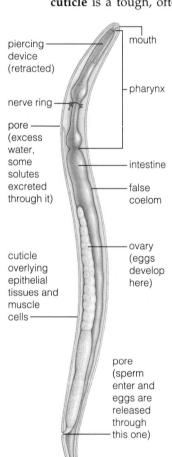

piercing device (retracted)

mouth

nerve ring

pharynx

pore (excess water, some solutes excreted through it)

intestine

false coelom

ovary (eggs develop here)

cuticle overlying epithelial tissues and muscle cells

pore (sperm enter and eggs are released through this one)

anus

Figure 26.17 Body plan of a female roundworm (*Paratylenchus*) that parasitizes plant roots.

Roundworms are cylindrical, bilateral, cephalized animals with a false coelom and a complete digestive system. Most cycle nutrients in communities; the parasites are notorious.

26.9 A ROGUE'S GALLERY OF WORMS

A number of parasitic flatworms and roundworms call the human body home. In any given year, for example, about 200 million people house blood flukes responsible for *schistosomiasis*. One of these, the Southeast Asian blood fluke *Schistosoma japonicum*, requires a human primary host, standing water in which larvae can swim, and an aquatic snail as an intermediate host. The flukes grow, become sexually mature, and mate inside a human host (Figure 26.18a). After being fertilized, the female's eggs leave the human body in feces and hatch into ciliated, swimming larvae (b) that burrow into a snail and multiply asexually (c). In time, many fork-tailed larvae develop (d). These leave the snail (e) and swim about until they contact human skin (f). They bore in and migrate to thin-walled intestinal veins, and the cycle begins anew. In infected humans, white blood cells that defend the body attack the masses of fluke eggs, and grainy masses form in tissues. In time, the liver, spleen, bladder, and kidneys deteriorate.

Some tapeworms parasitize humans. Different species use pigs, freshwater fish, or cattle as intermediate hosts. Humans become infected when they eat pork, fish, or beef that is raw, improperly pickled, or insufficiently cooked—and contaminated with tapeworm larvae (Figure 26.19).

Or consider a parasitic roundworm that causes thin, serpentlike ridges in human skin. For several thousand years, healers have been extracting the "serpents" by winding them out slowly, painfully, around a stick. The roundworms called pinworms and hookworms cause other problems. *Enterobius vermicularis*, a pinworm of temperate regions, parasitizes humans. It lives in the large intestine, but at night the centimeter-long females migrate to the anal region of the host and lay eggs. Their presence causes itching, and scratchings made in response transfer

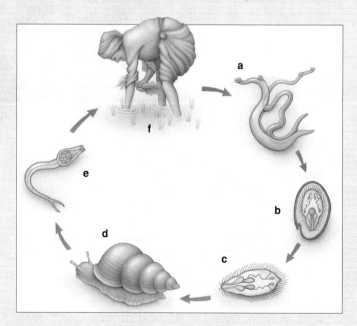

Figure 26.18 Life cycle of a dangerous blood fluke, *Schistosoma japonicum*.

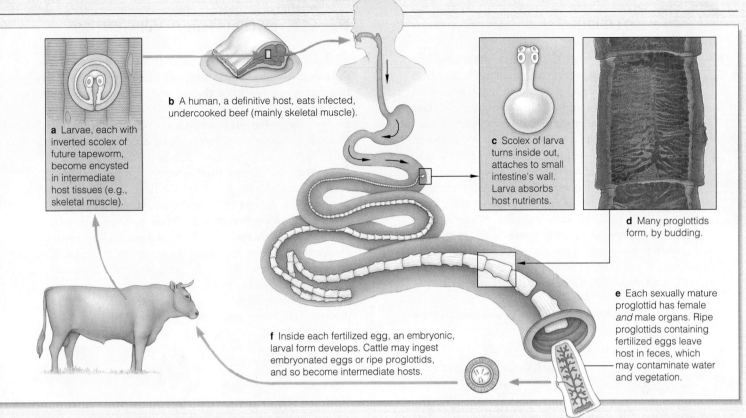

a Larvae, each with inverted scolex of future tapeworm, become encysted in intermediate host tissues (e.g., skeletal muscle).

b A human, a definitive host, eats infected, undercooked beef (mainly skeletal muscle).

c Scolex of larva turns inside out, attaches to small intestine's wall. Larva absorbs host nutrients.

d Many proglottids form, by budding.

e Each sexually mature proglottid has female *and* male organs. Ripe proglottids containing fertilized eggs leave host in feces, which may contaminate water and vegetation.

f Inside each fertilized egg, an embryonic, larval form develops. Cattle may ingest embryonated eggs or ripe proglottids, and so become intermediate hosts.

Figure 26.19 Life cycle of a beef tapeworm, *Taenia saginata*.

some eggs to hands, then to other objects. Newly laid eggs contain embryonic pinworms, but within a few hours they have developed into juveniles and are ready to hatch if another human inadvertently ingests them.

Hookworms are especially serious in impoverished areas of the tropics and subtropics. Adult hookworms live in the small intestine. After the toothlike devices or sharp ridges bordering their mouth cut into the intestinal wall, they feed on blood and other tissues and so compete with their host for nutrients. Adult females, about a centimeter long, can release a thousand eggs daily. These leave the body in feces, then hatch into juveniles. Walk barefoot, and a juvenile hookworm may penetrate the skin. Inside a host, the parasite travels the bloodstream to the lungs. There it works its way into the air spaces. After moving up the windpipe, the parasite moves into the gut when the host swallows. Soon it is in the small intestine, where it may mature and live for several years.

Another roundworm, *Trichinella spiralis*, causes painful, sometimes fatal symptoms. Adults live in the lining of the small intestine. Females release juveniles (Figure 26.20*a*), which work their way into blood vessels and travel to muscles. There they become encysted; they secrete a covering around themselves and enter a resting stage. Humans become infected mainly by eating insufficiently cooked meat from pigs or some game animals. It is not easy to detect the encysted juveniles when fresh meat is being examined, even in a slaughterhouse.

Figure 26.20*b* shows the results of prolonged, repeated infections by *Wuchereria bancrofti*, another roundworm.

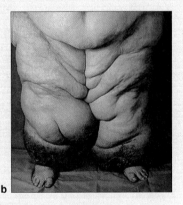

a   **b**

Figure 26.20 (**a**) Juveniles of a roundworm, *Trichinella spiralis*, living inside the muscle tissue of a host animal. (**b**) Legs of a woman infected by the roundworm *Wuchereria bancrofti*.

Adult worms become lodged in the lymph nodes, which filter lymph (excess tissue fluid) that normally flows into the bloodstream. In these nodes, the worms can obstruct the flow of lymph. When such an obstruction causes fluid to back up and accumulate in tissues, legs and other body regions undergo grotesque enlargement. This condition is called *elephantiasis*.

A mosquito is *Wuchereria*'s intermediate host. Females of this parasite produce active young that travel about at night, in the bloodstream. If a mosquito sucks blood from an infected person, the juveniles may enter the insect's tissues. In time they move near the insect's sucking device and enter a new host when it draws blood again.

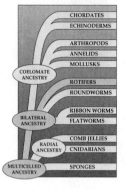

Like the roundworms just described, the **rotifers** (Rotifera) are bilateral, cephalized animals with a false coelom. All but about 5 percent live in freshwater, such as lakes, ponds, and even films of water on mosses and other plants. We typically find between 40 and 500 rotifers in a liter of pondwater; 5,000 were recorded on a few occasions. They eat bacteria and microscopic algae. Most types are not even a millimeter long, yet seldom have so many organs been packed in so little space. As Figure 26.21 shows, rotifers have a pharynx, an esophagus, digestive glands and a stomach, protonephridia, and usually an intestine and anus. Nerve cell bodies clustered in the head integrate body activities. Some rotifers have "eyes" (clusters of absorptive pigments). Two "toes" exude

Figure 26.21

A rotifer (*Philodina roseola*). Males are unknown in this species and many others. Females produce diploid eggs that become diploid females. The females of other species do the same, but they also can produce haploid eggs that develop into haploid males. If a haploid egg happens to be fertilized by a male, it develops into a female. The male rotifers appear only occasionally and are dwarfed and short-lived.

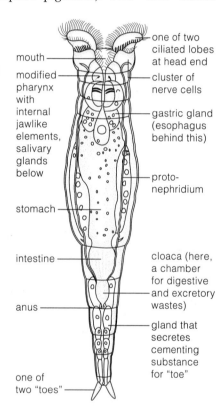

Labels: mouth; modified pharynx with internal jawlike elements, salivary glands below; stomach; intestine; anus; one of two "toes"; one of two ciliated lobes at head end; cluster of nerve cells; gastric gland (esophagus behind this); proto-nephridium; cloaca (here, a chamber for digestive and excretory wastes); gland that secretes cementing substance for "toe"

substances that attach free-living species to substrates at feeding time. A feature unique to rotifers is a crown of cilia at the head end that assists in swimming and in wafting food toward the mouth. Its rhythmic motions reminded early microscopists of a turning wheel; hence the name of the phylum ("rotifer" means wheel-bearer).

The rotifers are bilateral, cephalized animals with ciliated lobes at their head end and a false coelom that is packed with diverse organs.

Bilateral animals not much more complex than modern flatworms evolved in Cambrian times. Shortly after this, some species gave rise to two lineages of animals equipped with a coelom (Figure 26.1*b*). We call these great lineages the **protostomes** and the **deuterostomes**. Mollusks, annelids, and arthropods are all protostomes. Echinoderms and chordates are deuterostomes.

As a result of mutations, embryos of the animals in each lineage develop differently from fertilized eggs. For example, mitotic cell divisions cut the egg cytoplasm repeatedly, along prescribed planes, to form a tiny ball of cells (the early embryo). Protostomes undergo **spiral cleavage**, a pattern in which these early cuts are made at oblique angles relative to the genetically prescribed body axis. But deuterostomes undergo **radial cleavage**, a developmental pattern in which the early cuts are made parallel with and perpendicular to the axis:

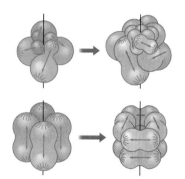

An early protostome embryo, consisting only of four cells, as it undergoes cleavages that are *oblique to* the original body axis:

An early deuterostome embryo, also at the four-cell stage, undergoing cleavages that are *parallel with* and *perpendicular to* the original body axis:

As another example, the very first opening to form at the surface of a protostome embryo becomes the mouth; an anus forms elsewhere. In a deuterostome embryo, the first opening becomes the anus; the second becomes the mouth. As a final example, a protostome coelom arises from spaces in the mesoderm, but a deuterostome coelom forms from outpouchings of the gut wall:

How the coelom forms in protostomes: pouch that will form mesoderm, enclose coelom; embryonic gut — coelom

How the coelom forms in deuterostomes: solid mass of mesoderm; embryonic gut

Such modifications to the embryonic stages of the two kinds of animals led to major differences in body plans.

Soon after the coelomate animals evolved in Cambrian times, two great lineages—the protostomes and deuterostomes—evolved through mutations that affected how their embryos develop, and this led to major differences in body plans.

A SAMPLING OF MOLLUSCAN DIVERSITY

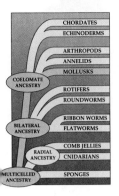

From children's books and explorations in gardens, few of us would have trouble recognizing a land snail when we see one (Figure 26.22*a*). Yet few of us know much about its 110,000 relatives in one of the largest of all animal groups, the phylum Mollusca. As their name implies, **mollusks** have fleshy soft bodies (*molluscus*, a Latin word, means soft). These bilateral animals have a small coelom. *Most* have a shell, or a reduced version of one, made of calcium carbonate and protein. These components are secreted from cells of a tissue that drapes like a skirt over the body mass. This tissue, the **mantle**, is unique to mollusks. The respiratory organs, gills of a type called ctenidia, contain thin-walled leaflets for gas exchange. *Most* mollusks have a fleshy foot. *Many* have a radula, a

their foot spreads out as they crawl. Many species have spirally coiled or conical shells. Coiling compacts the organs into a mass that can be balanced above the body, rather like a backpack. Other species have a reduced shell or none at all (Figure 26.22*a,b*).

Chitons are slow-moving or sedentary grazers with a dorsal shell divided into eight plates (Figure 26.22*c*). The bivalves, or animals with a "two-valved shell," include clams, scallops, oysters, and mussels (Figure 26.22*d*). Some bivalves are only a millimeter across. A few giant clams are over a meter across and weigh 225 kilograms (close to 500 pounds). Humans have been eating one type of bivalve or another since prehistoric times.

Cephalopods are highly active predators of the seas. They include the swiftest invertebrates (jet-propelled squids), the largest of all known invertebrates (the giant squid), and the smartest (octopuses, Figure 22.6*e*). For

Figure 26.22 A few mollusks. (**a**) One gastropod, a land snail. (**b**) Two busily mating sea slugs ("Mexican dancers"). Sea slugs are nudibranchs, a type of gastropod. Different gastropods creep, swim or float as they graze upon, prey upon, or parasitize other organisms. (**c**) From the intertidal zone of California's Monterey Bay, a chiton. With its broad foot, it creeps over and clings to rocks. (**d**) A scallop, one of the bivalves. Light-sensitive "eyes" (the small dark dots) fringe the two halves of its shell. Many bivalves, including a few pearl producers, have shells lined with iridescent mother-of-pearl. (**e**) An octopus, with eyes that resemble yours, although they form in a different way. Like other cephalopods, it has a well-developed nervous system.

tonguelike, toothed organ that shreds food destined for the gut. Mollusks with a well-developed head have eyes and tentacles, but not all have a head. Beyond these generalizations, there are no "typical" mollusks. They range from tiny snails in treetops to huge predators of the seas. In this section and the next, we sample four classes: chitons, gastropods, bivalves, and cephalopods.

The largest class, with 90,000 species of snails and slugs, is gastropods ("belly foots"), so named because

example, show an octopus an object with a distinctive shape and then give it a mild electric shock, and it will thereafter avoid that particular object. With respect to memory and learning abilities, octopuses and certain squids are the world's most complex invertebrates.

Mollusks are bilateral, soft-bodied, coelomate animals that vary tremendously in body details, size, and life-styles.

Maybe it was their fleshy, soft bodies—so forgiving of chance evolutionary changes in morphology—that gave the ancestors of mollusks the potential to diversify in so many ways and to radiate into so many habitats. Let's explore this idea by using a few characteristics of our representative mollusks. As a point of departure, start by studying the body plan shown in Figure 26.23.

Twisting and Detwisting of Soft Bodies

Notice, in Figure 26.23*a*, how evolution put an unusual twist in the soft snail body. Its anus dumps wastes near the mouth! As a gastropod embryo develops, a cavity between its mantle and the shell gets twisted 180° counterclockwise, and so does nearly all of the visceral mass (the gut, heart, gills, and other internal organs). This process, called **torsion**, occurs only in gastropods:

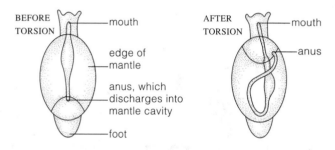

Such a drastic rearrangement of body parts could have come about through mutations that affected retractor muscles, which attach a gastropod embryo or larva to its shell. The muscles on the body's right side develop before those on the left. As they grow toward the front of the body, they drag the mantle cavity and organs with them. It takes them a few hours at most to do this.

By one hypothesis, torsion proved to be adaptive, for the head could withdraw into the mantle cavity in times of danger. However, in some 1985 experiments by J. Pennington and F. Chia, predators ate just as many torsioned larvae as "pre-torsioned" ones. Was torsion a bad evolutionary experiment? By putting the

gills, anus, and kidneys above the mouth, it certainly created a potentially awful sanitation problem. At the very least, the uptake of discharged wastes in ancestral torsioned species must have been distasteful.

In fact, we find evolutionary compensations for this state of affairs. Most gastropods now have enough cilia in this region to create currents that sweep the wastes away. Also, torsion is not as pronounced as it once was in some lineages. Nudibranchs have even undergone an apparent detorsion; the soft larval body twists, but then it untwists. At some point in their evolution, they also ended up losing most of their mantle cavity and all of their ctenidia. Most species have other outgrowths that function in gas exchange (Figures 26.22*b* and 26.23*c*).

Hiding Out, One Way or Another

If you were small, edible, and soft of body, an external shell would be a distinct advantage, as it is for chitons and clams. When a chiton is disturbed by predators or surf or exposed by a receding tide, it hunkers under its shell. Muscles in its foot pull the body mass down, and the mantle's edge around the shell's rim presses like a suction cup against a rock. That eight-plated shell is flexible. Pull a chiton from a rock, and it can roll up in a ball until it can unroll and become reattached elsewhere.

Besides having a shell, protection also can be had by hiding in sediments and other substances. A bivalve's head is not much to speak of, but its foot is usually large and specialized for burrowing. Bivalves burrowed

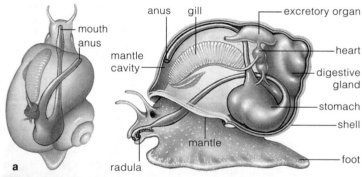

Figure 26.23 (**a**) Body plan of an aquatic snail, a gastropod. (**b**) Close-up of a radula, a feeding device of snails and some other mollusks. As the radula is rhythmically protracted and retracted, it rasps food and then draws it toward the gut on the retraction stroke. (**c**) The sea slug *Aplysia*, also called the sea hare. The two flaps above its dorsal surface are foot extensions that undulate and so help ventilate the mantle cavity. Like many other gastropods, *Aplysia* is hermaphroditic. It can function as a male, a female, or both.

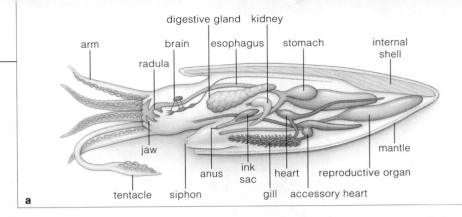

digestive gland kidney

arm   brain   esophagus   stomach   internal
                                                  shell
     radula

jaw

tentacle   siphon   anus   ink   heart   gill   accessory heart
                                  sac   reproductive organ

mantle

a

in sand or mud have a pair of siphons: extensions of the mantle edges, fused into tubes (Figure 26.24*a*). Water is drawn into the mantle cavity through one siphon and leaves through the other, carrying wastes. One wonders what preyed on the ancestors of geoducks of the Pacific Northwest, which have siphons more than a meter long.

On the Cephalopod Need for Speed

Some 500 million years ago the cephalopods, with their buoyant, chambered shells, were the supreme predators in Ordovician seas (Section 21.5). And yet, of a lineage having more than 7,000 ancestral species, the shell of all but one of the existing descendant species is reduced or gone (Figure 26.25). What happened? This evolutionary trend coincided with an adaptive radiation of the bony fishes—which preyed on cephalopods or were strong competitors for the same prey. During what may have been a long-term race for speed and wits, cephalopods lost their thick external shell and became streamlined and highly active. Of all mollusks, they now have the largest brain relative to body size and display the most

Figure 26.25 (**a**) Body plan (generalized) of a cuttlefish, a cephalopod. Its tentacles, thinner than the arms, are specialized for capturing prey. (**b**) A squid (*Dosidiscus*) and a diver inspecting each other. (**c**) A chambered nautilus, the only existing cephalopod that has an external shell. Being so active, cephalopods have great demands for oxygen. They are the only mollusks with a closed circulatory system. Their blood is pumped from a main heart to two gills, each with a booster (accessory) heart at its base that speeds blood flow, hence the uptake of oxygen (for muscle cells especially) and removal of carbon dioxide.

b

c

mouth   left mantle   retractor muscle

retractor
muscle

water flows
out through
exhalant
siphon

water flows
in through
inhalant
siphon

a   foot   palps   left gill   shell

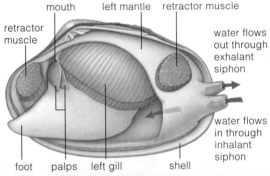

b

Figure 26.24 (**a**) Body plan of a clam, with half of its shell removed. In nearly all bivalves, gills serve in collecting food and in respiration. As water moves through the mantle cavity, mucus on the gills traps food. Cilia move the mucus and food to palps, where final sorting takes place before suitable bits are driven to the mouth. (**b**) A scallop escaping from a sea star by clapping its valves and producing a propulsive water jet.

complex behavior. Nerves connect the brain to muscles that can make quick responses to food or danger. Blood circulation and respiration are highly efficient (Figure 26.25). Except for the chambered nautilus, cephalopods can discharge dark fluid from an ink sac, maybe to confuse predators.

Jet propulsion became the name of the game. Cephalopods force a jet of water out of the mantle cavity and a funnel-shaped siphon. As mantle muscles relax, water is drawn into the cavity. As they contract, a water jet is squeezed out. When the mantle's free edge closes down on the head at the same time, a jet shoots out through the siphon. The brain controls the siphon's activity and so influences the direction of escape or pursuit.

Lively stories emerge when evolutionary theory is used to interpret the fossil record and the range of existing species diversity, as we have done for the mollusks.

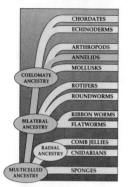

Maybe you've noticed earthworms after a downpour, when they wriggle out of their burrows to avoid drowning. Earthworms are among 15,000 or so species of bilateral, segmented animals known as the **annelids** (Annelida). Their relatives are a few kinds of leeches and the polychaetes, which are far more diverse but not nearly as well known (Figure 26.26). The phylum name means "ringed forms." But the "rings" are really a series of repeating body units, and this segmentation is pronounced. Also, except for leeches, nearly all segments have pairs or clusters of chitin-reinforced bristles on each side of the body. The bristles are also called *setae* or *chaetae*, but these are just formal names for "bristles." When pushed into soil, the bristles provide the traction required for crawling or burrowing. They have become broadened paddles in some swimming species. Earthworms, one of the oligochaetes, have few setae per body segment, and marine polychaete worms typically have many of them (*oligo-*, few; *poly-*, many).

Advantages of Segmentation

A segmented body has great evolutionary potential, for individual parts can undergo modification and become highly adapted for specialized tasks. Although most of an earthworm's segments are similar, the leeches have suckers at both ends, and polychaetes have an elaborate head and fleshy-lobed appendages known as parapods ("closely resembling feet"). By analyzing existing species, we catch glimpses of developments that led to increases in size and to more complex internal organs.

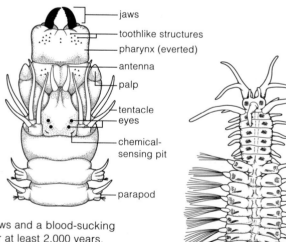

Figure 26.26 Representative annelids. (**a**) Most leeches have sharp jaws and a blood-sucking device. This one is shown before and after gorging on human blood. For at least 2,000 years, *Hirudo medicinalis*, a freshwater leech, has been used as a blood-letting tool to "cure" problems ranging from nosebleeds to obesity. Today leeches are still used, but more selectively. Thus, after surgeons reattach a severed ear, lip, or fingertip, leeches may be used to draw off pooled blood. A patient's body cannot do this on its own until severed blood circulation routes are reestablished.

(**b**) A less scary annelid—an earthworm. (**c,d**) From the polychaetes, examples of the dizzying variety of modifications that have evolved, starting from a segmented, coelomic body plan. The photograph shows a tube-dweller. Featherlike structures at its head end are coated with mucus. After the mucus has trapped bacteria and other bits of food, coordinated beating of cilia sweeps them to the mouth. Most polychaetes live in marine habitats. They actually are one of the most common types of animals along coasts. Many are predators or scavengers; others dine on algae.

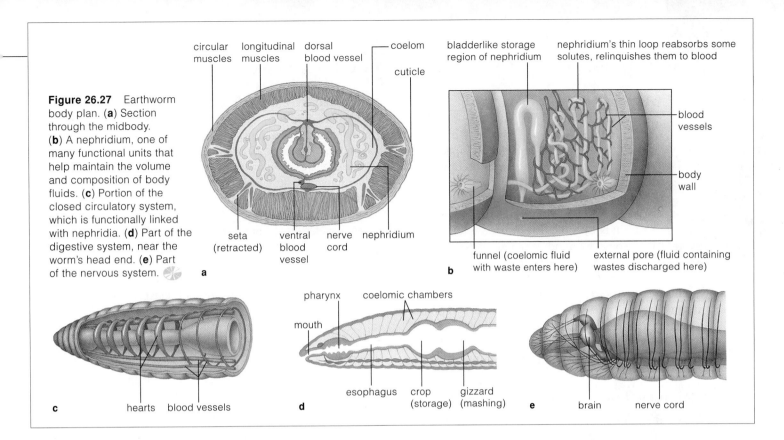

Figure 26.27 Earthworm body plan. (**a**) Section through the midbody. (**b**) A nephridium, one of many functional units that help maintain the volume and composition of body fluids. (**c**) Portion of the closed circulatory system, which is functionally linked with nephridia. (**d**) Part of the digestive system, near the worm's head end. (**e**) Part of the nervous system.

Labels for (a): circular muscles · longitudinal muscles · dorsal blood vessel · coelom · cuticle · seta (retracted) · ventral blood vessel · nerve cord · nephridium

Labels for (b): bladderlike storage region of nephridium · nephridium's thin loop reabsorbs some solutes, relinquishes them to blood · blood vessels · body wall · funnel (coelomic fluid with waste enters here) · external pore (fluid containing wastes discharged here)

Labels for (c): hearts · blood vessels

Labels for (d): pharynx · coelomic chambers · mouth · esophagus · crop (storage) · gizzard (mashing)

Labels for (e): brain · nerve cord

Annelid Adaptations—A Case Study

Earthworms are the usual textbook examples of annelids. As Figure 26.27 shows, partitions divide the body into a series of coelomic chambers. In most chambers we find repeats of muscles, blood vessels, branching nerves, and other organs. The gut extends through all the chambers, from mouth to anus. Like all annelids, an earthworm has a cuticle of secreted material that surrounds the body surface. It bends easily and is permeable to water as well as gases, which is one reason why annelids are restricted to aquatic habitats or moist habitats on land.

Earthworms are scavengers. They ingest moist soil and mud that contains decomposing plant material and other organic matter. Each worm ingests its own weight every twenty-four hours. Collectively, their burrowing and feeding activities aerate soil and lift nutrients to the surface, to the benefit of many plants.

As in other annelids, the fluid-cushioned coelomic chambers serve as a hydrostatic skeleton against which muscles act. Each segment's wall incorporates a layer of circular muscles (Figure 26.27a). When longitudinal muscles that span several segments are contracting, the circular ones relax, so that segments shorten and fatten. When the pattern reverses, the segments lengthen. While this is going on, bristles on different segments are protracting and retracting. When the first few segments lengthen, the body is extended forward. Bristles of the segments behind them plunge into the ground and hold the body in its extended position. As the first segments plunge *their* bristles into the ground, the ones behind them retract their bristles and are pulled forward. The alternating contractions and elongations proceed along the body's length and move the whole worm forward.

Figure 26.27b shows part of a system of **nephridia** (singular, nephridium), units that regulate the volume and composition of body fluids. In many annelids, cells of these units are similar to flame cells, which implies an evolutionary link between flatworms and annelids. More often, a nephridium starts as a funnel that collects excess fluid from one coelomic chamber. The funnel connects with a tubular part of the nephridium, which delivers fluid to a surface pore in the body wall of the next coelomic chamber in line.

The worm's head end has a rudimentary **brain**, an aggregation of nerve cell bodies that integrate sensory input and muscle responses for the whole body. Paired **nerve cords**, each a bundle of extensions of nerve cell bodies, lead away from the brain. They are pathways for rapid communication. In each segment, the paired nerve cords broaden into a **ganglion** (plural, ganglia), a cluster of nerve cell bodies that controls local activity.

Finally, as is true of most annelids, earthworms have a closed circulatory system, with blood confined in hearts and muscularized blood vessels. Contractions keep blood circulating in one direction. Smaller blood vessels service the gut, nerve cord, and body wall.

Annelids are bilateral, coelomate, segmented worms that have complex organ systems. Some species show the degree of specialization possible with a segmented body plan.

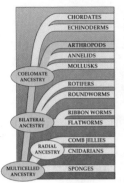

Arthropod Diversity

Evolutionarily speaking, "success" means having the greatest number of species, producing the most offspring, occupying the most habitats, effectively fending off predators and competitors, and having a capacity to exploit the greatest amounts and kinds of food. These are the features that come to mind when we attempt to characterize the **arthropods** (Arthropoda).

Over a million species (mostly insects) are known, and new ones are being discovered weekly. Of four major lineages, the trilobites are extinct (Section 21.5). The other three lineages are chelicerates (spiders and their kin), crustaceans (such as barnacles and crabs), and uniramians (centipedes, millipedes, and insects).

Adaptations of Insects and Other Arthropods

Six important adaptations contributed to the success of arthropods in general and insects in particular:

1. A hardened exoskeleton
2. Jointed appendages
3. Fused and modified segments
4. Specialized respiratory structures
5. Efficient nervous system and sensory organs
6. Division of labor in the life cycle

HARDENED EXOSKELETONS Arthropods have a cuticle of chitin, proteins, and surface waxes. It is sufficiently hardened, as by calcium carbonate deposits, to serve as a rigid, protective external skeleton—an exoskeleton. Such cuticles might have evolved as defenses against predation, but they took on added functions when arthropods invaded the land. They support a body deprived of water's buoyancy. Their waxy surface restricts evaporative water loss. Hard cuticles do restrict increases in size, but arthropods grow in spurts, by **molting**. At certain stages of their life cycle, they secrete a new, soft cuticle under their old one, which they shed (Figure 26.28). The body mass then increases by repeated, rapid cell division before the new cuticle hardens.

Figure 26.28 Molting, as demonstrated by a bright orange centipede backing out of its old exoskeleton.

JOINTED APPENDAGES If arthropods had a uniformly hardened cuticle, they wouldn't move much. However, the cuticle thins down at joints (places where different body parts abut). Muscles associated with the joints can make the thinned cuticle bend in specific directions and so move the attached body parts. A jointed exoskeleton was an evolutionary innovation that led to specialized appendages, including diverse wings and antennae as well as legs (arthropod means "jointed foot").

FUSED AND MODIFIED SEGMENTS The first arthropods were segmented, like the annelid stock that presumably gave rise to them. In most of their existing descendants, however, the serial repeats of the body wall and organs are masked. Segments in different parts of the body have become fused together and modified for far more specialized functions. For example, in the ancestors of insects, different segments fused to form three regions —head, thorax, and abdomen—which morphologically diverged from one another in astounding ways.

RESPIRATORY STRUCTURES Many aquatic arthropods depend on gills for gas exchange. Air-conducting tubes evolved among insects and other land-dwellers. Insect tracheas begin as pores on the body surface and branch into tubes that deliver oxygen directly to tissues. They support energy-consuming activities, such as flight.

SPECIALIZED SENSORY STRUCTURES Intricate eyes and other sensory organs contributed to arthropod success. Numerous species have a wide angle of vision and can process visual information from many directions.

DIVISION OF LABOR Moths, butterflies, beetles, flies, and many other species divide the job of surviving and reproducing among different stages of development. As their immature forms grow, body tissues are massively reorganized and body parts become remodeled. (Figure 1.4). Growth and major transformation of an immature form into the adult is called metamorphosis. Typically, immature stages such as caterpillars specialize in feeding and increasing in size, and the adult specializes mainly in dispersal and reproduction. A life cycle that turns on such a *division of labor* among different developmental stages is adaptive to environmental changes, including seasonal variation in food sources and water supplies.

As a group, the arthropods are exceptionally abundant and widespread, and they have enormously different life-styles.

Their success arises largely from their hardened, jointed exoskeletons; fused, modified body segments; specialized appendages; specialized respiratory, nervous, and sensory organs; and often a division of labor in the life cycle.

A LOOK AT SPIDERS AND THEIR KIN

The chelicerates originated in shallow seas early in the Paleozoic. The only surviving marine species are a few mites, sea spiders, and horseshoe crabs (Figure 26.29). The familiar chelicerates—scorpions, spiders, ticks, and chigger mites—are classified as arachnids. Most species of arachnids live on land, and we might say this about them: Never have so many been loved by so few.

Figure 26.29 Horseshoe crab. Its hard, shieldlike cover hides *five* pairs of legs, one of its defining features.

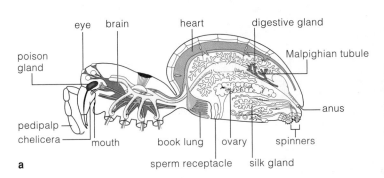

a

Figure 26.30 (**a**) Internal organization of a spider body. (**b**) Wolf spider. Like most spiders, it helps keep insect populations in check and its bite is harmless to humans. (**c**) Female black widow. Its bite can be painful and sometimes dangerous. (**d**) Brown recluse, with a violin-shaped mark on its forebody. Its bite can be severe to fatal.

Scorpions and spiders are predators; they sting or bite and may subdue prey with venom. When they sting or bite us, the reaction might be painful but is rarely serious. Spiders especially are beneficial in that they prey on great numbers of pestiferous insects. Bites of the blood-sucking ticks that parasitize deer, mice, and other vertebrates can cause maddening itches and often serious diseases. For example, the bites of some ticks transmit the bacterial agents of Rocky Mountain spotted fever or Lyme disease to humans (Section 22.4). Most mites are free-living scavengers.

Arachnids have segments fused into a forebody and hindbody. The forebody's jointed appendages include four pairs of legs, a pair of pedipalps with primarily sensory functions, and a pair of chelicerae that inflict wounds and discharge venom. The appendages of the hindbody spin out silk threads for webs and for egg cases. Most webs are netlike. One type of spider spins a vertical thread with a ball of sticky material at the end. It uses a leg to swing the ball at insects passing by! Inside the body is an *open* circulatory system, with a heart that pumps blood into tissues, then receives blood through small openings in its wall. Some blood travels through moist folds of book lungs. These respiratory organs resemble book pages, and they greatly increase the surface area available for gas exchange with the air. Figure 26.30*a* shows the arrangement of book lungs and other major organs inside the spider body.

The spiders, scorpions, and their relatives have a variety of appendages specialized for predatory or parasitic life-styles.

A LOOK AT THE CRUSTACEANS

Shrimps, lobsters, crabs, barnacles, pillbugs, and other crustaceans got their name because they have a hard yet flexible "crust" (an external skeleton), but so do nearly all arthropods. Only some of the 35,000 species live in freshwater or on land. The vast majority live in marine habitats, where they are so abundant they have been dubbed the insects of the seas. Lobsters and crabs are the "giants" of this subphylum; most crustaceans are less than a few centimeters long. All have major roles in food webs, and humans harvest many edible types.

The simplest crustaceans have many pairs of similar appendages for most of their length and may resemble their annelid ancestors. In other lineages, unspecialized appendages evolved into diverse structures of the sort shown in Figure 26.31. The strong claws of lobsters and crabs are used to collect food, intimidate other animals, and sometimes dig burrows. Feathery appendages of barnacles comb microscopic bits of food from the water.

Many crustaceans have sixteen to twenty segments; some have more than sixty. In crabs, lobsters, and some other crustaceans, the dorsal cuticle extends back from the head as a shield-like cover (carapace) over some or all of the segments. The head incorporates two pairs of antennae, a pair of mandibles (jawlike appendages),

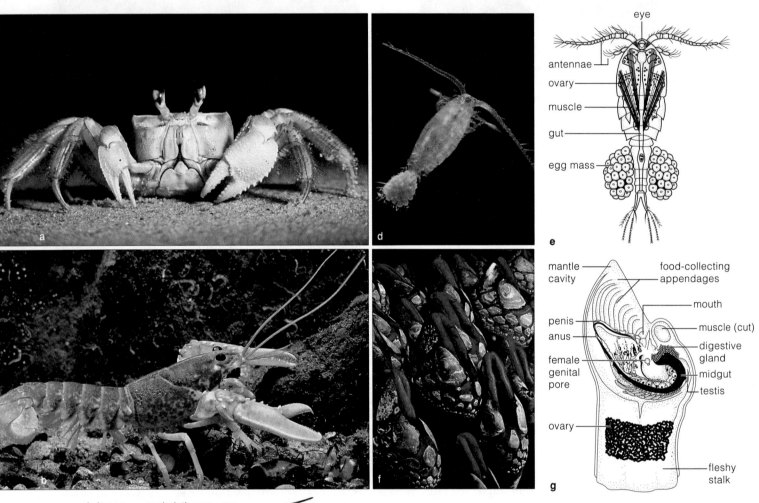

Figure 26.31 A sampling of crustaceans and their diverse life-styles. A crab (**a**). Photograph and body plan of (**b**,**c**) a lobster, (**d**,**e**) a copepod, and (**f**,**g**) stalked barnacles. Lobsters are secretive for most of their lives. Crabs skitter actively and openly across sand and rocks. The copepods are free-living filter feeders, predators, or parasites. This female has a pair of long antennae and is carrying her eggs around with her. As adults, barnacles cement themselves to one spot. The goose barnacles shown earlier in Figure 5.1 have no problem attaching to boat hulls, floating logs or bottles, and other available objects in the water. You might mistake barnacles for mollusks, but as soon as they open their hinged shell to filter feed, their jointed appendages—the hallmark of arthropods—are apparent.

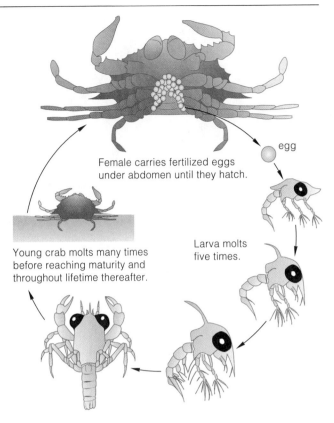

Figure 26.32 Life cycle of a crab. The larval and juvenile stages molt repeatedly during their growth.

Female carries fertilized eggs under abdomen until they hatch.

egg

Young crab molts many times before reaching maturity and throughout lifetime thereafter.

Larva molts five times.

and two pairs of maxillae (food-handling appendages). Crayfish, crabs, lobsters, shrimps, and their kin have five pairs of walking legs.

Figure 26.31*d* shows a copepod. Copepods are less than two millimeters long and are the most numerous animals in aquatic habitats, maybe even in the world. About 1,500 kinds parasitize various invertebrates and fishes. The majority—8,000 species—are consumers of phytoplankton, the "pastures" of aquatic habitats. Some also eat larval or small adult invertebrates, fish eggs, and fish larvae, which they grab with their pair of food-handling appendages. In turn, the copepods are food for different invertebrates, fishes, and baleen whales.

Of all the arthropods, only barnacles have a calcified "shell," a modified external skeleton that protects them from predators, drying out, and battering currents and surf. Adult barnacles cement themselves to rocks, wharf pilings, and similar surfaces (Figure 26.31*f,g*). A few kinds attach themselves only to the skin of whales.

Like other arthropods, crustaceans molt repeatedly to shed the exoskeleton during their life cycle. Figure 26.32 shows a crab's larval stages and increases in size.

Crustaceans differ greatly in the number and kind of their appendages. As for arthropods generally, they repeatedly replace their external skeleton by molting.

HOW MANY LEGS?

The **millipedes** and **centipedes** have a long, segmented body with many legs. Of course, millipedes don't have "a thousand," as their name implies. Most have about 100, although one exuberant individual grew 752. And centipedes have between 15 and 177 pairs of legs, not a nicely rounded number of "one hundred."

Millipedes have a rounded body (Figure 26.33*a*). As they develop, pairs of segments fuse, so what looks like a single segment in the adult has *two* pairs of legs. They typically bulldoze through soil and forest litter, mostly as scavengers of decaying vegetation. Centipedes have a flattened body, and all but two segments have a pair

Figure 26.33 (**a**) Millipede. (**b**) A Southeast Asian centipede.

of walking legs. All species are fast-moving, aggressive predators, outfitted with fangs and venom glands. They prey on insects, earthworms, and snails. The one shown in Figure 26.33*b* subdues small lizards, toads, and frogs. The house centipede (*Scutigera coleoptrata*) often lurks in buildings, where it actively preys on cockroaches, flies, and other pests. Athough helpful in this respect, its vaguely terrifying body keeps it from being welcomed.

The mild-mannered, scavenging millipedes and aggressive, predatory centipedes do not lend themselves to leg counts as they walk by.

A LOOK AT INSECT DIVERSITY

As a group, insects share the adaptations listed earlier in Section 26.15. Here we expand the list a bit. Insects have a head, a thorax, and an abdomen. The head has paired sensory antennae and paired mouthparts used in biting, chewing, sucking, or puncturing (Figure 26.34). The thorax has three pairs of legs and usually two pairs of wings. Most abdominal appendages are reproductive structures, such as egg-laying devices. An insect has a foregut, midgut (where most digestion proceeds), and hindgut (where water is reabsorbed). Insects get rid of waste material by **Malpighian tubules**, small tubes that connect with the midgut. Nitrogen-containing wastes from protein breakdown diffuse from blood into the tubules and are converted into harmless crystals of uric acid. The crystals are eliminated with feces. This system allows the land-dwelling insects to get rid of potentially toxic wastes without losing precious water.

We've already catalogued more than 800,000 species of insects. If we use sheer numbers and distribution as the yardstick, the most successful insects are small in size and have a staggering reproductive capacity. For example, you may find some of these species growing and reproducing in great numbers on a single plant that might be only an appetizer for another animal. By one estimate, if all the progeny of a single female fly were to survive and reproduce through six more generations, that fly would have more than 5 trillion descendants!

Besides this, the most successful insect species are winged. In fact, they are the *only* winged invertebrates. They can move among food sources that are too widely scattered to be exploited by other kinds of animals. The capacity for flight contributed to their success on land.

Finally, insect life cycles commonly proceed through stages that allow exploitation of different resources at different times. As an insect embryo develops, organs required for feeding and other vital activities form and become functional. Before an insect becomes an adult (the sexually mature form of the species), it proceeds through immature, post-embryonic stages. **Nymphs** and **pupae**, as well as larvae, are examples of the stages.

Like human infants, some insect nymphs are shaped like miniature adults; but unlike children, they undergo growth and molting (Figure 26.35*a*). Other insects go through post-embryonic stages of reactivated growth, tissue reorganization, and remodeling of body parts. Metamorphosis, recall, is the name for this resumption of growth and transformation into an adult form. Figure 26.35*b,c* shows how the transformation is more drastic in some species than in others.

The factors that contribute to insect success also make them our most aggressive competitors. Insects destroy crops, stored food, wool, paper, and timber. As they stealthily draw blood from us and from our pets, they often transmit pathogenic microorganisms. On the

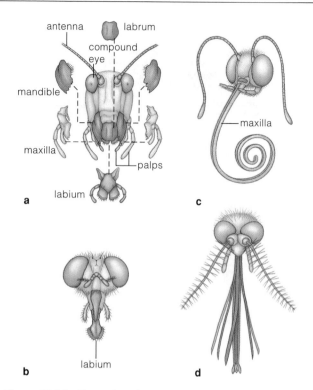

Figure 26.34 Examples of insect appendages. Headparts of (**a**) grasshoppers, a chewing insect; (**b**) flies, which sponge up nutrients with a specialized labium; (**c**) butterflies, which siphon up nectar with a specialized maxilla; and (**d**) mosquitoes, with piercing and sucking appendages.

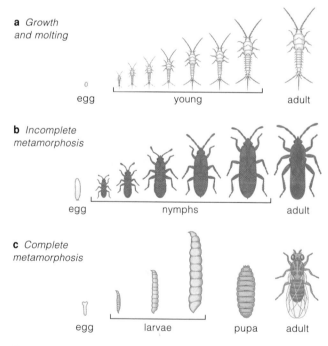

Figure 26.35 Examples of post-embryonic development. (**a**) Young silverfish are adults in miniature, changing little except in size and proportion as they mature into adults. (**b**) True bugs show *incomplete* metamorphosis, which involves gradual, partial change from the first immature form until the last molt. (**c**) Fruit flies show *complete* metamorphosis. Tissues of immature forms are destroyed and replaced before emergence of the adult.

Figure 26.36 Insects. (**a**) Stinkbugs (order Hemiptera), newly hatched. (**b**) A honeybee (order Hymenoptera) attracting its hive mates with a dance, as described in Section 51.4. (**c**) Ladybird beetles (order Coleoptera) swarming. These beetles are raised commercially and released as biological controls of aphids and other pests. Also in this order, the scarab beetle (**d**). With more than 300,000 species, Coleoptera is the largest order of the animal kingdom. (**e**) Luna moth (order Lepidoptera) of North America. Like most other moths and butterflies, its wings and body are covered with microscopic scales. (**f**) Dragonfly (order Odonata), which swiftly captures and eats insects in midflight.

(**g**) Duck louse (order Mallophaga). It eats bits of feathers and skin. (**h**) European earwig (order Dermaptera), a common household pest. (**i**) Flea (order Siphonaptera), with strong legs for jumping onto and off animal hosts. (**j**) Mediterranean fruit fly (order Diptera). Its larvae destroy citrus fruit and other crops.

bright side, many insects pollinate flowering plants in general and crop plants in particular. And many "good" insects attack or parasitize the ones we would rather do without. We now leave our survey of insects and other arthropods with Figure 26.36 and its representatives from some of the major orders of insects.

As a group, insects show immense variation on the basic arthropod body plan. Many have wings, and their life cycles have stages that allow exploitation of different and often widely scattered food sources. Many species produce great numbers of small individuals that pass through immature, post-embryonic stages before the adult form emerges.

THE PUZZLING ECHINODERMS

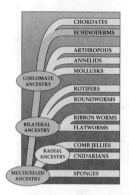

We turn, finally, to the second lineage of the coelomate animals, the deuterostomes. The major invertebrate members of this lineage are **echinoderms** (Echinodermata). The feather star, sea urchin, sea cucumber, and brittle stars shown in Figure 26.37 belong to this phylum. So do the sea lilies (crinoids), sand dollars, and sea biscuits. Sea lilies, which look a bit like stalked plants, flourished in Silurian times (Figure 21.14). About 13,000 echinoderm species are known from the fossil record, but most of these became extinct. Nearly all of the 6,000 or so existing species live in marine habitats.

An echinoderm body wall bears a number of spines, spicules, or plates made stiff with calcium carbonate. (Echinodermata means spiny skinned.) These structures serve defensive functions, as you might suspect if you have ever stepped barefoot on a sea urchin. Its spines trigger painful swelling if they break off under the skin. Most echinoderms also have a well-developed internal skeleton, which is composed of calcium carbonate and other substances secreted from specialized cells.

Oddly, the adult echinoderms are radial with some bilateral features. Most species even produce bilateral larvae during their life cycle. Did bilateral invertebrates give rise to the ancestors of echinoderms, which later picked up some radial features as they evolved? Maybe.

Adult echinoderms have no brain. However, their decentralized nervous system allows them to respond to information about food, predators, and so forth that is coming from different directions. For instance, any arm of a sea star that senses the shell of a tasty scallop

Figure 26.37 Representative echinoderms. (**a**) Feather star, with finely branched food-gathering appendages. (**b**) Sea urchin, which moves about on spines and tube feet. (**c**) Sea cucumber, with rows of tube feet along its body. (**d**) Brittle stars. Their arms (rays) make rapid, snakelike movements.

tube feet   spine

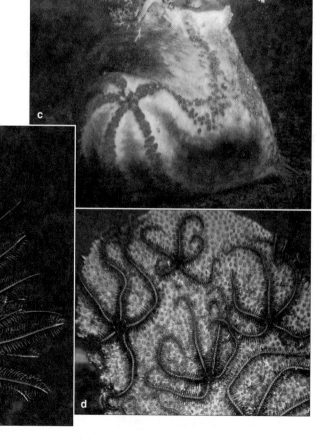

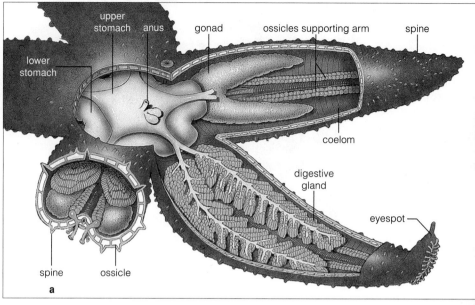

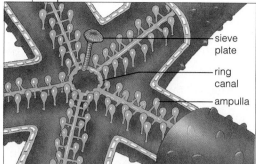

part of the water-vascular system

sieve plate

ring canal

ampulla

upper stomach
anus
gonad
ossicles supporting arm
spine
lower stomach
coelom
digestive gland
eyespot
spine
ossicle

a

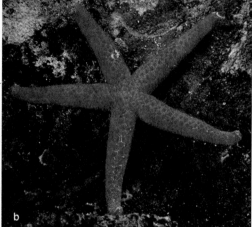

b

Figure 26.38 (**a**) Some key aspects of the radial body plan of a sea star. The water-vascular system, in combination with great numbers of tube feet, is the basis of locomotion. (**b,c**) Five-armed sea star, with closer views of tube feet.

can become the leader, directing the rest of the body to move in a direction suitable for prey capture.

Figures 26.37 and 26.38 show examples of tube feet. These fluid-filled, muscular structures have suckerlike adhesive disks. Sea stars use their tube feet for walking, burrowing, clinging to a rock, or gripping a clam or a snail about to become a meal. Tube feet are part of a **water-vascular system** unique to echinoderms. In sea stars, the system includes a main canal in each arm. Short side canals extend from them and deliver water to the tube feet. Each tube foot has an ampulla, a fluid-filled, muscular structure shaped rather like the rubber bulb on a medicine dropper. As an ampulla contracts, it forces fluid into the foot and causes it to lengthen.

Tube feet change shape constantly as muscle action redistributes fluid through the water-vascular system. Hundreds of tube feet may move at a time. After being released, each one swings forward, reattaches to the substrate, then swings backward and is released before swinging forward again. Their motions are splendidly coordinated, so sea stars glide rather than lurch along.

On their ventral surface, sea stars have a formidable feeding apparatus, such as the one shown here:

c

tube feet on the ventral surface of a sea star arm

Some eager types simply swallow their prey whole. Others push part of their stomach outside the mouth and around their prey, then start digesting their meal even before swallowing it. Sea stars get rid of coarse, undigested residues through the mouth. They do have a small anus, but this is of no help in getting rid of empty clam or snail shells.

With their curious traits, echinoderms are a suitable point of departure for this chapter. Even though we can identify broad trends in animal evolution, we should keep in mind that there are confounding exceptions to the perceived macroevolutionary patterns.

Echinoderms are coelomate animals with spines, spicules, or plates in the body wall. From the evolutionary perspective, they are a puzzling mix of bilateral and radial features.

---

1. Animals are multicelled, aerobic heterotrophs that ingest or parasitize other organisms. Nearly all have diploid body cells, arranged as tissues, organs, and organ systems. Animals reproduce sexually and often asexually. They go through embryonic development, and most are motile during at least part of the life cycle.

2. Animals range from structurally simple placozoans and sponges to vertebrates. The species of each phylum share certain characteristics.

   a. By comparing major animal phyla and integrating the information with the fossil record, biologists have identified major evolutionary trends among them.

   b. The revealing aspects of body plans are the type of symmetry, gut, and cavity (if any) between the gut and body wall; whether there is a head end; and whether the body is divided into a series of segments (Figure 26.39).

3. *Trichoplax*, the only known placozoan, is the simplest animal. It consists of little more than two layers of cells, with a fluid matrix in between.

4. A sponge body is at the cellular level of construction and has no symmetry. A sponge has several kinds of cells, but these are not organized as epithelia and other tissues of complex animals.

5. Cnidarians (such as jellyfishes, sea anemones, and hydras) have radial symmetry and are at the tissue level of construction. Cnidarians alone produce nematocysts (capsules with dischargeable threads, used mainly in prey capture).

6. Nearly all animals more complex than cnidarians show bilateral symmetry, and they form tissues, organs, and organ systems. Their gut may be saclike, as it is in flatworms, but it usually is complete, with an anus and a mouth. Like most animals, flatworms have a coelom or false coelom, which are cavities between the gut and body wall. A coelom has a special lining (peritoneum).

7. Two major lineages diverged shortly after flatworms evolved. One (protostomes) gave rise to the mollusks, annelids, and arthropods. The other (deuterostomes) gave rise to echinoderms and chordates.

8. All mollusks have a fleshy soft body, a mantle, and a shell or remnant of one. They vary greatly in size, body details, and life-styles.

9. Annelids (the earthworms, polychaetes, and leeches) have a segmented body, complex organs, and a series of coelomic chambers.

10. Collectively, arthropods are the most successful of all groups in terms of diversity, numbers, distribution, defenses, and capacity to exploit food resources.

   a. The arthropods, including arachnids, crustaceans, insects, all have hardened, jointed exoskeletons, as well as modified segments, specialized appendages, highly

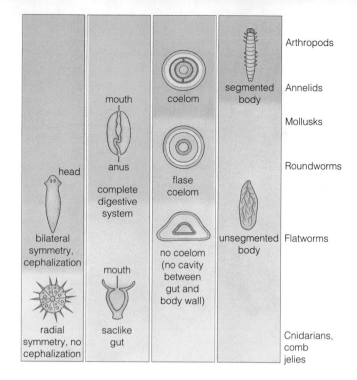

Figure 26.39 Summary of key trends in the evolution of animals, as identified by comparing body plans of major phyla. All of these features did not appear in every group.

specialized respiratory, nervous, and sensory organs, and (in insects only) wings.

   b. Arthropods develop by growth in size and molting or develop through a series of immature stages, such as larvae and nymphs. Many metamorphose; the tissues of immature forms undergo major reorganization and body parts are remodeled before the adult emerges.

11. Echinoderms have spines, spicules, or plates in their body wall. Evolutionarily, the phylum is puzzling. The larvae of most species form bilateral features, but they go on to develop into basically radial adults.

Review Questions

1. List the six main features that characterize animals. *26.1*

2. When attempting to discern evolutionary relationships among major groups of animals, which aspects of their body plans provide the most useful clues? *26.1*

3. What is a coelom? Why was it important in the evolution of certain animal lineages? *26.1*

4. Name some animals with a saclike gut. Evolutionarily, what advantages does a complete gut afford? *26.1, 26.4–26.6*

5. Choose a species of insect that lives in your neighborhood and describe some of the observable adaptations that underlie its success. *26.15, 26.19*

Self-Quiz (*Answers in Appendix IV*)

1. Which is *not* characteristic of the animal kingdom?
   a. multicellularity; cells form tissues, organs
   b. exclusive reliance on sexual reproduction
   c. motility at some stage of the life cycle
   d. embryonic development during the life cycle

Figure 26.40 *Telesto*, one of the soft branching corals.

2. Jellyfishes, sea anemones, and their relatives have _____ symmetry, and their cells form _____ .
   a. radial; mesoderm   c. radial; tissues
   b. bilateral; tissues   d. bilateral; mesoderm

3. In sheer numbers and distribution, _____ are the most successful animals.
   a. arthropods   c. snails and clams   e. vertebrates
   b. sponges   d. sea stars

4. Bilateral, segmented bodies and hardened exoskeletons occur among the _____ .
   a. arthropods   c. snails and clams   e. vertebrates
   b. sponges   d. sea stars

5. Which phylum contains members that are notorious for causing serious diseases in humans?
   a. cnidarians   c. segmented worms
   b. flatworms   d. chordates

6. More complex animals have a _____ between the gut and body wall.
   a. pharynx   c. coelom
   b. pseudocoelom   d. archenteron

7. Most animals that are more complex than cnidarians have _____ symmetry, and _____ forms in their embryos.
   a. radial; mesoderm   c. bilateral; mesoderm
   b. bilateral; endoderm   d. radial; endoderm

8. Match the terms with the appropriate groups.
   ___ sponges   a. spiny-skinned
   ___ cnidarians   b. vertebrates and kin
   ___ flatworms   c. flukes and tapeworms
   ___ roundworms   d. no tissue organization
   ___ rotifers   e. no males for some
   ___ mollusks   f. nematocysts, radial symmetry
   ___ annelids   g. hookworms, elephantiasis
   ___ arthropods   h. jointed exoskeleton
   ___ echinoderms   i. "belly-foots" and kin
   ___ chordates   j. segmented worms

Critical Thinking

1. A carnivorous sponge was discovered in an underwater cave in the Mediterranean Sea. Unlike other sponges, it has no pores or canals. Projecting from branching outgrowths of its body surface are hooklike spicules that act like Velcro to trap shrimp and other animals. The outgrowths envelop prey, which sponge cells then digest. Which components of the sponge body had to evolve for such a feeding strategy?

2. Tapeworms are hermaphroditic. What selective advantages might this feature offer, and in what kinds of environments?

3. People who eat raw oysters or clams harvested from sewage-polluted waters develop mild to severe gastrointestinal ailments.

Think about the feeding modes of these mollusks and develop a hypothesis to explain why mollusk eaters can get sick.

4. You are diving in calm, warm waters behind a tropical reef. You see something that looks like a bright-red and white plant, but it has a profusion of tiny tentacles (Figure 26.40). It is a branching soft coral with small individual polyps. You will never see such a coral growing on reef surfaces exposed to the open sea. Propose two possible explanations for this.

5. The animal shown in Figure 26.41*a* is packed with organs and has a crown of cilia at its head end. The marine worm in Figure 26.41*b* has a segmented body. Most of its segments are similar to one another, and each has bristles on the sides that were used to dig its burrow in sediments. To which groups do the two animals belong?

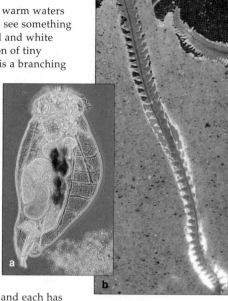

Figure 26.41 Go ahead, name the mystery animals.

Selected Key Terms

adult 26.3	ganglion 26.14	nymph 26.19
animal 26.1	gonad 26.5	organ 26.7
annelid 26.14	gut 26.1	organ system 26.7
arthropod 26.15	hermaphrodite 26.7	placozoan 26.2
bilateral	hydrostatic	planula 26.5
symmetry 26.1	skeleton 26.4	polyp 26.4
brain 26.14	invertebrate 26.1	proglottid 26.7
centipede 26.18	larva 26.3	protostome 26.11
cephalization 26.1	Malpighian	pupa 26.19
cnidarian 26.4	tubule 26.19	radial cleavage 26.11
coelom 26.1	mantle 26.12	radial symmetry 26.1
collar cell 26.3	medusa 26.4	ribbon worm 26.7
comb jelly 26.6	mesoderm 26.1	rotifer 26.10
contractile cell 26.4	metamorphosis 26.15	roundworm 26.8
cuticle 26.8	millipede 26.18	spiral cleavage 26.11
deuterostome 26.11	mollusk 26.12	sponge 26.3
echinoderm 26.20	molting 26.15	torsion 26.13
ectoderm 26.1	nematocyst 26.4	vertebrate 26.1
endoderm 26.1	nephridium 26.14	water-vascular
epithelium 26.4	nerve cell 26.4	system 26.20
flatworm 26.7	nerve cord 26.14	

Readings

Kozloff, E. 1990. *Invertebrates*. Philadelphia: Saunders.

Pearse, V., et al. 1987. *Living Invertebrates*. Palo Alto, California: Blackwell.

Pechenik, J. 1995. *Biology of Invertebrates*. Third edition. Dubuque: Wm. C. Brown.

Web Site See *http://www.wadsworth.com/biology* for practice quiz questions, hypercontents, BioUpdates, and critical thinking. The Wadsworth Biology Resource Center provides a wealth of information fully organized and integrated by chapter.

27 ANIMALS: THE VERTEBRATES

Making Do (Rather Well) With What You've Got

In 1798, a few naturalists were skeptically probing a specimen delivered to the British Museum in London, looking for signs that a prankster had slyly stitched the bill of an oversized duck onto the pelt of a small furry mammal. They didn't know it, but they were examining the remains of a platypus, a web-footed mammal about half the size of a housecat. Like the other mammals, the duck-billed platypus (*Ornithorhynchus anatinus*) has mammary glands and hair. And yet, like birds and reptiles, it has a cloaca, a single enlarged duct through which gametes, feces, and excretions from the kidneys pass. It lays shelled eggs, as birds and most reptiles do. Its young hatch pink and unfinished, as late embryonic stages too helpless to fend for themselves (Figure 27.1a). Its fleshy bill does look ducklike, and its broad, flat, furry tail does look like the one on a beaver (Figure 27.1b).

With its unusual traits, the platypus invites us to challenge preconceived notions of what constitutes "an animal." This particular combination of traits started coming together when the supercontinent Pangea was breaking up. As you will see, the platypus's ancestors happened to be stuck on a huge fragment that remained isolated from the other continents for 150 million years; eventually it became Australia. The geographic separation was a springboard for genetic divergence, and unique mutations in combination with selection pressures gave rise to the platypus body. Even though that body plan is much ridiculed, we scarcely can call it a failure. It has endured far longer than the one we humans inherited.

The platypus inhabits streams and lagoons in remote regions of Australia and Tasmania. During the day it hides in underground burrows. At night it slips into the water, where it hunts for small invertebrates. Even when it stays in water all night, its dense fur helps maintain body temperature, just as it does all day long when

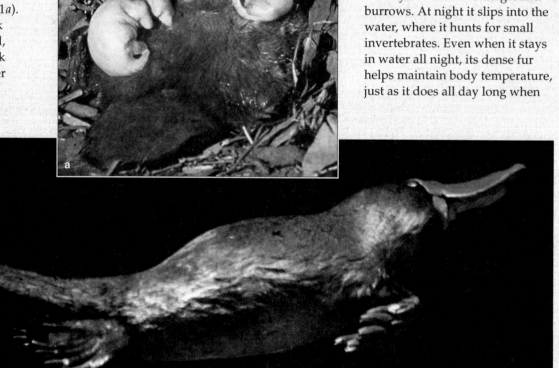

Figure 27.1 One of evolution's success stories—the platypus, (**a**) raising its incompletely formed offspring in a burrow and (**b**) underwater, eyes and ears shut, yet homing in on prey.

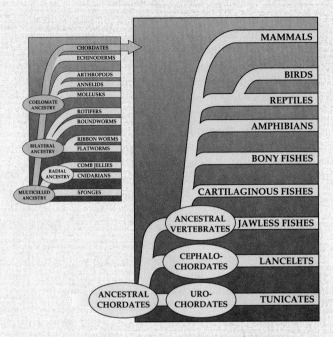

Figure 27.2 Family tree for vertebrates and other chordates.

the platypus is resting in its cool burrow. The broad, thick tail is a most excellent rudder during underwater maneuvers and a storehouse for energy-rich fat reserves. The broad, flattened bill is well-shaped for scooping up aquatic snails, mussels, worms, shrimps, and insect larvae. Horny pads lining the platypus jaws grind up and make short work of shelled and hard-cuticled meals. Back at the burrow, the platypus uses strong claws on its hind feet for digging.

Like a submarine, a platypus closes the hatches, so to speak, when it dives into the water. Its nostrils as well as a fleshy groove around the eyes and ears snap shut. No need for eyes or ears to zero in on prey; many of the 800,000 sensory receptors in its bill can detect tiny oscillations in water pressure as prey swim past. Other receptors detect changes in electric fields as weak as those initiated by the flicks of a shrimp tail.

The platypus is one of 4,500 kinds of mammals that now occupy the vertebrate branch of the animal family tree (Figure 27.2). Its vertebrate relatives include fishes, amphibians, reptiles, and birds, which are described in this chapter. As you consider each group, keep a key concept in mind. *Each animal is a mosaic of traits, many conserved from remote ancestors and others unique to its branch on the animal family tree.* This concept sets the stage for the chapter to follow. There you will trace the evolutionary history of the human species, starting with its mammalian and primate stocks.

KEY CONCEPTS

1. The chordate branch of the family tree for animals includes invertebrate and vertebrate species. All are bilateral animals with a supporting rod for the body (notochord), a dorsal nerve cord, a pharynx, and gill slits in the pharynx wall. These features typically appear in embryos or larvae; some or all may persist in the adults.

2. Existing invertebrate chordates include the tunicates and lancelets.

3. Of eight classes of vertebrates, seven have living representatives. These are jawless fishes, cartilaginous fishes, bony fishes, amphibians, reptiles, birds, and mammals. The other class, the placoderms, became extinct early in vertebrate history.

4. We can identify four major trends in the evolution of certain vertebrate lineages. First, structural support and movement came to depend less on the notochord and more on a backbone. Second, after jaws evolved, the nerve cord evolved into a spinal cord and brain. Third, during the invasion of land, gills became less important than lungs for gas exchange, a trend enhanced by the evolution of more efficient circulatory systems. Fourth, among the pioneers on land, fleshy fins with skeletal supports evolved into limbs, which evolved further among amphibians, reptiles, birds, and mammals.

Characteristics of Chordates

The preceding chapter left off with echinoderms, one of the most ancient lineages of the deuterostome branch of the animal family tree. Dominating this branch are their more recently evolved relatives, the **chordates** (phylum Chordata). Of these bilateral animals, only about 2,100 are, like echinoderms, invertebrates. The vast majority—about 47,000 species—are vertebrates.

Vertebrates are chordates with a backbone, either of cartilage or bone, and a brain located inside a protective chamber of skull bones. The "invertebrate chordates" share certain features with these dominant members of the phylum, but a backbone isn't one of them.

Four features are evident in chordate embryos, and in many species these features persist into adulthood. *First*, a **notochord**, a long rod of stiffened tissue (not cartilage or bone), helps support the body. *Second*, the nervous system has its foundation in a tubular, dorsal **nerve cord** that runs parallel to the notochord and gut. As the embryo develops, the anterior end of the nerve cord increases in mass and becomes modified to form the brain. *Third*, a **pharynx**, which is a muscularized tube, functions in feeding, respiration, or both. The wall of the chordate pharynx has distinctive slits. *Fourth*, a tail forms in embryos and extends past the anus.

Chordate Classification

Biologists group nearly all of the chordates into three subphyla. These are the Urochordata (tunicates and their kin), Cephalochordata (lancelets), and Vertebrata (vertebrates). There are eight classes of vertebrates:

Agnatha	*Jawless fishes*
Placodermi	*Jawed, armored fishes (extinct)*
Chondrichthyes	*Cartilaginous fishes*
Osteichthyes	*Bony fishes*
Amphibia	*Amphibians*
Reptilia	*Reptiles*
Aves	*Birds*
Mammalia	*Mammals*

Appendix I has an expanded classification scheme for the vertebrates. Unit VI provides details of their body plans and functions. Here we become acquainted with major trends in their evolution. We find clues to those trends among the invertebrate chordates.

The embryos of chordates alone have this combination of features: a notochord, a tubular dorsal nerve cord, a pharynx with slits in its wall, and a tail extending past the anus.

Tunicates

The 2,000 species of existing urochordates are baglike animals one to several centimeters long. More often they are called the **tunicates**, after the gelatinous or leathery "tunic" that adults secrete around themselves. Adults of the most common species, the "sea squirts," squirt water through a siphon when something irritates them.

Tunicates live in marine habitats, from the intertidal zone to surprising depths. Most adults remain attached to rocks and other suitably hard substrates. Some live solitary lives; others are colonial. Figure 27.3*a* shows an adult sea squirt. It develops from a bilateral, swimming larva that resembles a tadpole (Figure 27.3*b–d*). A larva, recall, is an immature stage between the embryonic and adult stages of an animal life cycle. The firm, flexible notochord of a tunicate larva is a series of cells that acts like a torsion bar. As muscles on one side of the tail or the other contract, the "bar" bends, then it springs back when the muscles relax. The strong, side-to-side motion propels the animal forward. Most fishes use muscles and the backbone for the same kind of motion.

Like some other animals, tunicates are **filter feeders**: they filter food from a current of water that is directed through part of their body. Water flows in through one siphon and passes through **gill slits**, or openings in the thin pharynx wall. Water flows out through a different siphon. Their pharynx also acts as a respiratory organ. Dissolved oxygen is more concentrated in the water than in the blood in vessels adjoining the pharynx, so it diffuses from water into the blood. Carbon dioxide's concentration gradient is such that it diffuses from blood into water leaving the pharynx.

Sea squirts undergo metamorphosis. By this process of development, recall, remodeling and reorganization of body tissues transform an immature stage into the adult. As a sea squirt larva metamorphoses, its tail and notochord disappear, and a tunic forms. The pharynx enlarges, and perforations in its wall get subdivided into many slits. The nerve cord regresses, so that all that remains is a greatly simplified nervous system.

Lancelets

About twenty-five species of fish-shaped, translucent animals called cephalochordates live just offshore, on seafloors around the world. Most are shorter than your little finger. Much of the time they are buried, almost up to their mouth, in sand and sediments. Their common name, **lancelets**, refers to the sharp tapering of their body at both ends. The lancelet body plan, shown in Figure 27.4*a*, clearly shows the four chordate features. Notice the segmented pattern of muscles on both sides of the notochord. Like tunicate larvae, lancelets use the

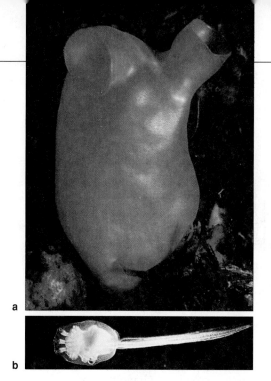

a

b

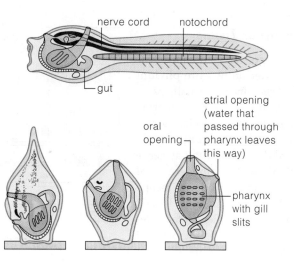

nerve cord   notochord

c Section through the tadpole-like larva.

gut

oral opening

atrial opening (water that passed through pharynx leaves this way)

pharynx with gill slits

d A new larva swims about for a brief period. Metamorphosis begins when its head attaches to a substrate. The notochord, tail, and most of the nervous system are resorbed (recycled to form new tissues). Slits in the pharynx wall multiply. Organs rotate until the openings through which water enters and leaves the pharynx are directed away from the substrate.

Figure 27.3   Example of a tunicate. (**a**) Adult form of a sea squirt. (**b,c**) Photograph and diagram of a sea squirt larva. (**d**) Metamorphosis of the larva into the adult.

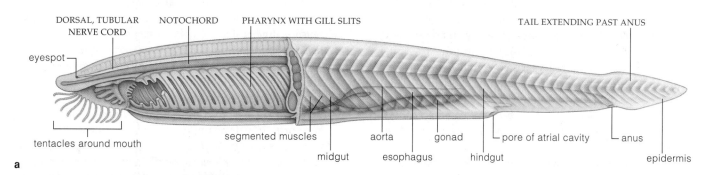

DORSAL, TUBULAR NERVE CORD   NOTOCHORD   PHARYNX WITH GILL SLITS   TAIL EXTENDING PAST ANUS

eyespot

tentacles around mouth

segmented muscles   midgut   esophagus   aorta   gonad   hindgut   pore of atrial cavity   anus   epidermis

a

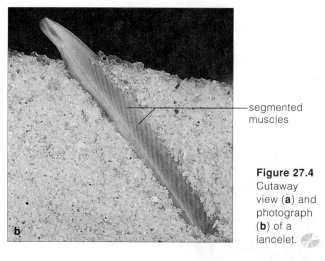

b

segmented muscles

Figure 27.4 Cutaway view (**a**) and photograph (**b**) of a lancelet.

Like tunicates, lancelets are filter-feeding animals. Cilia line their mouth cavity and create a current that draws water through it. The water then moves through the pharynx, where food becomes trapped in mucus. The trapped food is delivered to the rest of the gut. By itself, the collective beating of tiny cilia cannot deliver sufficient food to a filter-feeding animal. Delivery also depends on having a large food-trapping surface area. In lancelets, the pharynx provides the area. It is large, relative to the overall body length (Figure 27.4). Besides this, as many as 200 ciliated, food-trapping gill slits perforate its wall. As you will read in sections to follow, the pharynx turned out to be an organ with interesting evolutionary possibilities for the vertebrates.

notochord and muscles to produce swimming motions. Unlike tunicates, they have a closed circulatory system (but no red blood cells). A complex brain is nowhere in sight, but the anterior end of the dorsal nerve cord is expanded, and pairs of nerves extend into each muscle segment. The mode of respiration is simple yet effective for such a small body. Dissolved oxygen and carbon dioxide diffuse across the body's thin skin.

Tunicate larvae and the lancelets use their notochord and muscles for fishlike swimming movements. Their pharynx, a muscularized tube, has finely divided openings across its thin wall.

The tunicates and lancelets are filter feeders. They use their pharynx to filter microscopic food from a current of water that they draw in through the body. Tunicate larvae also use the pharynx as a respiratory organ, for gas exchange.

Puzzling Origins, Portentous Trends

Did vertebrates arise from tadpole-shaped chordates? It's possible, when you consider an obscure phylum called the hemichordates (*hemi-* meaning half, as in "halfway to chordates"). Evolutionarily, these marine invertebrates are midway between echinoderms and chordates. They don't have a notochord, but they do have a gill-slitted pharynx and a dorsal, tubular nerve cord. Their larvae resemble those of echinoderms *and* tunicates. What if mutations in an echinoderm lineage accelerated the rates at which sex organs developed? If those organs started functioning earlier, in tadpole-shaped *larvae*, then metamorphosis—and the original adult form—could be dispensed with. This idea is not far-fetched. A few existing tunicates look like larvae but have sex organs. Certain amphibians become sexually mature even though they are larvae in some respects, and they can reproduce, generation after generation.

Assuming tadpole-shaped chordates emerged, how did they evolve into vertebrates? The key trends started in fishes (Figures 27.5 and 27.6). One involved a shift from the notochord to reliance on a column of separate, hardened segments: **vertebrae** (singular, vertebra). This was the start of an endoskeleton (internal skeleton) that muscles could work against. *A vertebral column evolved in fast-moving predators, some of which were ancestral to all other vertebrates.*

In a related trend, part of the nerve cord expanded into a brain after **jaws** evolved. These skeletal elements arose through modification of the first in a series of structural elements that supported the gill slits. Jaws meant new feeding possibilities and greater competition

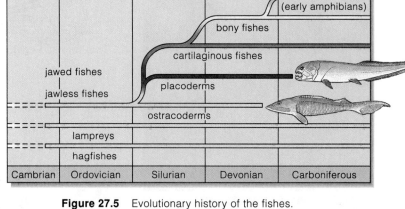

Figure 27.5 Evolutionary history of the fishes.

among predators. Fishes able to smell or see predators or food farther away would now have the advantage—provided that their brain also had evolved enough to process the new information. *A trend toward complex sensory organs and nervous systems did indeed begin in fishes, and it continued among land vertebrates.*

Another trend began when paired fins evolved. **Fins** are appendages that help propel, stabilize, and guide the body in water (compare Figure 27.9). In some lineages, ventral fins became fleshy and equipped with skeletal supports—the forerunners of limbs. *Paired, fleshy fins were the starting point for the legs, arms, and wings that evolved among amphibians, reptiles, birds, and mammals.*

Change in respiration was another trend. Think of the lancelet. Except when buried in sand, oxygen and carbon dioxide just diffuse across its body surface. In most vertebrate lineages, however, **gills** evolved. These respiratory organs have a moist, thin, and much-folded surface, they are richly endowed with blood vessels, and they offer a large surface area for gas exchange. For example, five to seven pairs of a shark's gill slits are continuous with gills extending from the pharynx to the body's surface. When a shark opens its mouth and closes its

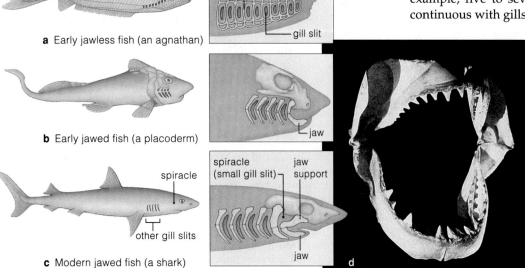

a Early jawless fish (an agnathan)

b Early jawed fish (a placoderm)

spiracle

other gill slits

c Modern jawed fish (a shark)

supporting structures

gill slit

jaw

spiracle (small gill slit)   jaw support

jaw

jaw

d

Figure 27.6 Comparison of gill-supporting structures in jawless fishes and jawed fishes. In the placoderms and other early jawed vertebrates, cartilage supported the mouth's rim. In modern jawed fishes, a gill slit between the jaws and a nearby supporting element serves as a spiracle, an opening through which water is drawn. In (**a**), the gill supports are just under the skin. In (**b**) and (**c**), they are internal to the gill surface. (**d**) Jaws of a modern shark.

external gill openings, the pharynx expands. Oxygen diffuses *from* water in the mouth into the gills, as carbon dioxide diffuses *into* that water. Muscles constrict the pharynx and so force oxygen-depleted, carbon dioxide-enriched water out through the gill slits.

As some fishes became larger and more active, gills became more efficient. But gills can't work out of water; they stick together unless water flows through them and keeps them moist. In fishes ancestral to the land vertebrates, pouches developed from the gut wall. The pouches evolved into **lungs**—internally moistened sacs for gas exchange. In a related trend, modifications to the heart enhanced the pumping of oxygen and carbon dioxide through the body. *Ancestors of land vertebrates relied less on gills and more on lungs. And more efficient circulatory systems accompanied the evolution of lungs.*

The First Vertebrates

Figure 27.5 gives a time frame for vertebrate evolution. Free-swimming species originated in Cambrian times and gave rise to two kinds of fishes—those without and those with jaws. One or the other kind probably gave rise to all vertebrate lineages that followed. The earliest jawless fishes (Agnatha) included **ostracoderms**. Figure 27.6*a* shows one of those bottom-dwelling filter feeders. Their skeleton probably consisted of a notochord and a protective covering for the enlarged brain. Armorlike plates on the body consisted of bony tissue and dentin, a hardened tissue still present in vertebrate teeth. The armor might have been useful against the giant pincers of sea scorpions, but it apparently wasn't much good against jaws. Ostracoderms disappeared when jawed fishes began their adaptive radiations.

Placoderms were among the first fishes with jaws and paired fins (Figure 27.6*b*). These bottom-dwelling scavengers or predators had a notochord reinforced with bony elements. The first gill-supporting structures were enlarged and had bony projections something like teeth, and they functioned as jaws. Before placoderms, feeding strategies had been limited to filtering, sucking, or rasping away at food material. When the placoderms started biting and tearing off large chunks of prey, they set off an evolutionary race of offensive and defensive adaptations that has continued to the present.

Placoderms diversified, then became extinct during the Carboniferous period. New kinds of predators, the cartilaginous and bony fishes, replaced them in the seas. We will consider the living descendants of the new predators after a look at existing jawless fishes.

The emergence of a vertebral column, jaws, paired fins, and lungs was pivotal in the evolution of certain vertebrates.

EXISTING JAWLESS FISHES

The **hagfishes** and the **lampreys** are living descendants of some early jawless fishes. All seventy-five or so of the known species have a cylindrical body, a cartilaginous skeleton, and no paired fins (Figure 27.7). Most are less than 1 meter (3.3 feet) long.

Hagfishes live together in groups on the sediments of continental shelves. They prey on polychaete worms or scavenge for weakened or dead organisms. Even though they lack jaws, hagfishes do eat well. They have sensory tentacles around the mouth and a tongue that rasps soft tissues from prey. They defend themselves by secreting copious amounts of sticky, smelly, slimy mucus from a series of glands along their body.

Lampreys are specialized predators that are almost parasites. Their suckerlike oral disk has horny, toothlike parts that rasp flesh from prey. Some types latch onto salmon, trout, and other commercially valuable fishes,

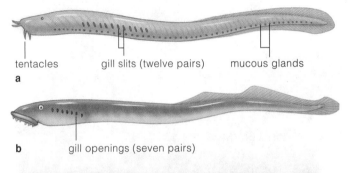

tentacles   gill slits (twelve pairs)   mucous glands
a

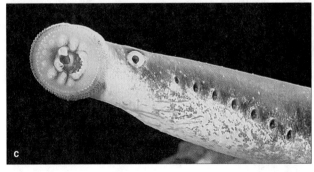

b   gill openings (seven pairs)

Figure 27.7 Body plan of (**a**) a hagfish and (**b**) a lamprey. (**c**) A lamprey's toothed oral disk, pressed against aquarium glass.

then suck out juices and tissues. Just before the turn of the century, lampreys began invading the Great Lakes of North America. Their introduction resulted in the collapse of populations of lake trout and other large, valued fishes, as it has done in other aquatic habitats.

Hagfishes and lampreys get along without jaws; they latch onto prey with their efficient, specialized mouthparts.

EXISTING JAWED FISHES

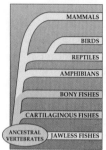

Few of us make a career of studying life beneath the surface of the seas and other bodies of water, so fishes are not widely recognized as being the world's dominant vertebrates. Their numbers exceed those of all other vertebrate groups combined. They also show far more diversity than the other groups; we have identified more than 21,000 species of bony fishes alone.

The form and behavior of a fish tell us something about the challenges it faces in water. For example, being about 800 times more dense than air, water resists fast movements. As an adaptation to this constraint, sharks and other predatory marine fishes are streamlined for pursuit (Figure 27.8a). Their long, trim body reduces friction; their tail muscles are organized for propulsive force and forward motion. By contrast,

some bottom-dwelling fishes, such as the rays, have a flattened body (Figure 27.8b). Theirs is not a high-speed body plan; it is good for hiding from predators or prey.

A motionless trout, suspended in shallow water, is another example of adaptation to water's density. Like many fishes, it maintains neutral buoyancy with a **swim bladder**, an adjustable flotation device that exchanges gases with blood. When a trout gulps air at the water's surface, it is adjusting the volume of its swim bladder.

Cartilaginous Fishes

Cartilaginous fishes (Chondrichthyes) include about 850 species of skates, sharks, and chimaeras (Figure 27.8). These marine predators have pronounced fins, a skeleton of cartilage, and five to seven gill slits on both sides of the pharynx. Most species have a few scales or many rows of them. **Scales** are small, bony plates at the body surface. Fish scales often protect the body without weighing it down.

The skates and rays are mainly bottom dwellers with flattened teeth suitable for crushing hard-shelled prey. Both have enlarged fins that extend onto the side of the head. The largest species, the manta ray, measures up to six meters from fin tip to fin tip. A venom gland in the tail of sting rays probably helps deter predators. Other rays have electric organs in the tail or fins that can stun prey with as much as 200 volts of electricity.

At fifteen meters from head to tail, some sharks are among the largest living vertebrates. As Figure 27.6d shows, sharks have formidable jaws. They continually shed and replace their sharp, triangular teeth (modified scales), which they use to grab prey and rip off chunks of flesh. The relatively few shark attacks on humans give the whole group a bad reputation. Surfboards with legs dangling over the side are a recent temptation for some sharks, which have been dining on invertebrates, fishes, and marine mammals for many millions of years.

The thirty or so species of chimaeras feed mostly on mollusks. With their bulky body and slender tail, they look a bit like a rat; hence the common name, ratfishes. They have a venom gland in front of the dorsal fin.

Bony Fishes

Bony fishes (Osteichthyes) are the most numerous and diverse vertebrates. They make up all but 4 percent of the existing species of modern fishes. Their ancestors arose during the Silurian and soon gave rise to three lineages: the ray-finned fishes, lobe-finned fishes, and lungfishes. Descendants of the early forms radiated into nearly all aquatic habitats.

Body plans vary greatly. Marine predators typically have a torpedo shape, a flexible body, and strong tail

Figure 27.8 Cartilaginous fishes: (**a**) shark, (**b**) blue-spotted reef ray, and (**c**) chimaera, sometimes called a ratfish.

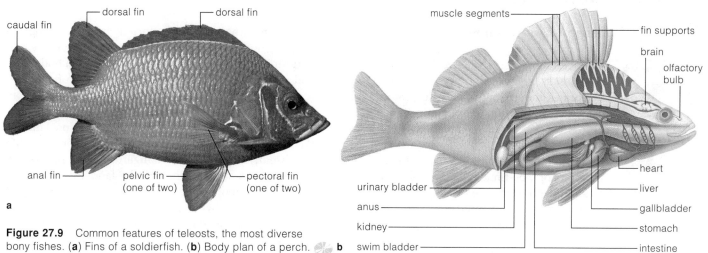

Figure 27.9 Common features of teleosts, the most diverse bony fishes. (**a**) Fins of a soldierfish. (**b**) Body plan of a perch.

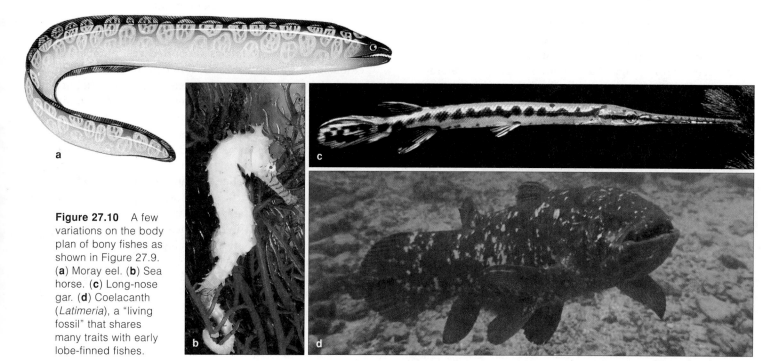

Figure 27.10 A few variations on the body plan of bony fishes as shown in Figure 27.9. (**a**) Moray eel. (**b**) Sea horse. (**c**) Long-nose gar. (**d**) Coelacanth (*Latimeria*), a "living fossil" that shares many traits with early lobe-finned fishes.

fins that aid in swift pursuit. Many reef dwellers are small finned and box shaped; they move easily through narrow spaces. Those with an elongated, flexible body, including the moray eel, wriggle through mud and slip into concealing crevices. Sea horses and many bottom-dwelling species have a cryptic body shape that helps conceal them from prey or predators. Figures 27.9 and 27.10 show examples of these varied body plans.

In ray-finned fishes, rays derived from the dermis (one of the skin's layers) support paired fins. Most of these fishes have highly maneuverable fins and light, flexible scales, both of which contribute to a capacity for complex movements. Their efficient respiratory system rapidly delivers oxygen to metabolically active tissues. One group of ray-finned fishes resembles their early

ancestors. It includes sturgeons and paddlefishes of the Mississippi River basin. Teleosts, the most abundant group, include salmon, tuna, rockfish, catfish, perch, minnows, moray eels, flying fish, sculpins, blennies, scorpionfish, and pikes. All are thin scaled or scaleless.

By contrast, only one species of lobe-finned fishes and three genera of lungfishes made it to the present. As their name suggests, a **lobe-finned fish** is unique in having paired fins that incorporate fleshy extensions from the body. As you will see next, neither it nor the lungfishes have changed much from their ancient stock.

Of all existing vertebrates, the bony fishes are the most spectacularly diverse and the most abundant.

AMPHIBIANS

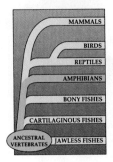

Origin of Amphibians

The lobe-finned fishes (Figure 27.10*d*) are living fossils, relics of a time when the first vertebrates moved onto land. Remember the earlier description of living conditions during the Devonian? Sea levels rose and fell repeatedly, and swamps fringing the coasts flooded and drained many times. Ancestors of lobe-finned fishes evolved in those trying times. They probably used their lobed fins to pull themselves from dried-up ponds to ones that were still habitable. What else made their pond-to-pond lurchings possible? They had sac-shaped outpouchings from the wall of the esophagus, which is a tube leading to the stomach. The sac began to be used as a way to supplement gas exchange. In other words, the ancestors of lobe-finned fishes had simple lungs.

Existing lungfishes provide more clues to how the ancestral forms might have made it through stressful times. They live in stagnant water but surface to gulp air. In dry seasons, when streams dwindle to mud, they encase themselves in a slathering of mud and slime that keeps them from drying out until the next rainy season.

Devonian lobe-finned fishes lurched over land only as a way of reaching more hospitable ponds. Yet the very act of traveling out of water favored the evolution

of stronger fins and more efficient lungs. Among the evolving forms were the ancestors of amphibians. An **amphibian** (Amphibia) is a vertebrate with a body plan and reproductive mode somewhere between fishes and reptiles. Most kinds have a largely bony endoskeleton, and they have four legs (or four-legged ancestors).

Early amphibians were spending time on land by the close of the Devonian (Figure 27.11*a*). For them, life in those drier habitats was dangerous—and promising. Temperatures shifted more on the land than in water, air didn't support the body as well as water did, and water was not always plentiful. However, air has far more oxygen, and the lungs continued to be modified in ways that enhanced oxygen uptake. Also, circulatory systems became better at rapidly distributing oxygen to cells throughout the body. Both modifications increased the energy base for more active life-styles.

New sensory information also challenged the early amphibians. Swamp forests supported vast numbers of edible insects and other invertebrate prey. Animals with good vision, hearing, and balance—the senses that are most advantageous on land—were favored. And brain regions concerned with those senses expanded.

All of the salamanders, frogs, toads, and caecilians alive today are descended from those first amphibians. None has escaped the water entirely. Even when gills or lungs are present, amphibians can use their thin skin as a respiratory surface (that is, for gas exchange). But respiratory surfaces must be kept moist, and skin dries easily in air. Some species still live their entire lives in water; others lay their eggs in water or produce aquatic larvae. Even species that have adapted to habitats on land must lay their eggs in moist places.

Figure 27.11 (**a**) *Ichthyostega*, one of the first of the Devonian amphibians. Fossils of this species have been recovered in Greenland. The skull, deep tail, and fins were decidedly fishlike. Unlike fishes, this species had four limbs adapted for moving on land, and a short neck intervened between its head and the rest of the body. Its vertebral column and rib cage were adapted to support the body's weight out of water. (**b,c**) Proposed evolution of skeletal elements inside the lobed fins of certain fishes into the limb bones of early amphibians.

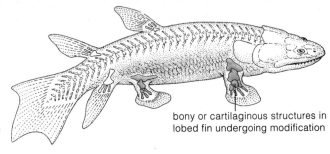

a

b
bony or cartilaginous structures in
lobed fin undergoing modification

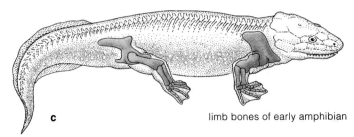

c
limb bones of early amphibian

Salamanders

Diverse salamanders and their kin, the newts, live in the world's north temperate regions and tropical parts of Central and South America. Most are less than fifteen centimeters long. And most have legs at right angles to the body, with forelimbs and hindlimbs about the same size (Figure 27.12a). Like fishes and early amphibians, salamanders bend from side to side when they walk:

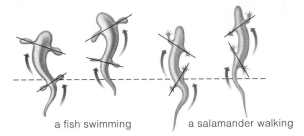

a fish swimming   a salamander walking

Probably the first four-legged vertebrates also walked this way. Both the larvae and adults are carnivores. Adults of some species retain several larval features. For example, the Mexican axolotl retains the external gills of the larval form, and the development of its teeth and bones is arrested at an early stage. As in some other species, its larvae are sexually precocious; they breed.

Frogs and Toads

With more than 3,000 species, frogs and toads are the most successful amphibians (Figure 27.12b,c). Their long

Figure 27.12 Amphibians. (**a**) Terrestrial stage in the life cycle of a red-spotted salamander. (**b**) A frog, splendidly jumping. (**c**) An American toad. (**d**) A caecilian.

hindlimbs and powerful muscles allow them to catapult into the air or barrel through the water. Most frogs flip a sticky-tipped, prey-capturing tongue from their mouth. An adult eats just about any animal it can catch; only its head size limits the size of prey. Skin glands of some species produce toxins, and poisonous types often have bright warning coloration. The skin of frogs, including the South African clawed frog (*Xenopus laevis*), harbors antibiotics against many pathogens in the water. In Unit VI, you will have opportunities to take a look at how frogs are put together and how they function.

Caecilians

The ancestors of caecilians lost their limbs and most of their scales. They gave rise to decidedly worm-shaped amphibians (Figure 27.12d). Nearly all of the 160 or so existing species burrow through soft, moist soil as they pursue insects and earthworms. A few live in shallow freshwater habitats. Adults of most species are blind.

Amphibians are halfway between fishes and reptiles in their body form and behavior. Regardless of how far they venture onto land, they have not fully escaped dependency on water.

THE RISE OF REPTILES

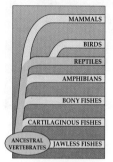

Reptiles (Reptilia) arose from amphibians during the late Carboniferous. Of all the vertebrates, they were the first to escape dependency on standing water. They did so mainly through four adaptations that clearly distinguish them from fishes and amphibians.

First, the reptiles have tough, dry, scaly skin that restricts loss of water from the body (Figure 27.13*a*). Second, fertilization is internal. A copulatory organ can deposit sperm into a female's body; sperm do not need free water to reach eggs. Third, reptilian kidneys are good at conserving water. Fourth, most species of reptiles produce **amniote eggs**, within which embryos develop to an advanced stage before being hatched or born into dry habitats. Amniote eggs contain membranes that retain water and protect or provide metabolic support to the embryo. Most of them have a leathery or calcified shell (Figure 27.13*b,c*). We will take a closer look at the structure and development of amniote eggs in Chapter 45.

Compared to amphibians, the early reptiles chased prey with far greater cunning and speed. With their well-muscled jawbones and formidable teeth, reptiles could seize and apply sustained, crushing force on prey. Their prey included other vertebrates as well as insects. Also, reptilian limbs generally were more efficient at supporting the trunk of the body on land. For many species, the nervous system increased in complexity. A reptile's brain is small compared to the rest of the body mass, but it governs complex forms of behavior that are unknown among amphibians. For example, the cerebral cortex is the most complex part of the forebrain. Here, information from diverse sensory organs is integrated and stored, and commands for responses are issued. The cerebral cortex evolved first among the reptiles.

b

c

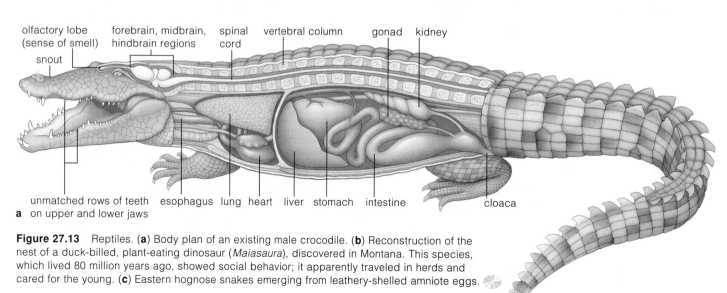

olfactory lobe (sense of smell)   forebrain, midbrain, hindbrain regions   spinal cord   vertebral column   gonad   kidney

snout

unmatched rows of teeth on upper and lower jaws   esophagus   lung   heart   liver   stomach   intestine   cloaca

a

Figure 27.13 Reptiles. (**a**) Body plan of an existing male crocodile. (**b**) Reconstruction of the nest of a duck-billed, plant-eating dinosaur (*Maiasaura*), discovered in Montana. This species, which lived 80 million years ago, showed social behavior; it apparently traveled in herds and cared for the young. (**c**) Eastern hognose snakes emerging from leathery-shelled amniote eggs.

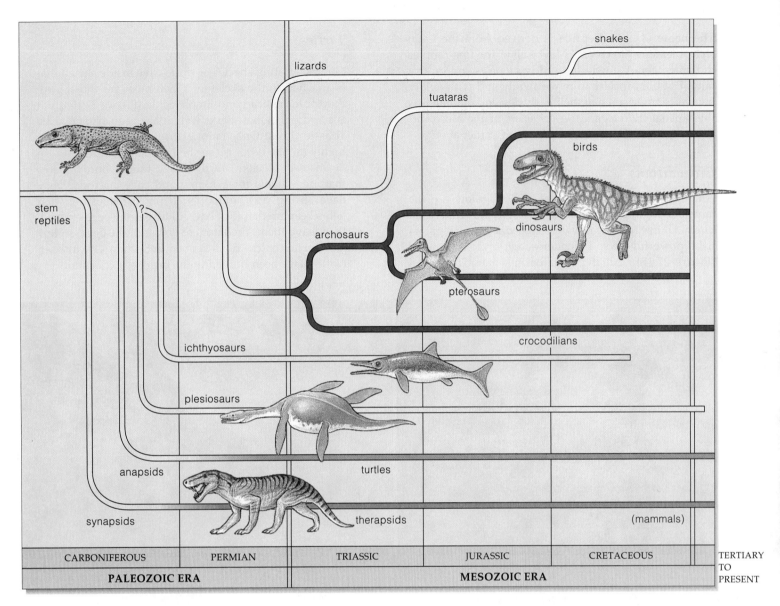

CARBONIFEROUS	PERMIAN	TRIASSIC	JURASSIC	CRETACEOUS	TERTIARY TO PRESENT
PALEOZOIC ERA		**MESOZOIC ERA**			

Figure 27.14 Evolutionary history of the reptiles.

The group of reptiles called crocodilians were the first vertebrates with a muscular, four-chambered heart fully separated into two halves. (Blood enters the first chamber of each half, and the second chamber pumps it out.) As Chapter 39 describes, such separation permits oxygen-rich blood to travel from the lungs to the rest of the body, and oxygen-poor blood from the body to the lungs, in two independent circuits. Also, gas exchange across the skin, so vital for amphibians, was abandoned by reptiles—nearly all of which depend on lungs.

The early reptiles underwent adaptive radiations that led to fabulously diverse forms. The group called dinosaurs evolved during the Triassic and for the next 125 million years were the dominant land vertebrates. The fossilized nests of one kind (*Maiasaura*) contain not only eggs but also juveniles a few months old, at most (Figure 27.13*b*). Such evidence suggests that at least some dinosaurs showed parental behavior; that is, they took care of their young during a period of dependency. (*Maiasaura* means "good mother lizard.")

When the Cretaceous ended abruptly, so did the last dinosaurs (Sections 21.6 and 21.7). As you can see from Figure 27.14, the reptilian groups that made it to the present are crocodilians, turtles, tuataras, snakes, and lizards.

Reptiles, with their tough, scaly skin, reliance on internal fertilization, water-conserving kidneys, and amniote eggs, were the first vertebrates to escape dependency on free-standing water in their habitats.

The move onto land also required major modifications in the nervous, circulatory, and respiratory systems.

MAJOR GROUPS OF EXISTING REPTILES

The name of class Reptilia is derived from the Latin *repto*, meaning "to creep." Maybe all of the first reptiles did creep about slowly in muddy swamps and on dry land. But the capacity to move swiftly and with agility evolved among some of their early descendants, such as the bipedal (two-legged) *Velociraptors* of the Mesozoic. Existing descendants race, lumber, and slither about.

Crocodilians

Among the modern crocodiles and alligators are the largest living reptiles, and the closest living relatives of birds. All live in or near water. Crocodiles and alligators have powerful jaws, a slender snout, and sharp teeth (Figure 27.15a). All live near or in water. The feared

Turtles

The 250 existing species of turtles live inside a shell that is attached to the skeleton. When most are threatened, they pull the head and limbs inside (Figure 27.15b,c). It is a body plan that works well; it has been around since Triassic times. Only among the sea turtles and other highly mobile types is the shell reduced in size.

Instead of teeth, turtles have tough, horny plates that are suitable for gripping and chewing food. They have strong jaws and often a fierce disposition that helps keep predators at bay. All turtles lay eggs on land, then leave them. Predators eat most of the eggs, so few new turtles hatch. As described in Section 41.9, the sea turtles have been hunted to the brink of extinction.

a

b

Figure 27.15 Reptiles. (**a**) Spectacled caiman. As is true of other crocodilians, its peglike upper teeth do not match up with peglike lower ones. The body plan and life-styles of crocodilians haven't changed much for nearly 200 million years. Their future is not rosy. Housing developments are encroaching on many of their habitats, and their belly skin is in demand for wallets, shoes, and handbags. (**b**) A heavy-shelled Galápagos tortoise. (**c**) Section through the skeleton and shell of a turtle, with head withdrawn and extended. (**d**) Frilled lizard, flaring a large ruff of neck skin as defensive behavior. (**e**) A rattlesnake in mid-strike. (**f**) Tuatara (*Sphenodon*).

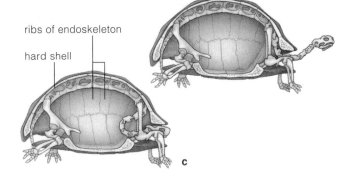

ribs of endoskeleton

hard shell

c

"man-eaters" of southern Asia and the Nile crocodiles weigh up to 1,000 kilograms. They drag a mammal or bird into the water, tear it apart by violently spinning about, then gulp down the torn chunks. Crocodilians are able to adjust body temperature by behavioral and physiological mechanisms, just as other reptiles and birds can do. Like birds, they also show complex social behavior, as when the parents guard nests and assist their hatchlings into the water.

Lizards and Snakes

About 95 percent of the existing reptiles, the lizards and snakes, are distant relatives of the dinosaurs. Most are small, but the Komodo monitor lizard is big enough to hunt young water buffalo. The longest modern snake would stretch across ten yards of a football field.

Most of the 3,750 kinds of lizards are insect eaters of deserts and tropical forests. They include aggressive, adhesive-toed geckos and iguanas; the photograph of the Unit VI introduction shows one of the more colorful types. Most lizards grab prey with small, peglike teeth. Chameleons capture prey with accurate flicks of their tongue, which is longer than their body (Section 33.4).

Being small themselves, most lizards are prey for many other animals. Some attempt to startle predators and intimidate rivals by flaring their throat fan (Figure 27.15d). Many give up their tail when a predator grabs them. The detached tail wriggles for a bit and may be distracting enough to permit a getaway.

During the early Cretaceous, some of the short-legged, long-bodied lizards gave rise to the elongated, limbless snakes. Most of the 2,300 existing species slither in S-shaped waves, much like salamanders. "Sidewinders"

make J-shaped movements and leave distinctive trails across loose sand and sediments. Some species have bony remnants of ancestral hindlimbs (Section 17.1).

Snakes have highly movable jaws; some swallow animals wider than they are. The pythons and boas coil around their prey and suffocate it into submission before eating. Fanged types, including rattlesnakes and coral snakes, bite and subdue prey with venom (Figure 27.15e). Snakes usually do not act aggressively toward humans. However, each year in the United States alone, rattlesnakes (one of the pit vipers) and other poisonous types bite 8,000 or so people and kill about 12 of them. In India, king cobras and other types bite 200,000 or so people and kill about 9,000 of them.

Even the most feared snakes are vulnerable during their life cycle; birds and other predators relish snake eggs. The female snakes store sperm and lay several clutches of fertilized eggs at intervals after they mate, which improves the odds that at least some will hatch.

Tuataras

Only two species of tuataras are still around, on small, windswept islands near New Zealand (Figure 27.15f). Their body plan has not changed much since Mesozoic times. The tuataras resemble lizards, but their lineage is more recent. At the top of their head is a third "eye," with retina, lens, and nerves to the brain. Being covered with skin, it can only register changes in daylength and light intensity. Does it have roles in hormonal controls over reproduction? Maybe (compare Section 37.8). The tuataras, like turtles, may live longer than sixty years. They engage in sex only after they are twenty years old.

Existing crocodilians are the closest relatives of dinosaurs and birds. Turtles, like tuataras, have changed little in body plan for millions of years. About 95 percent of all living reptiles are lizards or snakes.

BIRDS

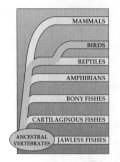

Of all organisms, only **birds** (Aves) grow feathers (Figure 27.16). **Feathers**, derived from skin, are lightweight structures that birds use in flight, as body insulation, or both. Judging from the fossil record, birds are descended from tiny reptiles that ran around on two legs 160 million years ago. *Archaeopteryx* was on or near the lineage leading to modern birds. Figure 17.7 shows how it had many reptilian traits as well as feathers and other avian traits.

In many respects, birds still resemble reptiles. For example, they have scales on their legs and a number of the same internal structures. They, too, lay their eggs and commonly engage in parental behavior, such as guarding the nest. One of Darwin's champions, Thomas Huxley, was the first to argue that birds are glorified reptiles. Today, many biologists do indeed classify birds as a branching from the reptilian lineage.

a

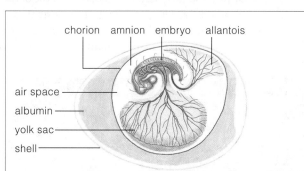

b

Figure 27.16 Characteristics of birds. (**a**) Flight. Of all living vertebrates, only birds and bats fly by flapping their wings. (**b**) All birds lay hard-shelled eggs. The photograph below this generalized sketch shows speckled eggs of a magpie.

(**c**) Feathers, the key defining characteristic of birds. The flamboyant plumage of this male pheasant is an outcome of sexual selection. This bird is a native of the Himalaya Mountains of India, and it is an endangered species. As is the case for many other kinds of birds, its jewel-colored feathers end up adorning humans—in this case, on the caps of native tribespeople.

(**d**) Many birds, including these Canada geese, show migratory behavior. *Animal migration* is a recurring pattern of movement between two or more locations in response to environmental rhythms. For example, seasonal change in daylength is a cue that acts on internal timing mechanisms (biological clocks), which in turn trigger physiological and behavioral changes. Such changes induce migratory birds to make round trips between distant regions that differ in climate. Canada geese spend the summer nesting in marshes and lakes in the northern United States and Canada. Their wintering grounds are in New Mexico and other parts of the southern United States.

c

d

There are almost 9,000 named species of birds. They show stunning variation in their body size, proportions, coloration, and capacity for flight. One of the smallest hummingbirds barely tips the scales at 2.25 grams (0.08 ounce). The largest existing bird, the ostrich, weighs about 150 kilograms (330 pounds). Ostriches cannot fly, but they are impressively long-legged sprinters (Figure 17.2). Many birds, such as warblers and other perching types, differ markedly in feather coloration and in their territorial songs. Bird songs and other complex social behaviors are topics of later chapters.

Flight demands high metabolic rates, which require a good flow of oxygen through the body. Just as you do, birds have a large, durable, four-chambered heart. The heart pumps oxygen-enriched blood to the lungs and to the rest of the body along separate routes, as it probably did in their reptilian ancestors. Also, a bird's respiratory system incorporates a unique ventilating apparatus that enormously enhances oxygen uptake. Bird respiration is described in Section 41.3.

Flight also demands an airstream, low weight, and a powerful downstroke that can provide lift (a force at right angles to the airstream). The bird wing, a modified forelimb, consists of feathers and lightweight bones attached to powerful muscles. The bones weigh very little (because of profuse air cavities in the bone tissue).

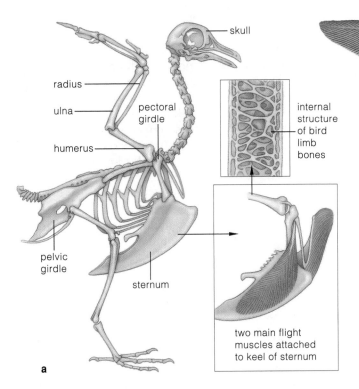

skull

radius

ulna — pectoral girdle

humerus

internal structure of bird limb bones

pelvic girdle

sternum

two main flight muscles attached to keel of sternum

a

barbule
barb
shaft

b

For example, the skeleton of a frigate bird, which has a seven-foot wingspan, weighs only four ounces. That is less than the feathers weigh! A bird's flight muscles attach to an enlarged breastbone, or sternum, and to upper limb bones adjacent to it (Figure 27.17). Contraction of the muscles creates the powerful downstroke required for flight. The wings, with their long flight feathers, serve as the airfoils. Usually, a bird spreads out long feathers on a downstroke and thus increases the size of the surface pushing against air (Figure 27.16a). On the upstroke, it folds the feathers somewhat, so that each wing presents the least possible resistance to the air.

Figure 27.17 (**a**) Body plan of a typical bird. Flight muscles attach to the large, keeled breastbone (sternum). (**b**) A bird wing is a complex system of lightweight bones and feathers. Feathers gain strength from a hollow central shaft and from tiny barbules that are interlocked in a latticelike array.

Of all animals, birds alone have feathers, which they use in flight, in heat conservation, and in socially significant communication displays.

THE RISE OF MAMMALS

Distinctly Mammalian Traits

Mammals are vertebrates with hair and mammary glands (hence the name of class Mammalia, from the Latin *mamma*, which means breast). Of all organisms, they are the only ones having these traits. Females feed their young with milk, a nutritious fluid that mammary glands secrete into ducts that lead to openings at the body's ventral or anterior surface (Figure 27.18*a*).

Mammals differ greatly in the amount, distribution, and type of hair. A few aquatic mammals, including the whales, lost most of their hair after their land-dwelling ancestors returned to the seas. But if you look closely, you see that even a whale snout has some "whiskers." These modified hairs serve sensory functions, just as they do in dogs or cats. More typically, mammals have a furry coat of *underhair* (a dense, soft, insulative layer that can trap heat) and coarser, longer *guard hairs* that protect the insulative layer from wear and tear (Figure 27.18*b*). When wet, the guard hairs of platypuses, otters, and other aquatic mammals become matted down like a protective blanket, and the underhair stays dry.

Mammals also are distinctive in that they care for the young for an extended period, and they also serve as models for their behavior. Young mammals are born with a capacity to learn and to repeat behaviors that have survival value. It is among mammals that we find the most stunning shows of **behavioral flexibility**—a capacity to expand on basic activities with novel forms of behavior—although the trait is far more pronounced in some species than in others. Behavioral flexibility coevolved with expansion of the brain, especially the cerebral cortex. Remember, this outermost layer of the forebrain receives, processes, and stores information from sensory structures, and it issues commands for complex responses. We find the most highly developed cerebral cortex among the mammals called primates. Chapter 28 starts with their story.

Unlike reptiles, which generally swallow their prey whole, most mammals secure, cut, and sometimes chew food before swallowing. They also differ from reptiles in **dentition** (that is, in the type, number, and size of teeth). Mammals have four distinctive types of upper and lower teeth that match up and work together to crush, grind, or cut food (Figure 27.18*c*). Their incisors (flat chisels or cones) nip or cut food. Horses and other grazing mammals have large incisors. Canines, with piercing points, are longest in meat-eating mammals. Premolars and molars (cheek teeth) are a platform with surface bumps (cusps); these crush, grind, and shear food. If a mammal has large, flat-surfaced cheek teeth, you can safely bet its ancestors evolved in places where fibrous plants were abundant foods. As you will see from the chapter that follows, fossilized jaws and teeth from early primates and species on the evolutionary road to modern humans offer many clues to their life-styles.

a

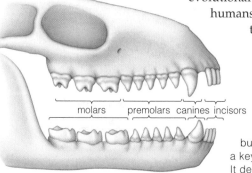

c

Figure 27.18 Three distinctly mammalian traits. (**a**) A human baby busily demonstrating a key defining feature: It derives nourishment from mammary glands. (**b**) A pair of juvenile raccoons displaying their fur coat. (**c**) Unlike the teeth of their reptilian ancestors, the upper and lower rows of mammalian teeth match up. (Compare Figure 27.15*a*.)

molars   premolars canines incisors

b

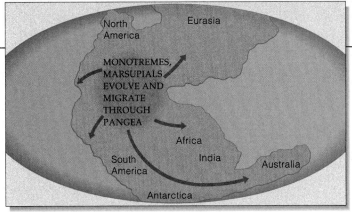

a About 150 million years ago (during the Jurassic)

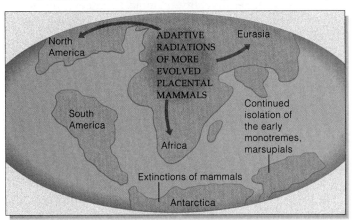

b Between 100 and 85 million years ago (Cretaceous times)

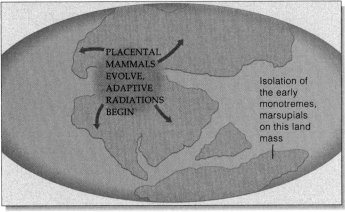

c About 20 million years ago (Miocene)

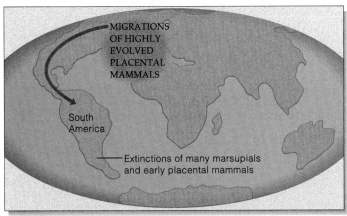

d About 5 million years ago (Pliocene)

Figure 27.19 Adaptive radiations of mammals.

Mammalian Origins and Radiations

More than 200 million years ago, during the Triassic, a genetic divergence from small, hairless reptiles called synapsids gave rise to the **therapsids**, the ancestors of mammals (Figure 27.14). And by Jurassic times, diverse plant-eating and meat-eating mammals called **therians** had evolved. Most were the size of a mouse, and they were endowed with hair and major changes in the jaws, teeth, and body form. For instance, their four limbs were positioned upright, under the body's trunk. This skeletal arrangement made it easier to walk erect, but a trunk higher from the ground was not as stable. At this time the cerebellum, a brain region dealing with the body's balance and spatial positioning, started expanding.

The therians coexisted with dinosaurs through the Cretaceous. When the last of the dinosaurs vanished, diverse adaptive zones opened up for three lineages of those previously inconspicuous mammals (Sections 19.5 and 21.8). New opportunities, differences in traits, and key geologic events put those lineages—the **monotremes** (egg-laying mammals), **marsupials** (pouched mammals), and **eutherians** (placental mammals)—on very different paths to the present. Compared with the monotremes and marsupials, which retained many archaic traits, the placental mammals had the competitive edge. They had higher metabolic rates, more precise ways of regulating body temperature, and a new way of nourishing their developing embryos. You will read about these traits in later chapters. For now, it is enough to know that the placental mammals radiated into many new adaptive zones, at the expense of their less competitive relatives.

As Figure 27.19a shows, the ancestors of monotremes and marsupials had entered the southern part of the supercontinent Pangea by the late Jurassic. Following the breakup of Pangea, the ones on the land mass that would become Australia were already isolated from the ancestors of placental mammals, which were evolving elsewhere (Figure 27.19b). In the future South America, monotremes were replaced by marsupials and early placental mammals, which evolved independently of their relatives on other continents. When a land bridge rejoined South and North America during the Pliocene, highly evolved placental mammals radiated southward and rapidly drove many of those previously isolated mammals to extinction. Only opossums and a few other species successfully invaded land in the other direction.

Mammals alone have hair and mammary glands. They have distinctive dentition, a highly developed nervous system, and a notable capacity for behavioral flexibility.

Much of mammalian history was a matter of luck, of species with particular traits being in the right or wrong places on a changing geologic stage at particular times.

Cases of Convergent Evolution

Evolutionarily distant, geographically isolated lineages have sometimes evolved in similar ways in similar habitats, so that they come to resemble each other in form and function. We call this **convergent evolution**, as defined in Section 20.4. With their interesting history, the three lineages of mammals offer classic examples.

The only living monotremes are two species of spiny anteaters and the duck-billed platypus described at the start of the chapter. One spiny anteater (*Tachyglossus*) is common in Australia; the other lives in the mountains of New Guinea. Both are small, burrowing mammals that feed mostly on ants. Like porcupines, they bristle with protective spines that are modified hairs (Figure 27.20a). The females lay eggs, like those of the platypus. Unlike their relative, they do not dig nests. A single egg is incubated, and the youngster suckles and completes development in a skin pouch that forms temporarily on the mother's ventral surface by muscle contractions.

Of the 260 existing species of marsupials, most are native to Australia and nearby islands; a few live in the Americas (Figure 27.20b,c). The tiny, blind, and hairless newborns suckle and complete development inside a permanent pouch on the female's ventral surface.

Every other existing descendant of the therians is a placental mammal. The **placenta** is a spongy tissue of maternal and fetal membranes. It forms in a pregnant

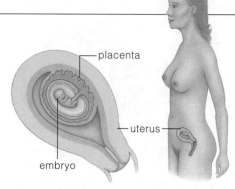

Figure 27.21 Location of the placenta in a human female.

female's uterus, the chamber in which embryos develop in relative freedom from harsh conditions in the outside world (Figure 27.21). Across the placenta, the female sends nutrients and oxygen to an embryo and removes its metabolic wastes. Placental mammals grow faster in the uterus than marsupials do in a pouch, and many are fully formed at birth. Figure 27.22 shows a few species; Appendix I lists the major groups. The body form and function, behavior, and ecology of the mammals will occupy us in later chapters. For now, simply reflect on Table 27.1, which lists some convergences among them.

Human Impact on Mammalian Diversity

In terms of species diversity, mammals don't even begin to approach, say, the mollusks or arthropods. Yet in terms of size, body form, functioning, and life-styles, the 4,500 known species are breathtaking. (In size alone they range from the 1.5-gram Kitti's hognosed bat to

Figure 27.20 (a) Spiny anteater (*Tachyglossus*), a monotreme. Marsupials: (b) From the southeastern United States, a female opossum and her young. (c) From Australia, a young kangaroo (joey) in its mother's pouch.

Table 27.1	Convergences Among Mammalian Groups	
Life Style	Home	Mammalian Family
Aquatic invertebrate eater	North America Central America Australia	Water shrew (Soricidae) Water mouse (Cricedidae) Platypus (Ornithorrhynchidae)
Land-dwelling carnivore	North America Australia	Wolf (Canidae) Tasmanian wolf (Thylacinedae)
Land-dwelling ant eater	South America Africa Australia	Giant anteater (Myrmecophagidae) Aardvark (Orycteropodidae) Spiny anteater (Tachyglossidae)
Ground-dwelling leaf, tuber eater	North America South America Eurasia	Pocket gopher (Geomyidae) Tuco-tuco (Ctenomyidae) Mole rat (Spalacidae)
Tree-dwelling leaf eater	South America Africa Madagascar Australia	Howler monkey (Cebidae) Colobus monkey (Cercopithecidae) Woolly lemur (Indriidae) Koala (Phascolarctidae)
Tree-dwelling nut, seed eater	Southeast Asia Africa Australia	Flying squirrel (Sciuridae) Flying squirrel (Anomaluridae) Flying squirrel (Phalangeridae)

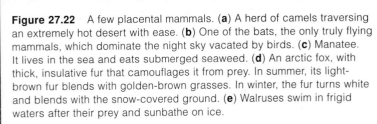

Figure 27.22 A few placental mammals. (**a**) A herd of camels traversing an extremely hot desert with ease. (**b**) One of the bats, the only truly flying mammals, which dominate the night sky vacated by birds. (**c**) Manatee. It lives in the sea and eats submerged seaweed. (**d**) An arctic fox, with thick, insulative fur that camouflages it from prey. In summer, its light-brown fur blends with golden-brown grasses. In winter, the fur turns white and blends with the snow-covered ground. (**e**) Walruses swim in frigid waters after their prey and sunbathe on ice.

whales, some of which weigh in at more than 100 tons.) They survived the dinosaurs and then one another, only to confront a modern challenge—a human population bent on hunting them or encroaching on their habitats, which commonly entails introducing novel species or favoring a few domesticated ones at the expense of wild stock. Think of it. In the 1800s, bison were nearly driven to extinction as part of a program to starve the Plains Indians into submission by killing off their major food source. Modern whaling industries based in Japan and elsewhere are exterminating the very mammals they hunt. About 300 mammalian species, including most whales, wild cats, otters, and primates other than humans, are currently on the endangered species list.

Convergent evolution has occurred among families of all three lineages of mammals that occupy similar habitats in different parts of the world. More recently, humans have introduced new challenges into the evolutionary equation.

SUMMARY

1. Nearly all chordate embryos have a notochord, a dorsal hollow nerve cord, a pharynx with gill slits (or hints of these), and a tail that extends past the anus. Some or all of these traits persist in the adult forms. Chordates with a backbone are known as vertebrates. Tunicates (including the sea squirts) and lancelets are invertebrate chordates.

2. The eight classes of vertebrates are jawless fishes, jawed armored fishes (now extinct), cartilaginous fishes, bony fishes, amphibians, reptiles, birds, and mammals.

3. The first vertebrates, jawless fishes that arose during the Cambrian era, included the ostracoderms. Their living descendants are lampreys and hagfishes. Jawed fishes also arose in the Cambrian. A dominant lineage, the placoderms, is extinct. Other lineages gave rise to the cartilaginous and bony fishes.

4. Four trends occurred during the evolution of certain vertebrate lineages:

   a. A vertebral column supplanted the notochord as a structural element against which muscles act. This led to swift predatory animals.

   b. Jaws evolved from gill-supporting elements. This led to increased predator-prey competition; it favored far more efficient nervous systems and sensory organs.

   c. In one lineage of bony fishes, paired fins evolved into fleshy lobes with structural elements (forerunners of paired limbs).

   d. In some fish lineages, respiration by gills came to be supplemented by lungs, which proved adaptive in the invasion of the land. In a related development, the circulatory system became more efficient at distributing oxygen through the body.

5. Amphibians were the first vertebrates to invade the land, but they never fully escaped the water. Their skin dries out, and aquatic stages persist in most life cycles.

6. Unlike amphibians, reptiles escaped dependence on standing free water. The key adaptations that allowed them to do so are their toughened, scaly skin, which restricts water loss from the body; a reliance on internal fertilization; more efficient, water-conserving kidneys; and amniote eggs, often leathery or shelled, that protect and metabolically support the embryos.

7. Reptiles, and the birds and mammals that descended from certain reptilian lineages, have efficient circulatory and respiratory systems and a highly developed nervous system and sensory organs.

8. Of all vertebrates, birds alone have feathers, which they use in flight, heat conservation, and social displays.

9. Of all animals, mammals alone have milk-producing mammary glands, and they have hair or thick skin that functions in insulation. They have distinctive dentition and a highly developed cerebral cortex, the brain's most complex region. Adults nurture their young through an extended period of dependency and learning.

Review Questions

1. List and describe features that distinguish chordates from other animals. 27.1

2. Name and describe the features of an organism from both groups of invertebrate chordates discussed in this chapter. 27.2

3. List four major trends that occurred during the evolution of at least some vertebrate lineages. 27.3

4. Which evolutionary modifications in fishes set the stage for the emergence of amphibians? 27.3, 27.6

5. List some of the characteristics that distinguish reptiles from amphibians. 27.7, 27.8

6. List some of the characteristics that distinguish birds from reptiles. 27.9

7. List some of the characteristics that distinguish each of the three mammalian lineages from the reptiles. 27.10, 27.11

Self-Quiz (Answers in Appendix IV)

1. Only _____ have a notochord, a tubular dorsal nerve cord, a pharynx with slits in the wall, and a tail extending past the anus.
   - a. echinoderms
   - b. tunicates and lampreys
   - c. vertebrates
   - d. both b and c
   - e. all are correct

2. Gill slits function in _____ .
   - a. respiration
   - b. circulation
   - c. food trapping
   - d. water regulation
   - e. both a and c

3. A shift from a reliance on _____ to reliance on _____ was pivotal in the evolution of all vertebrates.
   - a. the notochord; a backbone
   - b. filter feeding; jaws
   - c. gills; lungs
   - d. all are correct

4. The first vertebrates were _____ .
   - a. bony fishes
   - b. jawless fishes
   - c. jawed fishes
   - d. both a and b

5. Of all existing vertebrates, _____ are the most diverse.
   - a. cartilaginous fishes
   - b. bony fishes
   - c. amphibians
   - d. reptiles
   - e. birds
   - f. mammals

6. The bony fishes include _____ .
   - a. ray-finned fishes
   - b. lobe-finned fishes
   - c. lungfishes
   - d. all are correct

7. The only amphibians to entirely escape dependence on aquatic habitats are _____ .
   - a. salamanders
   - b. desert toads
   - c. caecilians
   - d. none is correct

8. Reptiles moved fully onto land owing to _____ .
   - a. tough skin
   - b. internal fertilization
   - c. good kidneys
   - d. amniote eggs
   - e. both b and d
   - f. all are correct

9. A four-chambered heart is characteristic of _____ .
   - a. bony fishes
   - b. amphibians
   - c. birds
   - d. mammals
   - e. both b and c
   - f. both c and d

10. _____ have highly efficient circulatory and respiratory systems, and a complex nervous system and sensory organs.
   a. Reptiles
   b. Birds
   c. Mammals
   d. all are correct

11. Birds use feathers in _____ .
   a. flight
   b. heat conservation
   c. social functions
   d. all are correct

12. Various mammals _____ .
   a. hatch
   b. complete their embryonic development in pouches
   c. complete their embryonic development in the uterus
   d. both b and c
   e. all are correct

13. Match the organisms with the appropriate features.
   _____ jawless fishes
   _____ cartilaginous fishes
   _____ bony fishes
   _____ amphibians
   _____ reptiles
   _____ birds
   _____ mammals

   a. complex cerebral cortex, thick skin or hair
   b. respiration by skin and lungs
   c. include coelacanths
   d. include hagfishes
   e. include sharks and rays
   f. complex social behavior, feathers
   g. first with amniote eggs

Critical Thinking

1. Describe the factors that might contribute to the collapse of native fish populations in a lake following the introduction of a novel predator, such as the lamprey.

2. Think about the flight muscles of birds and their demands for oxygen and ATP energy. What type of organelle would you expect to be profuse in these muscles? Explain your reasoning.

3. Kathie and Gary, two amateur fossil hunters, have unearthed the complete fossilized remains of a mammal. How can they determine whether their find is a herbivore, carnivore, or omnivore?

4. In Australia, many species of marsupials are competing with recently introduced placental mammals (such as rabbits) for resources—but they are not winning. Explain how it is that placental mammals that did not even evolve in the Australian habitats show greater fitness than the native mammals.

5. About seventy-five species of coral snakes live in the rain forests, grasslands, mountains, and desertlike regions of North, Central, and South America. Coral snakes of family Colubridae are harmless; those of family Elapidae are poisonous. In both families, the most poisonous types bite mostly to kill prey, yet moderately poisonous types can be extremely vicious. All coral snakes, including the deadliest species (*Micrurus*) in Figure 27.23, have distinctive color banding. When they are gliding along, even snake experts (herpetologists) may have trouble knowing which is which. Yet their relatives in other parts of the world have no such banding; neither do bigger snakes.

Thus, color banding in the two families can't be explained in terms of environmental differences, phylogeny, or lethality. What brought it about? According to a theory proposed by herpetologist R. Mertens, the ancestors of *Micrurus* migrated across a land bridge that connected Asia with North America more than 60 million years ago. Because predators learned to avoid them by associating the color banding with "taste trials," the ancestral snakes must *not* have been as deadly as some of their descendants. (If the snakes were *too* deadly, all trials might have ended in death.) Therefore, the protective function of the color banding must depend on unpleasant trials with only *moderately* poisonous snakes.

Among the harmless snakes (Colubridae), identical warning coloration also evolved. This was a case of *mimicry*, in which one

Figure 27.23 One of the extremely poisonous coral snakes (*Micrurus*).

group (mimics) came to bear deceptive resemblance to another group (models) that enjoy a survival advantage. For this to work, of course, mimics can't outnumber models. That said, explain the following numbers of trapped snakes that were brought to the Butantan Institut in São Paulo for the years shown (trappers did not distinguish among poisonous and nonpoisonous types):

	1950	1951	1952	1953
Micrurus:	59	41	58	56
Others:	218	246	265	284

Selected Key Terms

amniote egg 27.7	fin 27.3	notochord 27.1
amphibian 27.6	gill 27.3	ostracoderm 27.3
behavioral flexibility 27.10	gill slit 27.2	pharynx 27.1
bird 27.9	hagfish 27.4	placenta 27.11
bony fish 27.5	jaw 27.3	placoderm 27.3
cartilaginous fish 27.5	lamprey 27.4	reptile 27.7
chordate 27.1	lancelet 27.2	scale (fish) 27.5
convergent evolution 27.11	lobe-finned fish 27.5	swim bladder 27.5
dentition 27.10	lung 27.3	therapsid 27.10
eutherian 27.10	mammal 27.10	therian 27.10
feather 27.9	marsupial 27.10	tunicate 27.2
filter feeder 27.2	monotreme 27.10	vertebra 27.3
	nerve cord 27.1	vertebrate 27.1

Readings

Gould, S. J. (general editor). 1993. *The Book of Life*. New York: Norton. Splendid, easy-to-read essays, gorgeous illustrations.

Romer, A. S., and T. S. Parsons. 1986. *The Vertebrate Body*. Sixth edition. Philadelphia: Saunders.

Strickberger, M. 1996. *Evolution*. Second edition. Boston: Jones and Bartlett.

Wickler, W. 1968. *Mimicry in Plants and Animals*. New York: McGraw-Hill. Paperback.

Web Site See *http://www.wadsworth.com/biology* for practice quiz questions, hypercontents, BioUpdates, and critical thinking. The Wadsworth Biology Resource Center provides a wealth of information fully organized and integrated by chapter.

28

HUMAN EVOLUTION: A CASE STUDY

The Cave at Lascaux and the Hands of Gargas

Half a century ago, on a warm autumn day, four boys out for a romp stumbled onto a cave near Lascaux, a town in the Perigord region of France. What they discovered in that intricately tunneled cave stunned the world. Magnificent sketches, engravings, and paintings swept across the cave walls (Figure 28.1). The red, yellow, purple, and brown pigments were as vivid as if they had just been painted. Radiometric dating revealed that they were 17,000 to 20,000 years old. The prehistoric artists worked deep within the cave, where sunlight did not fade the images and neither winds nor water could wear them away. By the light of crude oil lamps, they captured the graceful, dynamic lines of bison, stags, stallions, ibexes, lions, a rhinoceros, and a heifer, since named the Great Black Cow.

Even earlier, art was committed to the walls of caves throughout southern France, northern Spain, and Africa, as when people made more than 150 imprints and outlines of hands in the cave of Gargas, in the Pyrenees, some 25,000 years ago.

Who were these artists? From their fossilized remains, we know they were people anatomically like us. From the way they planned and executed their art, we sense a level of abstract thinking that is unique to humans.

The quality of "humanness" did not emerge out of thin air. If we could go back 5 million years in time, to the place where the most recent ancestors of humans apparently originated, we might find ourselves in Africa, hiding in grasslands and dry woodlands of the sort shown in Figure 28.2. From there, our journey would take us back an additional 55 million years, to ancient tropical forests in which primates originated. We could not stop there. The mammalian ancestors of the primates evolved 250 million years ago, vertebrate ancestors of mammals evolved long before that, ancestors of vertebrates and all other animal groups evolved 900 million years ago—and so on back to the origin of the first living cells.

As we explore our own branch of the animal family tree, keep this greater evolutionary story in mind. *Our "uniquely" human traits emerged through modification of traits that evolved earlier, in ancestral forms.* From that perspective, the "ancient" cave paintings are the legacy of people who departed only yesterday, so to speak. The artists of Lascaux and Gargas are not remote from us. They *are* us.

Figure 28.1 Part of the human cultural heritage—prehistoric paintings from a cave in Lascaux, France, a unique outcome of a long history of biological evolution.

Figure 28.2 (*Below*) East Africa's Rift Valley, 3,200 kilometers long. Somewhere in this immense valley, sparsely wooded grasslands were the birthplace of the human species.

KEY CONCEPTS

1. In the primate branch of the mammalian lineage are the prosimians, tarsioids, and anthropoids (monkeys, apes, and humans). Apes and humans are hominoids. Only humans and some extinct species with a mosaic of apelike and humanlike traits are further classified as hominids.

2. When some early primates were adapting to life in the trees, their skull and eye sockets underwent modifications that led to increased reliance on daytime vision, compared to their night-foraging ancestors.

3. Hominoids evolved in Miocene times. In some of the apelike forms, certain skeletal modifications permitted an upright stance and bipedalism, or walking on two legs. Bipedalism freed the forelimbs for novel manipulative functions. Hands evolved, and among certain hominid descendants of Miocene apes, modifications in the bones and associated muscles led to a capacity to hold, carry, use, and make objects.

4. Starting in the Miocene, a long-term cooling trend led to seasonal changes in habitats and food sources. Forests gave way to grasslands with scattered stands of trees, and some hominids became scavengers of meat as well as foragers for fruits and nuts. The sizes and shapes of their jaws and teeth became adapted to processing new foods.

5. As food sources became scarcer, hominids underwent adaptive radiations through Africa and, later, into Europe and Asia. In some lineages, the brain increased in volume and became more complex. Eventually, the evolution of the brain became interlocked with cultural evolution.

6. Unlike earlier hominids, *Homo erectus* and *H. sapiens* displayed remarkable behavioral flexibility and creative experimentation with their environment, as when they started using fire. This characteristic allowed them to survive the challenges of dispersing into novel and often harsh environments around the world.

EVOLUTIONARY TRENDS AMONG THE PRIMATES

Having traversed the mammalian branch of the animal family tree in Chapter 27, we are now ready to follow it along the branchings leading to primates, then humans.

Primate Classification

The order **Primates** includes prosimians, tarsioids, and anthropoids (Table 28.1 and Appendix I). Figure 28.3 shows a few of these mammals. The prosimians are the oldest primate lineage (*pro*, before; *simian*, ape). They were **arboreal** (tree-dwelling), and they dominated the trees in North America, Europe, and Asia for millions of years before monkeys and apes evolved and almost displaced them. Tarsiers of Southeast Asia are the only living tarsioids. The traits of these small primates are somewhere between prosimians and anthropoids. All monkeys, apes, and humans are **anthropoids**. In their biochemistry and structure, apes are closer to humans than to monkeys. Apes, humans, and extinct species of the human lineage are grouped together, as **hominoids**.

By 5 million years ago, a divergence from the last shared ancestor of apes and humans was under way. All species that evolved on that separate road leading to humans are further classified as **hominids**.

Key Evolutionary Trends

Most primates live in tropical or subtropical forests, woodlands, or **savannas**, which are open grasslands with a few stands of trees, as in Figure 28.2. Like their ancestors, the vast majority are tree dwellers. Yet no one feature sets the primates apart from other mammals. Each lineage evolved in a distinct way and has its own defining traits. Five trends define the lineage leading to humans, which is our focus. They were set in motion when primates started adapting to life in the trees, and they contributed to the emergence of modern humans. *First*, there was less reliance on the sense of smell and more on daytime vision. *Second*, skeletal changes led to upright walking, which freed the hands for novel tasks. *Third*, changes in bones and muscles led to refined hand movements. *Fourth,* teeth became less specialized. *Fifth*, elaboration of the brain and changes in the skull led to truly complex behaviors. These developments became interlocked with one another and with cultural evolution.

ENHANCED DAYTIME VISION Early primates had an eye on each side of their head. Later ones had forward-directed eyes, an arrangement that is much better for sampling shapes and movements in three dimensions. Other modifications allowed their eyes to respond to

Figure 28.3 Representative primates. (**a**) A gibbon, with body and limbs adapted for swinging arm over arm in the trees. (**b**) A spider monkey, which is a four-legged climber, leaper, and runner. (**c**) Tarsiers, which are vertical climbers and leapers.

Table 28.1	Primate Classification
Taxon	Representatives
PROSIMIANS	
Lemuroids	Lemurs, lorises
TARSIOIDS	
Tarsioids	Tarsiers
ANTHROPOIDS	
Ceboids (New World monkeys)	Spider monkeys
Cercopithecoids (Old World monkeys)	Baboons
Hominoids:	
Hylobatids	Gibbon, siamang
Pongids	Orangutan, gorilla, chimpanzee
Hominids	Humans, extinct humanlike forms

Figure 28.4 Comparison of the skeletal organization and stance of a monkey, ape, and human. Monkeys climb and leap. Gorillas and chimps use forelimbs to climb and help support their weight, but most of the time they walk on all fours on the ground. Humans are two-legged walkers. The different modes of locomotion arose through modifications of the basic skeletal plan of mammals.

variations in color and in light intensity (dim to bright). Being able to interpret and respond quickly to these stimuli is advantageous when living in the trees.

UPRIGHT WALKING A primate's mode of locomotion depends on arm length and the shape and positioning of its shoulder blades, pelvic girdle, and backbone. With arm and leg bones about the same length, a monkey can climb, leap, and run on four legs, but not two (Figure 28.4). A gorilla has long arms and walks on all fours, including its knuckles. Only humans show **bipedalism**: they walk freely on two legs. Compared with monkeys and apes, the human backbone is shorter, S-shaped, and flexible. Bipedalism was a key innovation that evolved in the ancestors of hominids.

POWER GRIP AND PRECISION GRIP How did we get such splendid hands? The first mammals spread their toes apart to help support the body as they walked or ran on four legs. Primates still spread fingers or toes. Many can cup their fingers, as when monkeys lift food to the mouth. Among ancient tree-dwelling primates, handbone modifications allowed them to wrap their fingers around objects (*prehensile* movements) and to touch their thumb to the tip of each finger (*opposable* movements). In time, their hands became freed from load-bearing functions. Later, when hominids evolved, refinements led to the power grip and precision grip:

These hand positions gave early humans a capacity to make and use tools. They were a foundation for unique technologies and cultural development.

TEETH FOR ALL OCCASIONS Before hominids evolved, modifications in primate jaws and teeth accompanied a shift from eating insects to fruits and leaves, and on to a mixed diet. Later, rectangular jaws and long canines

came to be additional defining features of monkeys and apes. Along the road leading to humans, a bow-shaped jaw and smaller teeth of about the same length evolved.

BETTER BRAINS, BODACIOUS BEHAVIOR Among some early primates, an arboreal life-style required shifts in reproductive and social behavior. In many lineages, parents put more effort in fewer offspring. They formed stronger bonds with offspring, maternal care grew more intense, and the learning period increased (Figure 28.5). Expansions of the brain became interlocked with selection for increasingly complex behavior. New behavior promoted new neural connections in the brain regions that dealt with sensory inputs, a brain with more intricate wiring favored more new behavior, and so on. We find evidence of such interlocking in the parallel evolution of the human brain and culture. **Culture** is the sum total of behavior patterns of any social group, handed down from one generation to the next by learning and by symbolic behavior, especially language. A capacity for language arose among ancestral humans, through changes in the skull bones and brain.

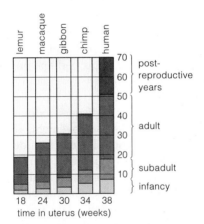

Figure 28.5 Trend toward longer life spans and longer dependency among primates.

Certain key adaptations evolved along the pathway from arboreal primates to modern humans: complex, forward-directed vision; bipedalism; refined hand movements; generalized dentition; and an interlocked elaboration of brain regions and cultural behavior.

Origins and Early Divergences

Primates evolved from mammals more than 60 million years ago, in tropical forests of the Paleocene. The first ones resembled small rodents or tree shrews (Figures 28.6 and 28.7). Like tree shrews, they probably had huge appetites and foraged at night for insects, seeds, buds, and eggs beneath the trees. They had a long snout and a good sense of smell, suitable for snuffling food and predators. They clawed their way up stems, although not with much speed or grace.

During the Eocene (between 54 and 38 million years ago), some primates were staying up in the trees. They had a shorter snout, enhanced daytime vision, a larger brain, and refined grasping movements. How did these traits evolve?

Consider the trees. An arboreal life-style had advantages: abundant food and safety from ground-dwelling predators. However, they also were habitats of uncompromising selection. Visualize an Eocene morning with dappled sunlight, boughs swaying in the breezes, colorful fruit hidden among the leaves, and perhaps predatory birds. A long, odor-sensitive snout would not have been of much use up in the trees, where air currents disperse odors. But a brain that could assess movement, depth, shape, and color would have been a definite plus. And so would a brain that could work fast when its owner was running and leaping (especially!) from branch to branch. The body's weight, the wind speed, and the distance and suitability of a destination had to be estimated all at the same time, and any adjustments for miscalculations had to be quick.

Figure 28.6 A nocturnal tree shrew of Indonesia.

By at least 36 million years ago, before the dawn of the Oligocene, tree-dwelling anthropoids had evolved in the forests. One form, the squirrel-sized *Catopithecus*, was on or very close to the evolutionary road that led to monkeys and apes. It had forward-directed eyes. Given its snoutless, flattened face and upper jaw with shovel-shaped front teeth, it must have used its hands to grab fruit and insects. Some of the early anthropoids lived in swamplands infested with formidable, predatory reptiles and rarely ventured to the ground. And maybe that is why it became imperative to think fast and grip strongly. Slip-ups were always possible; many primates still fall out of trees.

Between 25 and 5 million years ago (Miocene), apelike forms—*the first hominoids*—evolved, and they underwent an adaptive radiation into Africa, Europe, and southern Asia. Most forms became extinct. However, fossils as well as genetic analyses suggest that between 10 million and 5 million years ago, divergences from lineages of some Miocene apes led to gorillas and chimps, and to the first *hominids*.

The First Hominids

Most of the early hominids we know about lived in the East African Rift Valley. By 5 million years ago, at the dawn of the Pliocene, a long-term cooling trend was well under way as a result of drifting continents and changes in ocean circulation patterns. Figure 28.8 is one model of the geologic triggers for the trend. The once-

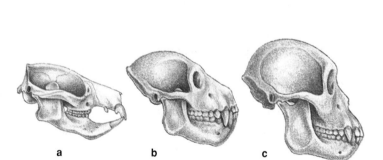

Figure 28.7 Comparison of skull shape and teeth of some early primates. (**a**) *Plesiadapis* (Paleocene) had rodentlike teeth. (**b**) *Aegyptopithecus* (Oligocene anthropoid) probably predates the divergence leading to Old World monkeys and apes. (**c**) Apelike dryopiths lived in the Miocene. *Plesiadapis* was as tiny as a tree shrew. *Aegyptopithecus* was monkey-sized, and some dryopiths, chimpanzee-sized.

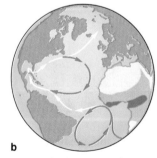

Figure 28.8 Long-term shift in an ancient climate. (**a**) Before the Isthmus of Panama formed, salinity levels of seawater were much the same around the world, and circulation patterns kept Arctic waters warmer. (**b**) After the Isthmus formed, North Atlantic waters became saltier and heavier; they sank before reaching the Arctic region. The Arctic ice cap formed and triggered a long-term trend toward a cooler, drier climate in Africa.

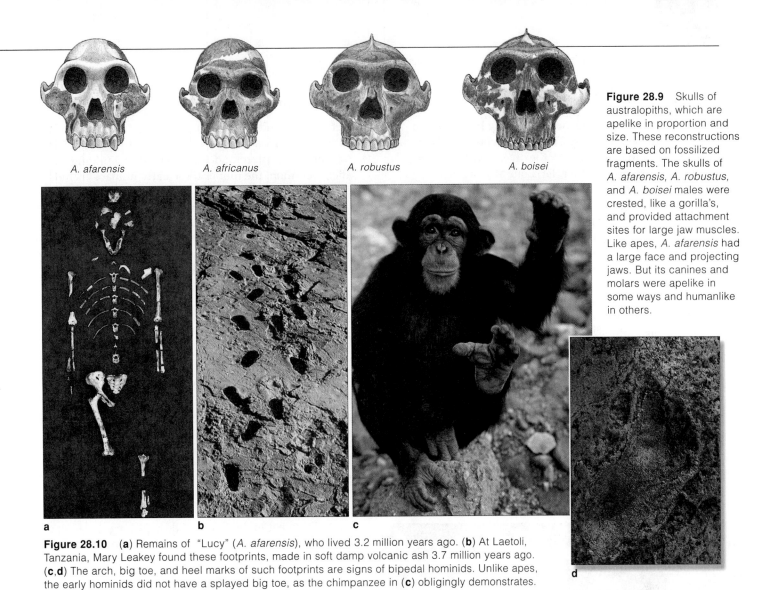

A. afarensis   A. africanus   A. robustus   A. boisei

Figure 28.9 Skulls of australopiths, which are apelike in proportion and size. These reconstructions are based on fossilized fragments. The skulls of *A. afarensis, A. robustus,* and *A. boisei* males were crested, like a gorilla's, and provided attachment sites for large jaw muscles. Like apes, *A. afarensis* had a large face and projecting jaws. But its canines and molars were apelike in some ways and humanlike in others.

a   b   c   d

Figure 28.10 (**a**) Remains of "Lucy" (*A. afarensis*), who lived 3.2 million years ago. (**b**) At Laetoli, Tanzania, Mary Leakey found these footprints, made in soft damp volcanic ash 3.7 million years ago. (**c,d**) The arch, big toe, and heel marks of such footprints are signs of bipedal hominids. Unlike apes, the early hominids did not have a splayed big toe, as the chimpanzee in (**c**) obligingly demonstrates.

vast tropical forests—with their bounty of edible soft fruits, leaves, and insects—gave way to dry woodlands and grasslands as the climate became more seasonal. The African savanna had emerged. The new foods were harder, and harder to find, and early hominids had two options: Move into the new adaptive zones or die out.

Fossil fragments 4 to 4.5 million years old indicate this was a "bushy" period of evolution; a confounding variety of hominids appeared (Figures 28.9 and 28.10). At present it is impossible to figure out their family tree, so they are simply grouped as **australopiths** (meaning southern apes). *Australopithecus anamensis* is the oldest of these. Like *A. afarensis* and *A. africanus,* it was gracile (slightly built). The ones called *A. boisei* and *A. robustus* were robust (muscular and heavily built).

With their large face, protruding jaws, and small skull (and brain) size, australopiths were apelike. Yet they differed from previous hominids. For example, their molars had thicker enamel and could grind harder foods. And they were good at walking upright. We know this from studies of fossilized hip and limb bones.

More telling, some australopiths left footprints. About 3.7 million years ago, *A. afarensis* individuals walked on newly fallen volcanic ash, which a light rain turned into quick-drying cement (Figure 28.10*b*). Bipedalism started to evolve earlier, when certain hominoids were forced to the ground late in the Miocene. They had mastered the power grip and precision grip in the trees. Rather than becoming specialized in running fast on all fours, they used their manipulative skills to advantage. They kept their hands free to carry offspring, and probably to carry precious food on their foraging expeditions.

Primates evolved from small, rodentlike mammals that moved into arboreal habitats 60 million years ago. During the Miocene, the first hominoids (apelike forms) evolved and radiated through Africa, Europe, and southern Asia.

Between 10 million and 5 million years ago, divergences from Miocene apes led to australopiths, the first hominids. They were apelike in many skeletal details, but they were humanlike in a crucial respect: They walked upright.

EMERGENCE OF EARLY HUMANS

Defining "Human"

What can fossilized fragments of the early hominids tell us about our own origins? The fossil record is still too sketchy for us to know how all the diverse australopiths were related to one another, let alone which ones may have been ancestral to humans. Besides, what *are* the traits that distinguish **humans** (members of the genus *Homo*) from other groups? Well, there's always the brain. In modern humans, it is the basis of great analytical and verbal skills, complex social behavior, and technological innovation. It easily sets us apart from apes, which have a skull volume and brain size much smaller than ours (Figure 28.11). Yet this feature alone cannot tell us when certain hominids made the transition to being human, because their brain size probably fell within the range for apes. They were makers of simple tools, but so are chimps and certain parrots. And it goes without saying that their behavior did not lend itself to fossilization.

And so we are left to speculate about a continuum of physical traits among a number of fossils—a skeleton adapted for bipedalism, manual dexterity, and larger brain volume; a smaller face; and smaller, more thickly enameled teeth. These traits, which originated late in the Miocene, were evident in what many consider to be the earliest humans, *Homo habilis* (meaning "handy man").

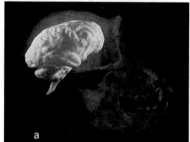
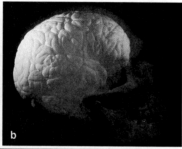

a   b

c

Homo habilis *and the First Stone Tools*

Between 2.5 and 1.6 million years ago, one or two forms of *H. habilis* lived in dry woodlands that punctuated the savannas of eastern and southern Africa. Judging from their dentition, the diet of these early humans included hard-shelled nuts and seeds as well as soft fruits, leaves, and insects. Supplies of these foods changed with the seasons. Most likely, *H. habilis* had to think ahead, to plan when to venture about to gather and possibly store foods that would help it survive the cold dry seasons.

H. habilis shared its habitat with saber-tooth cats and other formidable predators. The teeth of those cats could impale prey and shear off flesh, but they could not crush open the bones for marrow. Carcasses of their kills, with shreds of meat clinging to the marrow bones, were concentrated stores of nutrients in places that were nutrient-stingy. Although *H. habilis* was a forager, not a full-time carnivore, it opportunistically supplemented its diet by scavenging the carcasses (Figure 28.11c).

Fossil hunters have found numerous stone tools that date to the time of *H. habilis*. But they cannot say with certainty that *H. habilis* was the only species that made them. Possibly australopiths as well as *H. habilis* used sticks and other perishable tools before then, as modern apes do, but we have no way of knowing.

Maybe individuals on the road to modern humans started down a toolmaking road by picking up rocks to crack marrow bones. Maybe they started to scrape flesh from bones with small, sharp flakes that had fractured naturally from rocks. Eventually, early humans started *shaping* stone implements. Paleoanthropologist Mary Leakey was the first to discover evidence of toolmaking at Africa's Olduvai Gorge, which cuts through a great sequence of sedimentary rock layers. The most ancient tools at this site are crudely chipped pebbles that were buried in the deepest layers (Figure 28.12). They may have been used to smash marrow bones, dig for roots, and poke insects from tree bark. More recent layers have more complex tools. Large numbers of bones and tools have been found along the shores of ancient lakes that would have beckoned plenty of thirsty animals.

At such sites we find fossils of one form of *H. habilis* that was twice as brainy as the australopiths and that obviously ate well. There apparently was no selection pressure for more creativity in securing food resources; the stone tools of *H. habilis* did not change much for the next 500,000 years.

Figure 28.11 (**a**) Image of the brain of a modern chimpanzee, superimposed on a chimp skull. The brain of Lucy, one of the australopiths (*A. afarensis*), was only a bit larger than this and much smaller than (**b**) the modern human brain. (**c**) Artist's rendition of *Homo habilis* males in an East African woodland.

Figure 28.12 A few of the 37,000+ stone tools from Olduvai Gorge. (a) A crude chopper and (b) a more advanced form, with a joint and sharp edge. (c) Hand ax and (d) cleaver.

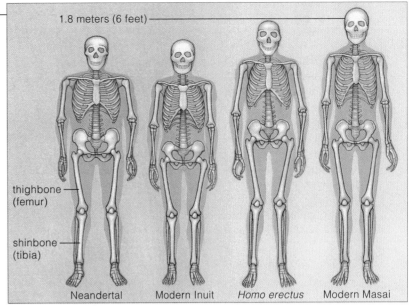

Figure 28.13 Climate and body build. Humans adapted to cold climates have a heat-conserving body (stockier, with shorter legs, compared with humans adapted to hot climates).

From Homo erectus *to* Homo sapiens

The ancestors of modern humans apparently stayed put in Africa until about 2 million years ago. At that time, genetic divergence from early members of *Homo* led to *Homo erectus*, a human species that the fossil record places on the evolutionary road to modern humans. Its name means "upright man." Although its forerunners also were upright, two-legged walkers, *H. erectus* did the name justice; its populations walked out of Africa, turned left into Europe, and right into Asia. It took some of them a long time to walk the 14,000 kilometers to China. *H. erectus* fossils from Southeast Asia and the former Soviet republic of Georgia are 1.8 million and 1.6 million years old. And the treks were strenuous. More than once, immense glaciers advanced down through northern Europe, southern Asia, and North America.

Whatever selection pressures triggered the adaptive radiations, this was a time of physical changes, as in skull size and leg length (Figure 28.13). It also was the time of cultural lift-off for the human lineage. *H. erectus* had a larger brain, it was a more creative toolmaker, and its social organization and communication skills must have been well developed to make such successful treks. From southern Africa on into England, different populations were using the same variety of hand axes and other tools designed to pound, scrape, shred, cut, and whittle. They withstood environmental challenges by building fires and using furs for clothing. Remains of fire use date from an ice age in the early Pleistocene.

Judging from fossils in Africa, **Homo sapiens** had evolved by 100,000 years ago. The origin and radiations of early *H. sapiens* are hotly debated topics (Section 28.4). Early *H. sapiens* had smaller teeth and jaws than *H. erectus*, and often it had a chin. Its facial bones were smaller, the skull was higher and rounder, and the brain was larger. Analysis of fossils indicates that these early forms may have had the capacity for complex language.

One group of early humans, the Neandertals, lived in Europe and the Near East from 200,000 to 35,000 years ago. Massively built and large brained, some of their populations were the first to adapt to the coldest regions (Figure 28.13). Their disappearance coincided with the appearance of anatomically modern humans in the same regions 40,000 to 30,000 years ago. There is no evidence that they warred or interbred with these later arrivals. We don't know yet what happened to them.

From 40,000 years ago to today, human evolution has been almost entirely cultural, not biological—and so we leave the story with these conclusions: Humans spread rapidly through the world by devising *cultural* means to deal with a broad range of environments. Compared with their predecessors, they developed rich and varied cultures. Even though hunters and gatherers persist in parts of the world, others moved from "stone-age" technology to the age of "high tech," attesting to the great plasticity and depth of human adaptations.

Cultural evolution has outpaced the biological evolution of the only remaining human species, *H. sapiens*. Today, humans everywhere rely on cultural innovation to adapt rapidly to a broad range of environmental challenges.

OUT OF AFRICA—ONCE, TWICE, OR . . .

If researchers are interpreting the fossil record of human evolution correctly, then it would seem that Africa was the cradle for us all. At this writing, at least, no one has found any fossils of humans that are older than 1 million years *except* in Africa. *H. erectus* coexisted for a time with earlier humans (*H. habilis*) before dispersing from the African savannas to the cooler grasslands, forests, and mountains of Europe and Asia. They apparently left Africa in waves between about 2 million and 500,000 years ago. Judging from recent examination of *H. erectus* fossils from Java, some populations may have survived, in relative isolation, until 27,000 to 53,000 years ago.

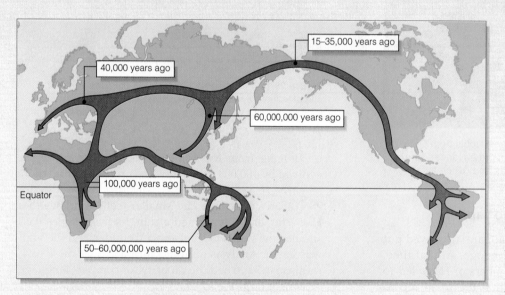

40,000 years ago

15–35,000 years ago

60,000,000 years ago

100,000 years ago

Equator

50–60,000,000 years ago

Figure 28.14 Dates when early *H. sapiens* populations were colonizing different parts of the world, based on fossil evidence. As shown here, the presumed dispersal routes (*brown* arrows) seem to support the African emergence model.

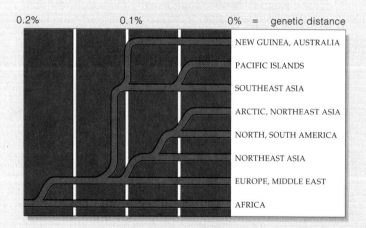

0.2%   0.1%   0% = genetic distance

NEW GUINEA, AUSTRALIA

PACIFIC ISLANDS

SOUTHEAST ASIA

ARCTIC, NORTHEAST ASIA

NORTH, SOUTH AMERICA

NORTHEAST ASIA

EUROPE, MIDDLE EAST

AFRICA

Figure 28.15 One proposed family tree for populations of modern humans (*Homo sapiens*) that are native to different regions of the world. Branch points indicate presumed genetic divergences. The tree is based on nucleic acid hybridization studies of many genes (including those for the ABO blood group) and immunological comparisons.

Where, on the larger geologic stage, do we place the origin of *H. sapiens*? *Here we find a good example of how the same body of evidence can be interpreted in different ways.* These interpretations are called the multiregional model and African emergence model for modern human origins. Both attempt to explain the world distribution of fossils of particular ages and the measured genetic distances among modern, existing human populations. For example, Figure 28.14 includes locations of *H. sapiens* fossils that have been dated to particular times. As another example, evidence from biochemical and immunological studies suggests the greatest genetic distance separates *H. sapiens* populations native to Africa from populations everywhere else; the next greatest distance separates Southeast Asia (and Australia) from everywhere else (Figure 28.15).

MULTIREGIONAL MODEL By this model, *H. erectus* populations had spread through much of the world by 1 million years ago. Those geographically separated groups were subject to different selection pressures, so their traits evolved in regionally distinctive ways. Then subpopulations ("races") of *H. sapiens* evolved from them in different places. Although they differed phenotypically, they did not evolve into separate species because gene flow continued among them, even to the present day. (Thus, for example, while the armies of Alexander the Great were sweeping eastward, they also contributed blue-eye genes from the Greeks to the allele pool of generally brown-eyed subpopulations in Africa, the Near East, and Asia.)

AFRICAN EMERGENCE MODEL This model does not dispute fossil evidence that populations of *H. erectus* evolved in distinctive ways in different regions. But it holds that *H. sapiens*—modern humans—originated in sub-Saharan Africa somewhere between 200,000 and 100,000 years ago. Only later did *H. sapiens* populations move out of Africa, then into other regions along the routes indicated by the Figure 28.14 map. In each region that *H. sapiens* populations settled, they replaced the archaic *H. erectus* populations that had preceded them. Only then did regional phenotypic differences become superimposed on the original *H. sapiens* body plan.

In support of this model, the oldest known *H. sapiens* fossils are indeed from Africa. Also, in Zaire, new finds of finely wrought barbed-bone harpoons and other tools suggest the African populations were as skilled at making tools as *Homo* populations known earlier from Europe.

SUMMARY

1. Like other mammals, humans and other primates have an internal skeleton, a complex brain and sensory organs fully enclosed in a skull, mammary glands (in females), and distinctive teeth. Adults nourish, protect, and serve as behavioral models for the young.

2. Primates include the prosimians (lemurs and related forms) tarsioids, and the anthropoids (which include monkeys, apes, and humans). Only apes and humans are hominoids. Only the anatomically modern humans (*H. sapiens*) and others of their lineage (from *A. afarensis* to *H. erectus*) are further classified as hominids.

3. The first primates were small, rodentlike mammals that evolved by 60 million years ago in tropical forests. Lineages evolving in the trees developed a power grip, a precision grip, and good daytime vision. Some lineages gave rise to anthropoids, including the ancestors of monkeys, apes, and humans, by 35 million years ago.

4. Apelike forms (the first hominoids) evolved between 25 and 13 million years ago, in the Miocene. In Africa, some had given rise to australopiths, the earliest known hominids, before 4 million years ago (Figure 28.16). The evolution of late Miocene apes and their hominid descendants has been correlated with strong selection pressures that emerged during a long-term trend from tropical climates to cooler, drier climates. The African savanna emerged, and the hominids that survived were the ones with physical changes that permitted upright walking (bipedalism), a more varied diet (changes in dentition), and more brain complexity to figure out how to gather scarcer and seasonally available food.

5. *H. habilis*, the earliest known members of the genus *Homo*, had evolved by 2.5 million years ago. It may have been the first stone toolmaker. It opportunistically supplemented its diet of nuts, fruits, and seeds with meat and marrow from carcasses.

6. *Homo erectus*, the presumed ancestor of modern human populations, evolved by 2 million years ago. *H. erectus* populations radiated out of Africa, into Asia and Europe. The oldest known fossils of early modern humans (*H. sapiens*) are from Africa; they are 100,000 years old. About 40,000 years ago, cultural evolution outstripped biological evolution of the human form.

7. Modern humans are adapted to a wide range of environments. This capacity resulted from evolutionary modifications in certain primate lineages. Starting with arboreal primate ancestors, there was less reliance on the sense of smell and more on enhanced daytime vision. Also among arboreal primates, manipulative skills increased as the hands began to be freed from load-bearing functions. Starting with Miocene apes, there was a shift from four-legged climbing to bipedalism, a shift from specialized to omnivorous eating habits, and increases in brain complexity and behavior.

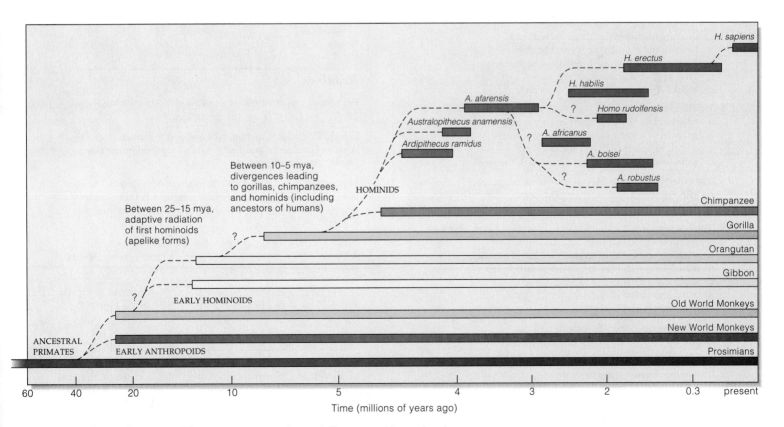

Figure 28.16 Summary of lineages on or near the evolutionary road to modern humans.

Review Questions

1. What is the difference between "hominoid" and "hominid"? Are we hominoids, hominids, or both? 28.1

2. List the presumed macroevolutionary trends among certain primate lineages that were ancestral to modern humans. 28.1

3. What environmental changes are correlated with the great adaptive radiation of apelike forms during the Miocene? 28.2

4. On the basis of the fossil record, where did the first humans (of genus *Homo*) originate? 28.3

5. Why is it difficult to determine how australopiths were related and which ones might have been ancestral to early humans? 28.2

6. Did *H. erectus* populations coexist for a time with *H. habilis*, which may have been ancestral to them? 28.3

7. Briefly describe some of the conserved physical traits that link anatomically modern humans with their mammalian ancestors, then with their primate ancestors. What are the characteristics that set modern humans apart from other primates? 28.3

8. Explain the difference between the multiregional model and African emergence model of modern human origins. 28.4

Self-Quiz *(Answers in Appendix IV)*

1. Primates include _____ .
   a. lemurs
   b. monkeys
   c. apes
   d. humans
   e. all of the above

2. Behavioral flexibility can be observed among _____ .
   a. mammals
   b. primates
   c. humans
   d. all of the above

3. _____ are hominoids; _____ are hominids.
   a. Lemurs and monkeys; apes and humans
   b. Apes and humans; australopiths and humans
   c. Monkeys; apes and australopiths
   d. Monkeys, apes, and humans; apes only

4. Hominids _____ .
   a. adapted to a wide range of environments
   b. adapted to a narrow range of environments
   c. had flexible bones that cracked easily
   d. were limber enough to swing through the trees

5. The oldest known primates date from the _____ .
   a. Miocene
   b. Paleocene
   c. Oligocene
   d. Pliocene

6. The adaptive radiations of Miocene apes and of early hominids may have been correlated with _____ .
   a. a long-term trend toward warmer global climates
   b. a long-term trend toward cooler global climates
   c. continental drift and changes in ocean circulation
   d. a and c are most likely
   e. b and c are most likely

7. The first known hominids were the _____ .
   a. Homos
   b. dryopiths
   c. cercopiths
   d. australopiths

8. The earliest known stone tools are _____ years old.
   a. 8 million
   b. 6 million
   c. 4.2 million
   d. 2.5 million

9. Match the primates with their descriptions.
   ____ prosimian
   ____ tarsioid
   ____ anthropoid
   ____ hominoid
   ____ hominid
   a. monkeys, apes, and humans
   b. lemurs
   c. humans and recent ancestors
   d. apes and humans
   e. tarsiers

Critical Thinking

1. Fossil evidence suggests that Neandertals and *H. erectus* populations coexisted in the same parts of the Middle East for 25,000 to 50,000 years. By one hypothesis, the Neandertals evolved in Europe, from *H. erectus* stock that had dispersed from Africa much earlier, then migrated to the Middle East. It appears that both groups used the same kinds of tools, and neither developed the techniques required for cave painting or for constructing decorative artifacts. Their extended coexistence implies that they did not compete with each other. However, the Neandertals became extinct and the other group gave rise to populations of anatomically modern humans. Does this scenario fit better with the multiregional model or African emergence model of modern human origins?

2. When it comes to human origins, different researchers "read" the fossil record in different parts of the world in different ways. Do you interpret this as an indication that the primate fossils they are citing might have nothing to do with human origins? Why or why not?

Selected Key Terms

anthropoid 28.1
arboreal (life-style) 28.1
australopith 28.2
bipedalism 28.1
culture 28.1
hominid 28.1
hominoid 28.1
Homo erectus 28.3
Homo habilis 28.3
Homo sapiens 28.3
human 28.3
Primates (order) 28.1
savanna 28.1

Readings

Beardsley, T. April 1996. "Out of Food?" *Scientific American.* 20–22. Update on *H. erectus* migrations.

Larick, R., and R. Ciochon. November–December 1996. "The African Emergence and Early Asian Dispersal of the Genus *Homo.*" *American Scientist.*

Monastersky, R. 3 August 1996. "Out of Arid Africa." *Science News* (150): 74–75. Account of debates on whether climatic changes sparked pulses of evolutionary change among the hominids.

Waters, T. May 1990. "Almost Human." *Discover.* 43–53. Account of how noted sculptor John Gurche is painstakingly reconstructing the heads and facial features of seven forms of early hominids and of a modern human.

Weiss, M., and A. Mann. 1990. *Human Biology and Behavior.* Fifth edition. New York: Harper Collins.

Web Site See *http://www.wadsworth.com/biology* for practice quiz questions, hypercontents, BioUpdates, and critical thinking. The Wadsworth Biology Resource Center provides a wealth of information fully organized and integrated by chapter.

APPENDIX I.   BRIEF CLASSIFICATION SCHEME

The classification scheme that follows is a composite of several that microbiologists, botanists, and zoologists use. The major groupings are agreed upon, more or less. There is not always agreement, however, on what to call a given grouping or where it might fit within the overall hierarchy. There are several reasons for the lack of total consensus.

First, the fossil record varies in its quality and in its completeness. Therefore, the phylogenetic relationship of one group to others is sometimes open to interpretation. Comparative studies at the molecular level are firming up the picture, but this work is still under way.

Second, ever since the time of Linnaeus, classification schemes have been based on the perceived morphological similarities and differences among organisms. Although some original interpretations are now open to question, we are so used to thinking about organisms in certain ways that reclassification often proceeds slowly. Traditionally, for example, birds and reptiles have been considered to be separate classes (Reptilia and Aves). And yet there are now compelling arguments for grouping the lizards and snakes together in one class, and the crocodilians, dinosaurs, and birds in a different class.

Third, researchers in microbiology, mycology, botany, zoology, and the other fields of biological inquiry have inherited a wealth of literature, based on classification schemes that were developed over time in each of those fields. Many see no good reason to give up the established terminology and thereby disrupt access to the past. Until recently, for example, microbiologists and botanists have been using *division*, and zoologists *phylum*, for taxa that are equivalent in the hierarchy of classification. Many still do. Opinions are still polarized with respect to the kingdom Protista, certain members of which could just as easily be grouped in the kingdoms of plants, fungi, or animals. Indeed, the term protozoan is a holdover from an earlier scheme in which amoebas and certain other single-celled organisms were ranked as simple animals.

Given the problems, why do we even bother imposing artificial frameworks on the history of life? We do this for the same reason that a writer might decide to break up the history of civilization into several volumes, a number of chapters, and many paragraphs. Both efforts are attempts to impart obvious structure to what might otherwise be an overwhelming body of knowledge and to enhance the retrieval of information from it.

Finally, bear in mind that we include this classification scheme primarily for your reference purposes. Besides being open to revision, it also is by no means complete. Numerous existing and extinct organisms of the so-called lesser phyla are not represented here. Our strategy is to focus mainly on organisms mentioned in the text. A few examples of organisms also are listed under the entries.

SUPERKINGDOM PROKARYOTA.   Prokaryotes. Almost all microscopic species with DNA concentrated in a region of cytoplasm, not inside a membrane-bounded nucleus. All are bacteria, either single cells or simple associations of cells. Autotrophs and heterotrophs (Table 22.2). Reproduce by prokaryotic fission, sometimes by budding and by bacterial conjugation. *Bergey's Manual of Systematic Bacteriology*, the authoritative reference in the field, calls this "a time of taxonomic transition." It groups bacteria mostly by numerical taxonomy (Section 22.6), not on phylogeny. The scheme presented here reflects strong evidence of evolutionary relationships for at least some bacterial groupings.

KINGDOM EUBACTERIA. Gram-negative and gram-positive forms. Peptidoglycan in cell wall. Photosynthetic autotrophs, chemosynthetic autotrophs, and heterotrophs.

PHYLUM GRACILICUTES.   Typical Gram-negative, thin wall. Autotrophs (photosynthetic and chemosynthetic) and heterotrophs. *Anabaena* and other cyanobacteria. *Escherichia, Pseudomonas, Neisseria, Myxococcus.*

PHYLUM FIRMICUTES.   Typical Gram-positive, thick wall. Heterotrophs. *Bacillus, Staphylococcus, Streptococcus, Clostridium, Actinomycetes.*

PHYLUM TENERICUTES.   Gram-negative, wall absent. Heterotrophs (saprobes, pathogens). *Mycoplasma.*

KINGDOM ARCHAEBACTERIA. Methanogens, halophiles, thermophiles. Strict anaerobes, distinct from other bacteria in cell wall, membrane lipids, ribosomes, and RNA sequences. *Methanobacterium, Halobacterium, Sulfolobus.*

SUPERKINGDOM EUKARYOTA. Eukaryotes. Single-celled and multicelled species. Cells start out life with a nucleus (encloses the DNA) and usually other membrane-bound organelles. Chromosomes with numerous proteins attached.

KINGDOM PROTISTA.   Diverse single-celled, colonial, and multicelled eukaryotic species, currently most easily classified by what they are *not* (not bacteria, fungi, plants, or animals). Autotrophs, heterotrophs, or both (Table 23.3). Reproduce sexually and asexually (by meiosis, mitosis, or both). Many related evolutionarily to plants, fungi, and possibly animals.

PHYLUM CHYTRIDIOMYCOTA.   Chytrids. Heterotrophs; saprobic decomposers or parasites. *Chytridium.*

PHYLUM OOMYCOTA.   Water molds. Heterotrophs. Decomposers, some parasites. *Saprolegnia, Phytophthora, Plasmopara.*

PHYLUM ACRASIOMYCOTA.   Cellular slime molds. Heterotrophs with free-living, phagocytic amoeboid cells and spore-bearing stages. *Dictyostelium.*

PHYLUM MYXOMYCOTA.   Plasmodial slime molds. Heterotrophs with free-living, phagocytic amoeboid cells and spore-bearing stages. Aggregate into streaming mass of cells that discard plasma membranes. *Physarum.*

PHYLUM SARCODINA.   Amoeboid protozoans. Heterotrophs, free-living or endosymbiotic, some pathogens. Soft-or shelled bodies, locomotion by pseudopods. The rhizopods (naked amoebas, foraminiferans), *Amoeba proteus, Entomoeba.* Also the actinopods (radiolarians, heliozoans).

PHYLUM MASTIGOPHORA.   Animal-like flagellated protozoans. Heterotrophs, free-living, many internal parasites. All with one to several flagella. *Trypanosoma, Trichomonas, Giardia.*

APICOMPLEXA.   Heterotrophs, many parasitic. Complex of rings, tubules, other structures at head end. Most familiar members called sporozoans. *Plasmodium, Toxoplasma.*

PHYLUM CILIOPHORA. Ciliated protozoans. Heterotrophs, predators or symbionts, some parasitic. All have cilia. Free-living, sessile, or motile. *Paramecium*, hypotrichs.

PHYLUM EUGLENOPHYTA. Euglenoids. Mostly heterotrophs, some autotrophs (photosynthetic). Flagellated. *Euglena*.

PHYLUM PYRRHOPHYTA. Dinoflagellates. Photosynthetic, mostly, but some heterotrophs. *Gymnodinium breve*.

PHYLUM CHRYSOPHYTA. Golden algae, yellow-green algae, diatoms. Photosynthetic. Some flagellated, others not. *Mischococcus, Synura, Vaucheria*.

PHYLUM RHODOPHYTA. Red algae. Photosynthetic, some parasitic. Nearly all marine, some freshwater. *Porphyra. Bonnemaisonia, Euchema*.

PHYLUM PHAEOPHYTA. Brown algae. Photosynthetic, nearly all in temperate or marine waters. *Macrocystis, Fucus, Sargassum, Ectocarpus, Postelsia*.

PHYLUM CHLOROPHYTA. Green algae. Mostly photosynthetic, some parasitic. Most freshwater, some marine or terrestrial. *Chlamydomonas, Spirogyra, Ulva, Volvox, Codium, Halimeda*.

KINGDOM FUNGI. Nearly all multicelled eukaryotic species. Heterotrophs; mostly saprobic decomposers, some parasites. Nutrition based on extracellular digestion of organic matter and absorption of nutrients by individual cells. Multicelled species form absorptive mycelium within substrates and structures that produce asexual spores (and sometimes sexual spores).

PHYLUM ZYGOMYCOTA. Zygomycetes. Zygosporangia (zygote inside thick wall) formed by sexual reproduction. Bread molds, related forms. *Rhizopus, Philobolus*.

PHYLUM ASCOMYCOTA. Ascomycetes. Sac fungi. Sac-shaped cells form sexual spores (ascospores). Most yeasts and molds, morels, truffles. *Saccharomycetes, Morchella, Neurospora, Sarcoscypha. Claviceps, Ophiostoma*.

PHYLUM BASIDIOMYCOTA. Basidiomycetes. Club fungi. Most diverse group. Produce basidiospores inside club-shaped structures. Mushrooms, shelf fungi, stinkhorns. *Agaricus, Amanita, Puccinia, Ustilago*.

IMPERFECT FUNGI. Sexual spores absent or undetected. The group has no formal taxonomic status. If better understood, a given species might be grouped with sac fungi or club fungi. *Verticillium, Candida, Microsporum, Histoplasma*.

LICHENS. Mutualistic interactions between fungal species and a cyanobacterium, green alga, or both. *Usnea, Cladonia*.

KINGDOM PLANTAE. Multicelled eukaryotes. Nearly all photosynthetic autotrophs with chlorophylls *a* and *b*. Some parasitic. Nonvascular and vascular species, generally with well-developed root and shoot systems. Nearly all adapted in form and function to survive dry conditions in land habitats; a few in aquatic habitats. Sexual reproduction predominant; also asexual reproduction by vegetative propagation.

PHYLUM RHYNIOPHYTA. Earliest known vascular plants; muddy habitats. Extinct. *Cooksonia, Rhynia*.

PHYLUM PROGYMNOSPERMOPHYTA. Progymnosperms. Ancestral to early seed-bearing plants; extinct. *Archaeopteris*.

PHYLUM PTERIDOSPERMOPHYTA. Seed ferns. Fernlike gymnosperms; extinct. *Medullosa*

PHYLUM CHAROPHYTA. Stoneworts.

PHYLUM BRYOPHYTA. Bryophytes: mosses, liverworts, hornworts. Seedless, nonvascular, haploid dominance. *Marchantia, Polytrichum, Sphagnum*.

PHYLUM PSILOPHYTA. Whisk ferns. Seedless, vascular. No obvious roots, leaves on sporophyte. *Psilotum*.

PHYLUM LYCOPHYTA. Lycophytes, club mosses. Seedless, vascular. Leaves, branching rhizomes, vascularized roots and stems. *Lycopodium, Selaginella*.

PHYLUM SPHENOPHYTA. Horsetails. Seedless, vascular. Some sporophyte stems photosynthetic, others nonphotosynthetic, spore-producing. *Equisetum*.

PHYLUM PTEROPHYTA. Ferns. Largest group of seedless vascular plants (12,000 species), mainly tropical, temperate habitats.

PHYLUM CYCADOPHYTA. Cycads. Type of gymnosperm (vascular, bears "naked" seeds). Tropical, subtropical. Palm-shaped leaves, simple cones on male and female plants. *Zamia*.

PHYLUM GINKGOPHYTA. Ginkgo (maidenhair tree). Type of gymnosperm. Seeds with fleshy outer layer. *Ginkgo*.

PHYLUM GNETOPHYTA. Gnetophytes. Only gymnosperms with vessels in xylem and double fertilization (but endosperm does not form). *Ephedra, Welwitchia*.

PHYLUM CONIFEROPHYTA. Conifers. Most common and familiar gymnosperms. Generally cone-bearing with needle-like or scale-like leaves.
Family Pinaceae. Pines, firs, spruces, hemlock, larches, Douglas firs, true cedars. *Pinus*.
Family Cupressaceae. Junipers, cypresses. *Juniperus*.
Family Taxodiaceae. Bald cypress, redwoods, Sierra bigtree, dawn redwood. *Sequoia*.
Family Taxaceae. Yews.

PHYLUM ANTHOPHYTA. Angiosperms (flowering plants). Largest group of vascular seed-bearing plants. Only organisms that produce flowers, fruits.
Class Dicotyledonae. Dicotyledons (dicots). Some families of several different orders are listed:
Family Nymphaeaceae. Water lilies.
Family Papaveraceae. Poppies.
Family Brassicaceae. Mustards, cabbages, radishes.
Family Malvaceae. Mallows, cotton, okra, hibiscus.
Family Solanaceae. Potatoes, eggplant, petunias.
Family Salicaceae. Willows, poplars.
Family Rosaceae. Roses, apples, almonds, strawberries.
Family Fabaceae. Peas, beans, lupines, mesquite.
Family Cactaceae. Cacti.
Family Euphorbiaceae. Spurges, poinsettia.
Family Cucurbitaceae. Gourds, melons, cucumbers, squashes.
Family Apiaceae. Parsleys, carrots, poison hemlock.
Family Aceraceae. Maples.
Family Asteraceae. Composites. Chrysanthemums, sunflowers, lettuces, dandelions.
Class Monocotyledonae. Monocotyledons (monocots). Some families of several different orders are listed:
Family Liliaceae. Lilies, hyacinths, tulips, onions, garlic.
Family Iridaceae. Irises, gladioli, crocuses.
Family Orchidaceae. Orchids.
Family Arecaceae. Date palms, coconut palms.
Family Cyperaceae. Sedges.
Family Poaceae. Grasses, bamboos, corn, wheat, sugarcane.
Family Bromeliaceae. Bromeliads, pineapples, Spanish moss.

KINGDOM ANIMALIA. Multicelled eukaryotes, nearly all with tissues, organs, and organ systems and with motility during at least part of the life cycle. Heterotrophs; predators (herbivores, carnivores, omnivores), parasites, detritivores. Reproduce sexually (and asexually in many species). Continuous stages of embryonic development.

PHYLUM PLACOZOA. Marine. Simplest known animal. Two cell layers, no mouth, no organs. *Trichoplax*.

PHYLUM MESOZOA. Ciliated, wormlike parasites, about the same level of complexity as *Trichoplax*.

PHYLUM PORIFERA. Sponges. No symmetry, tissues, or organs.

PHYLUM CNIDARIA. Radial symmetry, tissues, nematocysts.
Class Hydrozoa. Hydrozoans. *Hydra, Obelia, Physalia*.
Class Scyphozoa. Jellyfishes. *Aurelia*.
Class Anthozoa. Sea anemones, corals. *Telesto*.

PHYLUM CTENOPHORA. Comb jellies. Modified radial symmetry.

PHYLUM PLATYHELMINTHES. Flatworms. Bilateral, cephalized; simplest animals with organ systems. Saclike gut.
Class Turbellaria. Triclads (planarians), polyclads. *Dugesia*.
Class Trematoda. Flukes. *Schistosoma*.
Class Cestoda. Tapeworms. *Taenia*.

PHYLUM NEMERTEA.   Ribbon worms.

PHYLUM NEMATODA.   Roundworms. *Ascaris, Trichinella.*

PHYLUM ROTIFERA.   Rotifers.

PHYLUM MOLLUSCA.   Mollusks.

Class Polyplacophora. Chitons.

Class Gastropoda. Snails (periwinkles, whelks, limpets, abalones, cowries, conches, nudibranchs, tree snails, garden snails), sea slugs, land slugs.

Class Bivalvia. Clams, mussels, scallops, cockles, oysters, shipworms.

Class Cephalopoda. Squids, octopuses, cuttlefish, nautiluses. *Loligo.*

PHYLUM BRYOZOA.   Bryozoans (moss animals).

PHYLUM BRACHIOPODA.   Lampshells.

PHYLUM ANNELIDA.   Segmented worms.

Class Polychaeta. Mostly marine worms.

Class Oligochaeta. Mostly freshwater and terrestrial worms, but many marine. *Lumbricus* (earthworms).

Class Hirudinea. Leeches.

PHYLUM TARDIGRADA.   Water bears.

PHYLUM ONYCHOPHORA.   Onychophorans. *Peripatus.*

PHYLUM ARTHROPODA.

Subphylum Trilobita. Trilobites; extinct.

Subphylum Chelicerata. Chelicerates. Horseshoe crabs, spiders, scorpions, ticks, mites.

Subphylum Crustacea. Shrimps, crayfishes, lobsters, crabs, barnacles, copepods, isopods (sowbugs).

Subphylum Uniramia.

Superclass Myriapoda. Centipedes, millipedes.

Superclass Insecta.

Order Ephemeroptera. Mayflies.

Order Odonata. Dragonflies, damselflies.

Order Orthoptera. Grasshoppers, crickets, katydids.

Order Dermaptera. Earwigs.

Order Blattodea. Cockroaches.

Order Mantodea. Mantids.

Order Isoptera. Termites.

Order Mallophaga. Biting lice.

Order Anoplura. Sucking lice.

Order Homoptera. Cicadas, aphids, leafhoppers, spittlebugs.

Order Hemiptera. Bugs.

Order Coleoptera. Beetles.

Order Diptera. Flies.

Order Mecoptera. Scorpion flies. *Harpobittacus.*

Order Siphonaptera. Fleas.

Order Lepidoptera. Butterflies, moths.

Order Hymenoptera. Wasps, bees, ants.

PHYLUM ECHINODERMATA.   Echinoderms.

Class Asteroidea. Sea stars.

Class Ophiuroidea. Brittle stars.

Class Echinoidea. Sea urchins, heart urchins, sand dollars.

Class Holothuroidea. Sea cucumbers.

Class Crinoidea. Feather stars, sea lilies.

Class Concentricycloidea. Sea daisies.

PHYLUM HEMICHORDATA.   Acorn worms.

PHYLUM CHORDATA.   Chordates.

Subphylum Urochordata. Tunicates, related forms.

Subphylum Cephalochordata. Lancelets.

Subphylum Vertebrata. Vertebrates.

Class Agnatha. Jawless vertebrates (lampreys, hagfishes).

Class Placodermi. Jawed, heavily armored fishes; extinct.

Class Chondrichthyes. Cartilaginous fishes (sharks, rays, skates, chimaeras).

Class Osteichthyes. Bony fishes.

Subclass Dipnoi. Lungfishes.

Subclass Crossopterygii. Coelacanths, related forms.

Subclass Actinopterygii. Ray-finned fishes.

Order Acipenseriformes. Sturgeons, paddlefishes.

Order Salmoniformes. Salmon, trout.

Order Atheriniformes. Killifishes, guppies.

Order Gasterosteiformes. Seahorses.

Order Perciformes. Perches, wrasses, barracudas, tunas, freshwater bass, mackerels.

Order Lophiiformes. Angler fishes.

Class Amphibia. Mostly tetrapods; embryo enclosed in amnion.

Order Caudata. Salamanders.

Order Anura. Frogs, toads.

Order Apoda. Apodans (caecilians).

Class Reptilia. Skin with scales, embryo enclosed in amnion.

Subclass Anapsida. Turtles, tortoises.

Subclass Lepidosaura. *Sphenodon,* lizards, snakes.

Subclass Archosaura. Dinosaurs (extinct), crocodiles, alligators.

Class Aves. Birds. (In more recent schemes, dinosaurs, crocodilians, and birds are often grouped in the same category.)

Order Struthioniformes. Ostriches.

Order Sphenisciformes. Penguins.

Order Procellariiformes. Albatrosses, petrels.

Order Ciconiiformes. Herons, bitterns, storks, flamingoes.

Order Anseriformes. Swans, geese, ducks.

Order Falconiformes. Eagles, hawks, vultures, falcons.

Order Galliformes. Ptarmigan, turkeys, domestic fowl.

Order Columbiformes. Pigeons, doves.

Order Strigiformes. Owls.

Order Apodiformes. Swifts, hummingbirds.

Order Passeriformes. Sparrows, jays, finches, crows, robins, starlings, wrens.

Class Mammalia. Skin with hair; young nourished by milk-secreting glands of adult.

Subclass Prototheria. Egg-laying mammals (duckbilled platypus, spiny anteaters).

Subclass Metatheria. Pouched mammals or marsupials (opossums, kangaroos, wombats).

Subclass Eutheria. Placental mammals.

Order Insectivora. Tree shrews, moles, hedgehogs.

Order Scandentia. Insectivorous tree shrews.

Order Chiroptera. Bats.

Order Primates.

Suborder Strepsirhini (prosimians). Lemurs, lorises.

Suborder Haplorhini (tarsioids and anthropoids).

Infraorder Tarsiiformes. Tarsiers.

Infraorder Platyrrhini (New World monkeys).

Family Cebidae. Spider monkeys, howler monkeys, capuchin.

Infraorder Catarrhini (Old World monkeys and hominoids).

Superfamily Cercopithecoidea. Baboons, macaques, langurs.

Superfamily Hominoidea. Apes and humans.

Family Hylobatidae. Gibbon.

Family Pongidae. Chimpanzees, gorillas, orangutans.

Family Hominidae. Existing and extinct human species (*Homo*) and australopiths.

Order Carnivora. Carnivores.

Suborder Feloidea. Cats, civets, mongooses, hyenas.

Suborder Canoidea. Dogs, weasels, skunks, otters, raccoons, pandas, bears.

Order Proboscidea. Elephants; mammoths (extinct).

Order Sirenia. Sea cows (manatees, dugongs).

Order Perissodactyla. Odd-toed ungulates (horses, tapirs, rhinos).

Order Artiodactyla. Even-toed ungulates (camels, deer, bison, sheep, goats, antelopes, giraffes).

Order Edentata. Anteaters, tree sloths, armadillos.

Order Tubulidentata. African aardvarks.

Order Cetacea. Whales, porpoises.

Order Rodentia. Most gnawing animals (squirrels, rats, mice, guinea pigs, porcupines).

Metric-English Conversions

Length

English		Metric
inch	=	2.54 centimeters
foot	=	0.30 meter
yard	=	0.91 meter
mile (5,280 feet)	=	1.61 kilometer

To convert	multiply by	to obtain
inches	2.54	centimeters
feet	30.00	centimeters
centimeters	0.39	inches
millimeters	0.039	inches

Weight

English		Metric
grain	=	64.80 milligrams
ounce	=	28.35 grams
pound	=	453.60 grams
ton (short) (2,000 pounds)	=	0.91 metric ton

To convert	multiply by	to obtain
ounces	28.3	grams
pounds	453.6	grams
pounds	0.45	kilograms
grams	0.035	ounces
kilograms	2.2	pounds

Volume

English		Metric
cubic inch	=	16.39 cubic centimeters
cubic foot	=	0.03 cubic meter
cubic yard	=	0.765 cubic meters
ounce	=	0.03 liter
pint	=	0.47 liter
quart	=	0.95 liter
gallon	=	3.79 liters

To convert	multiply by	to obtain
fluid ounces	30.00	milliliters
quart	0.95	liters
milliliters	0.03	fluid ounces
liters	1.06	quarts

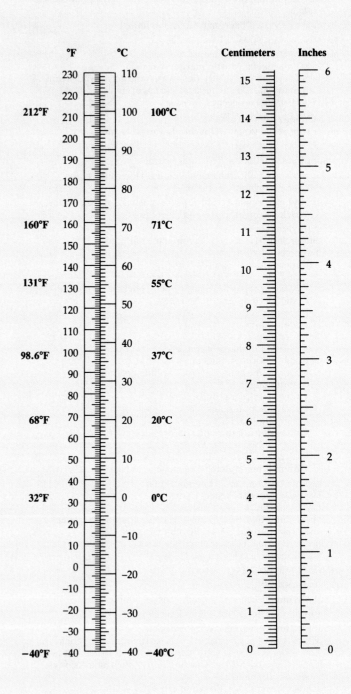

APPENDIX III. ANSWERS TO SELF-QUIZZES

CHAPTER 21
1. c
2. e
3 b
4. d
5. d
6. d
7. b, d, e, a, c

CHAPTER 22
1. c
2. c
3. b
4. c
5. d
6. d
7. e
8. d
9. d, e, b, c, a

CHAPTER 23
1. b
2. d
3. a
4. c
5. a
6. a
7. d
8. b

CHAPTER 24
1. b
2. c
3. b
4. a
5. a
6. e, c, d, a, f, b

CHAPTER 25
1. d
2. b
3. c
4. c
5. b
6. c, e, g, h, f, a, b, d

CHAPTER 26
1. b
2. c
3. a
4. a
5. b
6. c
7. c
8. d, f, c, g, e, i, j, h, a, b

CHAPTER 27
1. d
2. e
3. d
4. b
5. b
6. d
7. d
8. f
9. f
10. d
11. d
12. e
13. d, e, c, b, g, f, a

CHAPTER 28
1. d
2. b
3. c
4. c
5. b
6. c, e, g, h, f, a, b, d

A GLOSSARY OF BIOLOGICAL TERMS

ABO blood typing Method of using two surface proteins (A, B, or both) of red blood cells to characterize an individual's blood. O signifies the absence of both proteins.

abortion The spontaneous expulsion or the dislodging of an embryo or a fetus from the uterus.

abscisic acid (ab-SISS-ik) Plant hormone that promotes stomatal closure, bud dormancy, and seed dormancy.

abscission (ab-SIH-zhun) [L. *abscindere*, to cut off] The dropping of leaves, flowers, fruits, or other plant parts due to hormonal action.

absorption Of most animals, movement of nutrients, fluid, and ions across the gut lining and into the internal environment.

accessory pigment Light-trapping pigment molecule; it contributes to photosynthesis by extending the range of usable wavelengths beyond those absorbed by the chlorophylls.

acid [L. *acidus*, sour] A substance that releases hydrogen ions when dissolved in water.

acid rain The falling to Earth of rain (or snow) that contains sulfur and nitrogen oxides. Also called wet acid deposition (as opposed to dry acid deposition of airborne particles of sulfur and nitrogen oxides).

acoelomate (ay-SEE-luh-mate) Of some of the invertebrates, having no fluid-filled cavity between the gut and body wall.

acoustical signal Sounds that are normally used as a form of communication between animals of the same species.

actin (AK-tin) One of the motor proteins with roles in contraction. Interacts with myosin in muscle cells.

action potential Abrupt, brief reversal in the steady voltage difference across the plasma membrane (the resting membrane potential) of a neuron and other excitable cells.

activation energy The minimum amount of collision energy necessary to drive reactant molecules to an activated state (the transition state) at which a given chemical reaction will proceed spontaneously.

activator A regulatory protein having a role in a positive control system that promotes gene transcription.

active site A crevice in the surface of an enzyme molecule where a specific reaction is catalyzed, or made to proceed far faster than it would spontaneously.

active transport A pumping of one or more specific solutes through the interior of a transport protein that spans the lipid bilayer of a cell membrane. The solute is transported against its concentration gradient. An energy boost, as from ATP, activates the protein.

adaptation [L. *adaptare*, to fit] Of evolution, being adapted (or becoming more adapted) to a given set of environmental conditions. Of a sensory neuron, decreasing frequency of action potentials (or their cessation) even if a stimulus is maintained at constant strength.

adaptive behavior A behavior that promotes propagation of an individual's genes and that tends to increase in frequency in the population over time.

adaptive radiation A burst of divergences from a single lineage that gives rise to many new species, each adapted to an unoccupied or new habitat or to using a novel resource.

adaptive trait Any aspect of form, function, or behavior that helps an individual survive and reproduce under prevailing conditions.

adaptive zone A way of life available for organisms that are physically, ecologically, and evolutionarily equipped to live it, such as "catching insects in the air at night."

adenine (AH-de-neen) A purine; a nitrogen-containing base in certain nucleotides.

adenosine diphosphate (ah-DEN-uh-seen die-FOSS-fate) ADP, an organic compound that can make energy transfers in cells; typically formed by hydrolysis of ATP.

adenosine phosphate Any of a number of relatively small organic compounds, some of which function as chemical messengers within and between cells, and others (such as ATP) that function as energy carriers.

ADH Antidiuretic hormone. Hypothalamic hormone that induces water conservation as required during control of extracellular fluid volume and solute concentrations.

adhesion protein A protein that helps cells of the same type locate one another during development, adhere, and remain in the proper position in the proper tissue.

adipose tissue A type of connective tissue having an abundance of fat-storing cells and blood vessels for transporting fats.

ADP Adenosine diphosphate. A nucleotide coenzyme that accepts unbound phosphate or a phosphate group to become ATP.

ADP/ATP cycle In cells, a mechanism of ATP renewal. During a phosphate-group transfer, ATP reverts to ADP, then it forms again by phosphorylation of ADP.

aerobic respiration (air-OH-bik) [Gk. *aer*, air, + *bios*, life] The main pathway of ATP formation, for which oxygen is the final acceptor of electrons stripped from glucose or another organic compound. It proceeds from glycolysis through the Krebs cycle and electron transport phosphorylation. For each glucose molecule, a typical net yield is 36 ATP.

age structure The number of individuals in each of several or many age categories for a population.

agglutination (ah-glue-tin-AY-shun) In this defensive response, antibodies circulating in blood act against foreign cells (such as transfused red blood cells of the wrong type) and cause them to clump together.

aging A range of processes, including the breakdown of cell structure and function, by which the body gradually deteriorates. All multicelled species that show extensive cell differentiation undergo aging.

AIDS Short for acquired immunodeficiency syndrome. A set of chronic disorders that arises following infection by the human immunodeficiency virus (HIV), which destroys key cells of the immune system.

alcohol An organic compound that has one or more hydroxyl groups (—OH) and readily dissolves in water. Sugars are examples.

alcoholic fermentation Anaerobic pathway of ATP formation. Pyruvate from glycolysis is degraded to acetaldehyde, which accepts electrons from NADH to form ethanol with a net yield of two ATP. NAD$^+$ is regenerated.

aldosterone (al-DOSS-tuh-rohn) Hormone secreted by the adrenal cortex that helps regulate sodium reabsorption.

allantois (ah-LAN-twahz) [Gk. *allas*, sausage] One of four extraembryonic membranes that functions in respiration and in storing the metabolic wastes of embryos of reptiles, birds, and some mammals. In humans, it gives rise to blood vessels for the placenta and to the urinary bladder.

allele (uh-LEEL) At a given gene locus on a chromosome, one of two or more slightly different molecular forms of a gene that arise through mutation and that code for different versions of the same trait.

allele frequency The abundance of each kind of allele in the population as a whole.

allergen Normally harmless substance that provokes inflammation, excessive mucus secretion, and other defense responses.

allergy A response to an allergen.

allopatric speciation [Gk. *allos*, different, L *patria*, native land] Speciation that follows the end of gene flow between populations or subpopulations of a species as a result of their geographic isolation from each other.

allosteric control (AL-oh-STARE-ik) Form of control over a metabolic reaction or pathway that operates by binding a specific substance at a control site on a specific enzyme.

altruism (al-true-ISS-tik) Self-sacrificing behavior; an individual behaves in a way that helps others but decreases its own chances of reproductive success.

alveolus (ahl-VEE-uh-lus), plural **alveoli** [L. *alveus*, small cavity] One of the cupped, thin-walled outpouchings of respiratory bronchioles. A site where oxygen diffuses from air in the lungs to blood, and carbon dioxide diffuses from blood to the lungs.

amino acid (uh-MEE-no) A small organic molecule with a hydrogen atom, an amino group, an acid group, and an R group bonded covalently to a central carbon atom; the subunit of polypeptide chains.

ammonification (uh-moan-ih-fih-KAY-shun) A process by which certain soil bacteria and fungi break down nitrogenous wastes and remains of organisms; part of the nitrogen cycle.

amnion (AM-nee-on) Of land vertebrates, one of four extraembryonic membranes; the boundary layer of a fluid-filled sac that allows the embryo to grow in size, move freely, and be protected from sudden impacts and temperature shifts.

amniote egg An egg, often with a leathery or calcified shell, that has extraembryonic membranes, including the amnion.

amphibian A type of vertebrate somewhere between fishes and reptiles in body plan and reproductive mode; salamanders, frogs and toads, and caecilians are existing groups.

anaerobic pathway (an-uh-ROW-bik) [Gk. *an*, without, + *aer*, air] Metabolic pathway in which a substance other than oxygen serves as the final acceptor of electrons that have been stripped from substrates.

analogous structures (ann-AL-uh-gus) [Gk. *analogos*, similar to one another] Body parts that once differed in evolutionarily distant lineages, then converged in structure and function as the lineages responded to similar environmental pressures.

anaphase (AN-uh-faze) Of anaphase I of meiosis, the stage when each homologous chromosome separates from its partner and both move to opposite spindle poles. Of mitosis and of anaphase II of meiosis, sister chromatids of each chromosome separate and move to opposite poles.

aneuploidy (AN-yoo-ploy-dee) Having one more chromosome or one less relative to the parental chromosome number.

angiosperm (AN-gee-oh-spurm) [Gk. *angeion*, vessel, and *spermia*, seed] A flowering plant.

animal A multicelled heterotroph that feeds on other organisms, that is motile for at least part of the life cycle, that develops by a series of embryonic stages, and that usually has tissues, organs, and organ systems.

annelid A type of invertebrate classified as a segmented worm; an oligochaete (such as an earthworm), leech, or polychaete.

annual A flowering plant that completes its life cycle in one growing season.

anther [Gk. *anthos*, flower] A pollen-bearing part of a stamen.

antibiotic One of many metabolic products of certain microorganisms that can kill their bacterial competitors for nutrients in soil.

antibody [Gk. *anti*, against] One of a diverse array of antigen-binding receptors. Only B cells make antibody molecules and position them at their surface or secrete them.

anticodon A sequence of three nucleotide bases in a tRNA molecule that can base-pair with a codon in an mRNA molecule.

antigen (AN-tih-jen) [Gk. *anti*, against, + *genos*, race, kind] A molecular configuration that white blood cells recognize as foreign and that triggers an immune response. Most antigens are proteins at the surface of pathogens or tumor cells.

antigen-MHC complex Processed antigen fragments bound with a suitable MHC molecule and displayed at the cell surface; basis of antigen recognition that promotes lymphocyte cell divisions.

antigen-presenting cell Any cell displaying antigen-MHC complexes at its surface.

aorta (ay-OR-tah) [Gk. *airein*, to lift, heave] Main artery of systemic circulation; carries oxygenated blood away from the heart to all body regions except the lungs.

apical dominance Inhibitory influence of a terminal bud on growth of lateral buds.

apical meristem (AY-pih-kul MARE-ih-stem) [L. *apex*, top, + Gk. *meristos*, divisible] A mass of self-perpetuating cells responsible for primary growth at root and shoot tips.

apoptosis (APP-oh-TOE-sis) Of multicelled organisms, a form of cell death; molecular signals activate weapons of self-destruction already stockpiled in target cells. It occurs when a cell has completed its prescribed function or becomes altered, as by infection or cancerous transformation.

appendicular skeleton (ap-en-DIK-yoo-lahr) Bones of the limbs, hips, and shoulders.

Archaebacteria A kingdom of prokaryotes; encompasses methanogens, halophiles, and thermophiles, all of which differ from the eubacteria in chemical composition and in cell wall and membrane characteristics.

archipelago An island chain some distance away from a continent.

area effect Larger islands tend to support more species than smaller ones at equivalent distances from sources of colonizing species.

arteriole (ar-TEER-ee-ole) A blood vessel between an artery and capillary; a control point where the blood volume delivered to a given body region can be adjusted.

artery A large-diameter, rapid-transport blood vessel with a thick, muscular wall that smooths out the pulsations in blood pressure caused by heart contractions.

arthropod An invertebrate with a hardened exoskeleton, specialized body segments, and jointed appendages. Spiders, crabs, and insects are examples.

artificial selection Selection of traits among individuals of a population that occurs in an artificial environment, under contrived, manipulated conditions.

asexual reproduction Any of a number of modes of reproduction by which offspring arise from a single parent and inherit the genes of that parent only.

atmosphere A volume of gases, airborne particles, and water vapor that envelops the Earth; 80 percent of its mass is distributed within seventeen miles of the Earth's surface.

atmospheric cycle A biogeochemical cycle in which the atmosphere is the largest reservoir of an element. The carbon cycle and nitrogen cycle are examples.

atom The smallest particle unique to a given element; it has one or more positively charged protons, electrons, and (except for hydrogen), neutrons.

atomic number The number of protons in the nucleus of each atom of an element; the number differs for each element.

ATP Adenosine triphosphate (ah-DEN-uh-seen try-FOSS-fate). A nucleotide of adenine, ribose, and three phosphate groups that acts as an energy carrier. Its phosphate-group transfers drive nearly all energy-requiring metabolic reactions.

australopith (OHSS-trah-low-pith) [L. *australis*, southern, + Gk. *pithekos*, ape] Any of the earliest known hominids; a primate species on or near the evolutionary road that led to modern humans.

autoimmune response Misdirected immune response in which lymphocytes mount an attack against normal body cells.

autonomic nervous system (auto-NOM-ik) All nerves from the central nervous system to the smooth muscle, cardiac muscle, and glands of the viscera (internal organs and structures) of the vertebrate body.

autosome Any of the pairs of chromosomes that are the same in both males and females of the species.

autotroph (AH-toe-trofe) [Gk. *autos*, self, + *trophos*, feeder] Organism that synthesizes its own organic compounds using carbon dioxide (as the carbon source) and energy from the physical environment (such as sunlight energy). *Compare* heterotroph.

auxin (AWK-sin) A plant hormone that influences growth, such as stem elongation.

axial skeleton (AX-ee-uhl) Of a vertebrate skeleton, the skull, backbone, ribs, and breastbone (sternum).

axon A cylindrical extension from the cell body of a neuron, often with finely branched endings, that is specialized for the rapid propagation of action potentials.

B lymphocyte (B cell) The only white blood cell that produces antibodies, then positions them at the cell surface or secretes them as weapons in immune responses.

bacterial conjugation A transfer of plasmid DNA from one bacterial cell to another.

bacterial flagellum Of many bacterial cells, a whiplike motile structure that does not contain a core of microtubules.

bacteriophage (bak-TEER-ee-oh-fahj) [Gk. *baktērion*, small staff, rod, + *phagein*, to eat] Category of viruses that infect bacterial cells.

balancing selection All forms of selection that are maintaining two or more alleles for a trait in a population. When the resulting genetic variation persists, this is a case of balanced polymorphism.

Barr body In cells of female mammals, one of two X chromosomes that was randomly condensed so that its genes are inactivated.

basal body A centriole which, after giving rise to microtubules of a flagellum or cilium, remains attached to its base in the cytoplasm.

base Any substance that accepts hydrogen ions when dissolved in water.

base pair Two nucleotide bases that are located in two adjacent strands of DNA or RNA and that are hydrogen-bonded to each other.

base sequence The particular order in which one nucleotide base follows the next in a strand of DNA or RNA. The order is unique in at least some regions for each species.

basophil Fast-acting white blood cell that secretes histamine and other substances to maintain an inflammatory response.

behavior, animal A response to external and internal stimuli that requires sensory, neural, endocrine, and effector components. Behavior has a genetic basis and can evolve; it also can be modified through learning.

benthic province All sediments and rocky formations of the ocean bottom.

biennial (bi-EN-yul) A flowering plant that lives through two growing seasons.

bilateral symmetry Body plan in which the left and right halves of an animal are mirror-images of each other.

binary fission Of flatworms and some other animals, a mode of asexual reproduction; growth by way of mitotic cell divisions is followed by division of the whole body into two parts of the same or different sizes. Not the same as prokaryotic fission, by which bacteria reproduce.

biogeochemical cycle The movement of an element from the environment to organisms, then back to the environment.

biogeographic realm [Gk. *bios*, life, + *geographein*, to describe the Earth's surface] One of six major land divisions, each with distinguishing plants and animals; it retains its general identity because of climate and geographic barriers to gene flow.

biological clock Internal time-measuring mechanism that helps adjust an organism's daily activities, seasonal activities, or both in response to environmental cues.

biological magnification The increasing concentration of a nondegradable or slowly degradable substance in body tissues as it is passed along food chains.

biological species concept A species is one or more populations of individuals that are interbreeding under natural conditions and producing fertile offspring, and that are reproductively isolated from other such populations. The concept applies only to sexually reproducing species.

bioluminescence A flashing of light that emanates from an organism when excited electrons of luciferins (highly fluorescent substances) return to a lower energy level.

biomass Combined weight of all organisms at a given trophic level in an ecosystem.

biome A broad, vegetational subdivision of a biogeographic realm; shaped by climate, topography, and composition of regional soils.

biosphere [Gk. *bios*, life, + *sphaira*, globe] All regions of the Earth's waters, crust, and atmosphere in which organisms live.

biosynthetic pathway A metabolic pathway by which organic compounds necessary for life are synthesized.

biotic potential Of population growth for a given species, the maximum rate of increase per individual under ideal conditions.

bipedalism Habitually walking on two feet, as by ostriches and humans.

bird The only vertebrate that produces feathers and that has strong resemblances and evolutionary connections to reptiles.

blastocyst (BLASS-tuh-sist) [Gk. *blastos*, sprout, + *kystis*, pouch] Outcome of cleavage of a fertilized mammalian egg; a surface layer of blastomeres around a cavity filled with their secretions (the blastocoel), and an inner cell mass (several blastomeres huddled together against the cavity's inner surface).

blastomere One of the small, nucleated cells produced during the cleavage stage of animal development.

blastula (BLASS-chew-lah) Outcome of one type of cleavage pattern; blastomeres formed by the successive cuts surround a fluid-filled cavity (blastocoel).

blood A fluid connective tissue composed of water, solutes, and formed elements (blood cells and platelets); it carries substances to and from cells and helps maintain an internal environment favorable for cell activities.

blood pressure Fluid pressure, generated by heart contractions, that circulates blood.

blood-brain barrier Mechanism that exerts some control over which solutes enter the cerebrospinal fluid and thus helps protect the brain and spinal cord.

bone Mineral-hardened connective tissue of bones; one of the organs of the vertebrate skeleton that help move the body and its parts, protect of other organs, store minerals, and (in some) produce blood cells.

bottleneck A severe reduction in population size, as brought about by intense selection pressure or a natural calamity.

Bowman's capsule The cup-shaped portion of a nephron that receives water and solutes being filtered from the blood in the kidneys.

brain The most complex integrating center of most nervous systems; it receives, processes, and integrates sensory input and issues coordinated commands for response by muscles and glands.

brain stem The vertebrate nervous tissue that evolved first and that still persists in the hindbrain, midbrain and forebrain.

bronchiole A component of the finely branched bronchial tree inside each lung.

bronchus, plural **bronchi** (BRONG-CUSS, BRONG-kee) [Gk. *bronchos*, windpipe] Tube-like branchings of the trachea that lead into the lungs of most vertebrates.

brown alga A photoautotrophic protistan with a notable abundance of xanthophyll pigments; such algae occupy marine habitats.

bryophyte A nonvascular land plant, such as a moss.

bud An undeveloped shoot of meristematic tissue, primarily; often covered and protected by scales (modified leaves).

buffer system A partnership between a weak acid and the base that forms when it dissolves in water. The two work as a pair to counter slight shifts in pH.

bulk A volume of fiber and other undigested material that absorption processes in the small intestine cannot decrease.

bulk flow In response to a pressure gradient, the movement of more than one kind of molecule in the same direction in the same medium (as in blood, sap, or air).

C4 pathway A pathway of photosynthesis in which carbon dioxide is fixed twice, in two different cell types. Carbon dioxide accumulates in the leaf and helps counter photorespiration. The first compound formed is the 4-carbon oxaloacetate.

Calvin-Benson cycle Cyclic reactions that are the *synthesis* part of the light-independent reactions of photosynthesis. In plants, RuBP or some other compound to which carbon has been affixed undergoes rearrangements; a sugar phosphate forms and the RuBP is regenerated. The cycle runs on ATP and NADPH from the light-dependent reactions.

CAM plant A plant that conserves water by opening stomata only at night, when it fixes carbon dioxide by way of a C4 pathway.

cambium (KAM-bee-um), plural **cambia** One of two types of meristems responsible for secondary growth (increases in stem and root diameter). Vascular cambium gives rise to secondary xylem and secondary phloem; cork cambium gives rise to periderm.

camouflage Adaptations in body color, form, or patterning, and in behavior that function in predator avoidance; they help an organism hide in the open (blend with its surroundings) and escape detection.

cancer A malignant tumor; its cells show gross abnormalities in the plasma membrane and cytoplasm, skewed growth and division, and weakened capacity for adhesion within the parent tissue (leading to metastasis). Unless eradicated, cancer is lethal.

capillary, blood [L. *capillus*, hair] A thin-walled vessel that functions in the exchange of carbon dioxide, oxygen, and some other substances between blood and interstitial fluid, which bathes living cells.

capillary bed A diffusion zone, consisting of a great number of capillaries, where substances are exchanged between blood and interstitial fluid.

carbohydrate [L. *carbo*, charcoal, + *hydro*, water] A molecule that consists of carbon, hydrogen, and oxygen in a 1:2:1 ratio (there are exceptions). All cells use carbohydrates as structural materials, energy reservoirs, and transportable forms of energy. The monosaccharides, oligosaccharides, and polysaccharides are three classes.

carbon cycle A biogeochemical cycle in which carbon moves from the atmosphere (its largest reservoir), through the ocean and organisms, then to the atmosphere.

carbon dioxide fixation Of photosynthesis, the first enzyme-mediated step of the light-independent reactions. Carbon (from CO_2) is affixed to RuBP or another compound for entry into the Calvin-Benson cycle.

carcinogen (kar-SIN-uh-jen) A substance or an agent, such as ultraviolet radiation, that can trigger cancer.

cardiac cycle (KAR-dee-ak) [Gk. *kardia*, heart, + *kyklos*, circle] The sequence of muscle contraction and relaxation for one heartbeat.

cardiac pacemaker Sinoatrial (SA) node; the basis of the normal rate of heartbeat. The self-excitatory cardiac muscle cells that spontaneously generate rhythmic waves of excitation over the heart chambers.

cardiovascular system Of most animals, an organ system of blood, one or more hearts, and blood vessels that functions in the rapid transport of substances to and from cells.

carnivore [L. *caro, carnis*, flesh, + *vovare*, to devour] An animal that eats other animals; a type of heterotroph.

carotenoid (kare-OTT-en-oyds) A light-sensitive, accessory pigment that transfers absorbed energy to chlorophylls. Different types absorb violet and blue wavelengths and transmit red, orange, and yellow.

carpel (KAR-pul) The female reproductive part of a flower; sometimes called a pistil. The lower portion of a single carpel (or of a structure composed of two or more) is an ovary. The upper portion has a stigma (a pollen-capturing surface tissue) and often a style (slender extension of the ovary wall).

carrying capacity The maximum number of individuals in a population (or species) that can be sustained indefinitely by a given environment.

cartilage A type of connective tissue with solid yet pliable intercellular material that resists compression.

Casparian strip A waxy band that is an impermeable barrier between the walls of abutting cells making up the endodermis (and exodermis, if present) inside roots.

cDNA Any DNA molecule copied from a mature mRNA transcript by way of reverse transcription.

cell [L. *cella*, small room] The smallest living unit; an organized unit that can survive and reproduce on its own, given suitable DNA instructions and environmental resources—notably energy and raw materials.

cell count The number of cells of a given type in a microliter of blood.

cell cycle Events by which a cell increases in mass, roughly doubles its number of cytoplasmic components, duplicates its DNA, then undergoes nuclear and cytoplasmic division. It extends from the time a new cell is produced until it completes division.

cell differentiation Developmental process in which different cell populations activate and suppress a fraction of their genes in different ways and so become specialized in composition, structure, and function.

cell junction Of multicelled organisms, a point of contact that links two adjoining cells physically, functionally, or both.

cell plate A disklike structure that forms from remnants of a microtubular spindle when a plant cell divides; it develops into a crosswall that partitions the cytoplasm.

cell theory A theory in biology stating that (1) all organisms are composed of one or more cells, (2) the cell is the smallest unit that retains a capacity for independent life, and (3) all cells arise from preexisting cells.

cell wall A semirigid, permeable structure that helps a cell hold its shape and resist rupturing if internal fluid pressure rises.

central nervous system The brain and spinal cord of vertebrates.

central vacuole A fluid-filled organelle in mature, living plant cells that stores amino acids, sugars, ions, and toxic wastes. As it enlarges, it forces increases in cell surface area that improve nutrient uptake.

centriole (SEN-tree-ohl) A cylinder of triplet microtubules that gives rise to microtubules of cilia and flagella.

centromere (SEN-troh-meer) [Gk. *kentron*, center, + *meros*, a part] A small, constricted region of a chromosome having attachment sites for the microtubules that move the chromosome during nuclear division.

cephalization (sef-ah-lah-ZAY-shun) [Gk. *kephalikos*, head] During the evolution of bilateral animals, the concentration of sensory structures and nerve cells in a head.

cerebellum (ser-ah-BELL-um) [L. diminutive of *cerebrum*, brain] Hindbrain region with reflex centers for maintaining posture and smoothing out limb movements.

cerebral cortex Thin surface layer of the cerebral hemispheres. Some parts receive sensory input, others integrate information and coordinate suitable responses.

cerebrospinal fluid Clear extracellular fluid surrounding and cushioning the brain and spinal cord.

cerebrum (suh-REE-bruhm) Forebrain region that first evolved to integrate olfactory input and select motor responses to it. In mammals, it evolved into the most complex integrating center.

channel protein A transport protein that acts as a channel through which specific ions and other water-soluble substances cross the plasma membrane. Some channels remain open; others are gated, and these open and close in controlled ways.

chemical bond A union between the electron structures of two or more atoms or ions.

chemical synapse (SIN-aps) [Gk. *synapsis*, union] A small cleft between a presynaptic neuron and a postsynaptic cell (another neuron, a muscle cell, or a gland cell) that is bridged by neurotransmitter molecules released from the presynaptic neuron.

chemiosmotic theory (kim-ee-OZ-MOT-ik) An electrochemical gradient across a cell membrane drives ATP formation. Hydrogen ions accumulate in a compartment formed by the membrane. The combined force of the H^+ concentration and electric gradients propels ions through transport proteins (ATP synthases) spanning the membrane. By enzyme action at these proteins, ADP and inorganic phosphate combine to form ATP.

chemoreceptor (KEE-moe-ree-sep-tur) A sensory receptor that detects chemical energy (ions or molecules) dissolved in the fluid that bathes it.

chemosynthetic autotroph (KEE-moe-sin-THET-ik) One of a few kinds of bacteria able to synthesize its own organic compounds by using carbon dioxide as the carbon source and certain inorganic substances (such as sulfur) as the energy source.

chlorofluorocarbon (KLORE-oh-FLOOR-oh-car-bun), or **CFC** One of the odorless, invisible compounds of chlorine, fluorine, and carbon, widely used in commercial products, that are contributing to the thinning of the ozone layer above the Earth's surface.

chlorophyll (KLOR-uh-fills) [Gk. *chloros*, green, + *phyllon*, leaf] A light-sensitive pigment that absorbs violet-to-blue and red wavelengths but that transmits green. The main pigments in all but one small group of photoautotrophs. Certain chlorophylls donate electrons to the light-dependent reactions of photosynthesis.

chloroplast (KLOR-uh-plast) An organelle that specializes in photosynthesis in plants and photosynthetic protistans.

chordate An animal having a notochord, a dorsal hollow nerve cord, a pharynx, and gill slits in the pharynx wall for at least part of the life cycle.

chorion (CORE-ee-on) Of placental mammals, one of four extraembryonic membranes; it becomes a key component of the placenta. Absorptive structures (villi) develop at its surface and enhance the rapid exchange of substances between the embryo and mother.

chromatid (CROW-mah-tid) Of a duplicated eukaryotic chromosome, one of two DNA molecules (and associated proteins) that remain attached at their centromere region until separated by mitosis or meiosis; after this, each is a separate chromosome.

chromosome (CROW-moe-some) [Gk. *chroma*, color, + *soma*, body] Of eukaryotes, a DNA molecule with many associated proteins. Of prokaryotes, a DNA molecule without a comparable profusion of proteins.

chromosome number The sum total of chromosomes in cells of a given type. *See* haploidy; diploidy.

cilium (SILL-ee-um), plural **cilia** [L. *cilium*, eyelid] Of eukaryotic cells, a short, hairlike projection with an internal, regular array of microtubules. Cilia can serve as motile or sensory structures or help create currents of fluids. Typically more profuse than flagella.

circadian rhythm (ser-KAYD-ee-un) [L. *circa*, about, + *dies*, day] A cycle of physiological events that is completed every twenty-four hours or so independently of environmental change.

circulatory system An organ system having a muscular pump (heart, most often), blood vessels, and blood; the system transports materials to and from cells and often helps stabilize body temperature and pH.

cladogram [Gk. *clad-*, branch] Evolutionary tree diagram that arranges groups by branch points to show their relative relationships. Groups closer together share a more recent common ancestor than those farther apart.

classification system A way of organizing and retrieving information about species.

cleavage Third stage of animal embryonic development. Mitotic cell divisions divide the volume of egg cytoplasm into a number of smaller, nucleated cells (blastomeres). The number of cells increases, but the original volume of egg cytoplasm does not.

cleavage furrow A ringlike depression that forms during cytoplasmic division of an animal cell and that defines the cleavage plane for the cell. Microfilaments attached to the plasma membrane contract and draw it inward to cut the cell in two.

cleavage reaction A molecule splits into two smaller ones. Hydrolysis is an example.

climate Prevailing weather conditions for an ecosystem, such as temperature, humidity, wind speed, cloud cover, and rainfall.

climax community A self-perpetuating, stable array of species in equilibrium with one another and with their habitat.

climax pattern model Idea that one climax community may extend into another along gradients of environmental conditions, such as variations in climate, topography, and species interactions.

cloaca Of some vertebrates, the last part of a gut that receives feces, urine, and sperm or eggs; of some invertebrates, an excretory, respiratory, or reproductive duct.

cloned DNA Multiple, identical copies of restriction fragments that have been inserted into plasmids or some other cloning vector.

club fungus A fungus with reproductive structures having microscopic, club-shaped cells that produce and bear spores.

cnidarian A radial invertebrate at the tissue level of organization and the only organism to produce nematocysts. Two body forms (medusae and polyps) are common.

coal A nonrenewable source of energy that formed more than 280 million years ago from submerged, undecayed plant remains.

codominance A pair of nonidentical alleles that specify two phenotypes are expressed at the same time in heterozygotes.

codon One of the base triplets in an mRNA molecule, the linear sequence of which corresponds to a linear sequence of amino acids in a polypeptide chain. Of 64 codons, 61 specify different amino acids, and 3 of these also are start signals for translation; 1 serves as a stop signal for translation.

coelom (SEE-lum) [Gk. *koilos*, hollow] A cavity, lined with peritoneum, between the gut and body wall of most animals.

coenzyme A nucleotide; an enzyme helper that accepts electrons and hydrogen atoms stripped from substrates at a reaction site and transfers them elsewhere.

coevolution The joint evolution of two or more closely interacting species; when one species evolves, the change affects selection pressures operating between the two, so the other also evolves.

cofactor A metal ion or coenzyme; it helps an enzyme catalyze a reaction or transfers electrons, atoms, or functional groups from one substrate to another.

cohesion Capacity to resist rupturing when placed under tension (stretched).

cohesion theory of water transport Theory that water moves up through plants due to hydrogen bonding among water molecules confined as narrow columns in xylem. The collective cohesive strength of the bonds allows water to be pulled up in response to transpiration (evaporation from leaves).

collenchyma (coll-ENG-kih-mah) A simple plant tissue that offers flexible support for primary growth, as in lengthening stems.

colon (CO-lun) The large intestine.

commensalism [L. *com*, together, + *mensa*, table] An ecological interaction between species that directly benefits one but does not affect other much, if at all.

communication display A pattern of behavior, often ritualized with intended changes in the function of common behavior patterns, that serves as a social signal.

communication signal A social cue encoded in stimuli that holds unambiguous meaning for individuals of the same species. Specific odors, sounds, coloration and patterning, postures, and movements are examples.

community All populations living in the same habitat. Also, a group of organisms with similar life-styles in a habitat, such as a community of birds.

companion cell A specialized parenchyma cell that helps load organic compounds into conducting cells of phloem.

comparative morphology [Gk. *morph*, form] Study of comparable body parts of adults or embryonic stages of major lineages.

competitive exclusion Theory that species that require identical resources cannot coexist indefinitely.

complement system A set of about twenty proteins circulating in inactive form within vertebrate blood; different kinds induce lysis of pathogens, promote inflammation, and stimulate phagocytes to act during both nonspecific defenses and immune responses.

compound A substance consisting of two or more elements in unvarying proportions.

concentration gradient A difference in the number of molecules or ions of a substance between adjoining regions. Energy inherent in their constant molecular motion makes them collide and careen outward from the region of higher to lower concentration. Barring other forces, all substances tend to diffuse down their concentration gradient.

condensation reaction Through covalent bonding, two molecules combine to form a larger molecule, often with the formation of water as a by-product.

cone cell In a vertebrate eye, a photoreceptor that responds to intense light and contributes to sharp daytime vision and color perception.

conifer A pollen- and seed-bearing plant of the dominant group of gymnosperms; mostly evergreen, woody trees and shrubs with needle-like or scale-like leaves.

conjugation, bacterial Of bacteria only, a mechanism by which a donor cell transfers plasmid DNA to a recipient cell.

connective tissue proper A category of animal tissues, all having mostly the same components but in different proportions. They incorporate fibroblasts and other cells, the secretions of which form fibers (mostly of collagen and elastin) and a ground substance of modified polysaccharides.

consumer [L. *consumere*, to take completely] A heterotroph that obtains energy and carbon by feeding on the tissues of other organisms. Herbivores, carnivores, and parasites are examples.

continuous variation Of a population, a more or less continuous range of small differences in a given trait among all of its individuals.

contractile vacuole (kun-TRAK-till VAK-you-ohl) [L. *contractus*, to draw together] Of some protistans, such as a paramecium, an organelle that takes up excess water in the cell body, then contracts; the contractile force is enough to expel the water outside the cell through a pore to its surface.

control group Of an experimental test, a group used to evaluate possible side effects of a test involving an experimental group. Ideally, the control group is identical to the experimental group in all respects except for the variable being studied.

cork cambium A lateral meristem that gives rise to a corky replacement for the epidermis of woody plant parts.

corpus callosum (CORE-pus ka-LOW-sum) A band of axons (200 million in humans) that functionally link two cerebral hemispheres.

corpus luteum (CORE-pus LOO-tee-um) A glandular structure that develops from cells of a ruptured ovarian follicle and secretes progesterone and estrogen.

cortex [L. *cortex*, bark] In general, a rindlike layer such as the kidney or adrenal cortex. In vascular plants, the ground tissue that makes up most of the primary plant body, supports plant parts, and stores food.

cotyledon A seed leaf, which develops as part of the embryo of monocots and dicots; cotyledons provide nourishment for the seedling at the time of germination and initial growth.

courtship display A pattern of ritualized social behavior between potential mates. It may include frozen postures as well as movements that are exaggerated and yet simplified. It may include visual signals, such as body parts that are conspicuously enlarged, distinctively colored or patterned, or some combination of these.

covalent bond (koe-VAY-lunt) [L. *con*, together, + *valere*, to be strong] A sharing of one or more electrons between atoms or groups of atoms. If electrons are shared equally, the bond is nonpolar. If shared unequally, it is polar (slightly positive at one end, slightly negative at the other).

continuous variation Of the individuals of a population, a range of small differences in one or more traits.

cross-bridge formation Of a muscle cell, a reversible interaction between its many actin and myosin filaments that is the basis of contraction.

crossing over During prophase I of meiosis, the breakage and exchange of corresponding segments between nonsister chromatids of a pair of homologous chromosomes; a form of genetic recombination that breaks up old combinations of alleles and puts new ones together in chromosomes.

culture The sum of behavior patterns of a social group, passed between generations by learning and by symbolic behavior, especially language.

cuticle (KEW-tih-kull) A body covering. Of land plants, a transparent cover of waxes and lipid-rich cutin deposited on the outer surface of epidermal cell walls. Of annelids, a thin, flexible surface coat. Of arthropods, a hardened, lightweight cover with protein and chitin components that functions as an exoskeleton.

cyclic AMP (SIK-lik) A nucleotide; cyclic adenosine monophosphate. It functions in intercellular communication, as when it is a second messenger (a cytoplasmic mediator of a cell's response to signaling molecules).

cyclic pathway of ATP formation Ancient photosynthetic pathway occurring at the plasma membrane of some bacteria and at the thylakoid membrane of chloroplasts. A photosystem embedded in the membrane gives up electrons to a transport system, which gives them back to the photosystem. The electron flow sets up concentration and electric gradients across the membrane that drive ATP formation at nearby membrane sites.

cyst Of many microorganisms, a resistant resting stage with thick, tough outer layers that typically forms in response to adverse conditions; of skin, any abnormal, fluid-filled sac without an external opening.

cytochrome (SIGH-toe-krome) [Gk. *kytos*, hollow vessel, + *chrōma*, color] Iron-containing protein molecule; a component of the electron transport systems used in photosynthesis and aerobic respiration.

cytokinesis (SIGH-toe-kih-NEE-sis) [Gk. *kinesis*, motion] Cytoplasmic division; the splitting of a parent cell into daughter cells.

cytokinin (SIGH-tow-KY-nin) Any of the class of plant hormones that stimulate cell division, promote leaf expansion, and retard leaf aging.

cytological marker One or more observable, unusual differences between chromosomes of the same type.

cytomembrane system [Gk. *kytos*, hollow vessel] Organelles functioning as a system to modify, package, and distribute newly formed proteins and lipids. Endoplasmic reticulum, Golgi bodies, lysosomes, and a variety of vesicles are its components.

cytoplasm (SIGH-toe-plaz-um) [Gk. *plassein*, to mold] All cellular parts, particles, and semifluid substances enclosed within the plasma membrane except for the nucleus (or nucleoid, in bacterial cells).

cytoplasmic localization When cleavage divides an animal zygote, each resulting blastomere receives a localized portion of maternal messages in the egg cytoplasm.

cytosine (SIGH-toe-seen) A pyrimidine; one of the nitrogen-containing bases in nucleotides.

cytoskeleton The internal "skeleton" of eukaryotic cells. Its microtubules and other components structurally support the cell and organize and move its internal components. The cytoskeleton also helps free-living cells move through their environment.

cytotoxic T cell A T lymphocyte that uses touch-killing to eliminate infected body cells or tumor cells. When it contacts targets, it delivers cell-killing chemicals into them.

decomposer [partly fr. L. *dis-*, to pieces, + *companere*, arrange] Of ecosystems, a heterotroph that gets energy and carbon by chemically breaking down the remains, products, or wastes of other organisms and helps cycle nutrients back to producers. Certain fungi and bacteria are examples.

deforestation The removal of all trees from a large tract of land, such as the Amazon Basin and the Pacific Northwest.

degradative pathway A metabolic pathway by which organic compounds are broken down in stepwise reactions that lead to products of lower energy.

deletion Loss of a chromosome segment.

denaturation (deh-NAY-chur-AY-shun) Of any molecule, the loss of three-dimensional shape following disruption of hydrogen bonds and other weak bonds.

dendrite (DEN-drite) [Gk. *dendron*, tree] A short, slender extension from the cell body of a neuron; commonly an input zone.

denitrification (DEE-nite-rih-fih-KAY-shun) Conversion of nitrate or nitrite by certain bacteria to gaseous nitrogen (N_2) and a small amount of nitrous oxide (N_2O).

density-dependent control A factor that limits population growth by reducing the birth rate, increasing the rates of death and dispersal, or all of these. Predation, parasitism, disease, and competition for resources are examples.

density-independent factor A factor that tends to cause a population's death rate to increase independently of its density. Storms and floods are examples.

dentition (den-TIH-shun) The type, size, and number of an animal's teeth.

derived trait A novel feature that evolved only once and is shared only by descendants of the ancestral species in which it evolved.

dermal tissue system All the tissues that cover and protect the surfaces of a plant.

dermis The layer of skin underlying the epidermis; consists primarily of dense connective tissue.

desert A biome that typically forms where the potential for evaporation greatly exceeds rainfall and vegetation cover is limited.

desertification (dez-urt-ih-fih-KAY-shun) Conversion of a grassland or an irrigated or rain-fed cropland to a desertlike condition, with a drop in agricultural productivity of 10 percent or more.

detrital food web A network of food chains in which energy flows mainly from plants through arrays of detritivores and decomposers.

detritivore (dih-TRY-tih-vorez) [L. *detritus*; after *deterere*, to wear down] A heterotroph that consumes decomposing particles of organic matter. Earthworms, crabs, and nematodes are examples.

deuterostome (DUE-ter-oh-stome) [Gk. *deuteros*, second, + *stoma*, mouth] A bilateral animal for which the first indentation that forms in the early embryo develops into an anus. An echinoderm or a chordate.

development Of multicelled organisms, the programmed emergence of specialized, morphologically different body parts.

diaphragm (DIE-uh-fram) [Gk. *diaphragma*, to partition] Muscular partition between the thoracic and abdominal cavities; its contraction and relaxation contribute to breathing. Also, a contraceptive device used temporarily to prevent sperm from entering the uterus during sexual intercourse.

dicot (DIE-kot) [Gk. *di*, two, + *kotylēdōn*, cup-shaped vessel] A dicotyledon. In general, a flowering plant characterized by seeds having embryos with two cotyledons (seed leaves); net-veined leaves; and floral parts arranged in fours, fives, or multiples of these.

diffusion Net movement of like molecules (or ions) down their concentration gradient. In the absence of other forces, the energy inherent in molecules makes them move constantly and collide at random. Their collisions are most frequent where they are most crowded together; thus they show a net outward movement from regions of higher to lower concentration.

digestive system An internal sac or tube from which ingested food is absorbed into the internal environment.

dihybrid cross An experimental cross in which true-breeding F_1 offspring inherit two gene pairs, each consisting of two nonidentical alleles.

diploidy (DIP-loyd-ee) The presence of two of each type of chromosome (that is, pairs of homologous chromosomes) in the interphase nucleus of somatic cells and germ cells. *Compare* haploidy.

directional selection A mode of natural selection by which the range of variation for some trait shifts in a consistent direction in response to directional change in the environment or to new environmental conditions.

disaccharide (die-SAK-uh-ride) [Gk. *di*, two, + *sakcharon*, sugar] A simple carbohydrate; one of the oligosaccharides consisting of two covalently bonded sugar monomers.

disease Outcome of an infection when the body's defenses cannot be mobilized fast enough; the pathogen's activities interfere with normal body functions.

disruptive selection A mode of natural selection by which forms of a trait at both ends of a range of variation are favored and intermediate forms are selected against.

distal tubule The tubular portion of a nephron farthest from the glomerulus; a region of water and sodium reabsorption.

distance effect Only species adapted for long-distance dispersal are potential colonists of islands far from their home range.

diversity, organismic Sum total of all the variations in form, function, and behavior that have accumulated in different lineages. Variations in traits generally are adaptive to prevailing conditions or were adaptive to conditions that existed in the past.

DNA Deoxyribonucleic acid (dee-OX-ee-RYE-bow-new-CLAY-ik). For all cells and many viruses, a nucleic acid that is the molecule of inheritance. It consists of two nucleotide strands twisted together helically and held together by numerous hydrogen bonds. The nucleotide sequence encodes instructions for synthesizing proteins and, ultimately, new individuals of a particular species.

DNA amplification Any of several methods by which a DNA library is copied again and again to yield multiple, identical copies of DNA fragments (cloned DNA).

DNA-DNA hybridization *See* nucleic acid hybridization.

DNA fingerprint A unique array of RFLPs, inherited in a Mendelian pattern from each parent, that gives each individual a unique identity.

DNA library A collection of DNA fragments produced by restriction enzymes and later incorporated into plasmids.

DNA ligase (LYE-gaze) An enzyme that seals together the new base-pairings during DNA replication; also used by technologists to seal base-pairings between DNA fragments and cut plasmid DNA.

DNA polymerase (poe-LIM-uh-raze) An enzyme that assembles a new strand on a parent DNA strand during replication; also takes part in DNA repair.

DNA probe A short DNA sequence that is synthesized from radioactively labeled nucleotides. Part of the probe is designed to base-pair with part of a gene under study.

DNA repair Following an alteration in the base sequence of a DNA strand, a process that may restore the original sequence, as carried out by DNA polymerases, DNA ligases, and other enzymes.

DNA replication Of cells, the process by which hereditary material is duplicated for distribution to daughter nuclei. Occurs prior to mitosis and meiosis in eukaryotic cells and during prokaryotic fission in bacterial cells.

dominance hierarchy A social organization in which some members of the group have adopted a subordinate status to others.

dominant allele In a diploid cell, an allele that masks the expression of its partner on the homologous chromosome.

dormancy [L. *dormire*, to sleep] A hormone-mediated time of inactivity during which metabolic activities idle. Perennials, seeds, many spores, cysts, and some animals go through dormancy.

double fertilization Of flowering plants only, the fusion of one sperm nucleus with the egg nucleus (to produce a zygote), *and* the fusion of a second sperm nucleus with nuclei of the endosperm mother cell, which gives rise to a nutritive tissue (endosperm).

doubling time The length of time it takes for a population to double in size.

drug addiction Chemical dependence on a drug following habituation and tolerance of it; in time the drug assumes an "essential" biochemical role in the body.

dry shrubland A biome that typically forms where annual rainfall is less than 25 to 60 centimeters; short, multibranched woody shrubs (e.g., chaparral) predominate.

dry woodland A biome that typically forms where annual rainfall is about 40 to 100 centimeters; there may be tall trees, but these do not form a dense canopy.

duplication A repeat of the same linear stretch of an individual's DNA in the same chromosome or in a different one.

ecdysone A hormone with major influence over the development of many insects.

echinoderm A type of invertebrate that has calcified spines, needles, or plates on the body wall. Although radially symmetrical, it has some bilateral features. Sea stars and sea urchins are examples.

ecology [Gk. *oikos*, home, + *logos*, reason] Study of the interactions of organisms with one another and with their physical and chemical environment.

ecosystem [Gk. *oikos*, home] An array of organisms and their physical environment, all of which are interacting through a flow of energy and a cycling of materials.

ecosystem modeling An analytical method, based on computer programs and models, of predicting unforeseen effects of specific disturbances to an ecosystem.

ectoderm [Gk. *ecto*, outside, + *derma*, skin] The first-formed, outermost primary tissue layer of animal embryos; forerunner of cell lineages that give rise to nervous system tissues and the integument's outer layer.

effector Of homeostatic systems, a muscle (or gland) that responds to signals from an integrator, such as the brain, by producing movement (or chemical change) that helps adjust the body to changing conditions.

effector cell A differentiated cell of one of the subpopulations of lymphocytes that form during an immune response; it acts at once to engage and destroy the antigen-bearing agent that triggered the response.

egg A type of mature female gamete; also called an ovum.

El Niño A recurring, massive eastward displacement of warm surface waters of the western equatorial Pacific, which displaces cooler waters off the South American coast. Causes global disruptions in climate.

electromagnetic spectrum The entire range of wavelengths, from the forms of radiant energy less than 10^{-5} nanometer long to radio waves more than 10 kilometers long.

electron A negatively charged unit of matter, with both particulate and wavelike properties, that occupies one of the orbitals around the atomic nucleus. Atoms can gain, lose, or share electrons with other atoms.

electron transfer The donation of one or more electrons stripped from one molecule to another molecule.

electron transport phosphorylation (FOSS-for-ih-LAY-shun) Final stage of aerobic respiration, when electrons from reaction intermediates flow through a membrane transport system that gives them up to oxygen. The flow sets up electrochemical gradients that drive ATP formation at other sites in the membrane.

electron transport system Organized array of enzymes and cofactors, bound in a cell membrane, that accept and donate electrons in series. When it operates, hydrogen ions flow across the membrane, and the flow drives ATP formation and other reactions.

element A substance that cannot be broken down to substances with different properties.

embryo (EM-bree-oh) [Gk. *en*, in, + probably *bryein*, to swell] Of animals generally, a stage formed by cleavage, gastrulation, and other early developmental events. Of seed plants, the young sporophyte, from the first cell divisions after fertilization until germination.

embryonic induction A change in the developmental fate of an embryonic tissue, as brought about by exposure to a gene product released from an adjacent tissue.

embryo sac Common name of the female gametophyte of flowering plants.

emerging pathogen A deadly pathogen, either a newly mutated strain of an existing species or one that evolved long ago and is only now taking great advantage of the increased presence of human hosts.

emulsification Of the chyme in the small intestine, a suspension of droplets of fat coated with bile salts.

end product A substance present at the end of a metabolic pathway.

endangered species A species at the brink of extinction owing to the extremely small size and severely limited genetic diversity of its remaining populations.

endergonic reaction (en-dur-GONE-ik) A chemical reaction having a net gain in energy.

endocrine gland A ductless gland that secretes hormones, which usually enter interstitial fluid and then the bloodstream.

endocrine system System of cells, tissues, and organs, functionally linked to the nervous system, that exerts control by way of its hormones and other chemical secretions.

endocytosis (EN-doe-sigh-TOE-sis) Transport of a substance into a cell by a vesicle, the membrane of which is a patch of plasma membrane that forms around the substance and sinks into the cytoplasm. Phagocytes also engulf prey or pathogens this way.

endoderm [Gk. *endon*, within, + *derma*, skin] The innermost primary tissue layer of animal embryos; gives rise to the inner lining of the gut and organs derived from it.

endodermis A sheetlike wrapping of single cells around the vascular cylinder of a root that functions in controlling the uptake of water and dissolved nutrients.

endometrium (EN-doh-MEET-ree-um) [Gk. *metrios*, of the womb] Innermost lining of the uterus, consisting of connective tissues, glands, and blood vessels.

endoplasmic reticulum or **ER** (EN-doe-PLAZ-mik reh-TIK-yoo-lum) An organelle that begins at the nucleus and curves through the cytoplasm. In rough ER (with many ribosomes on its cytoplasmic side), many new polypeptide chains acquire specialized side chains. Smooth ER (with no attached ribosomes) is a site of lipid synthesis.

endoskeleton [Gk. *endon*, within, + *skleros*, hard, stiff] An internal framework of bone, cartilage, or both in chordates. Together with skeletal muscle, supports and protects other body parts, helps maintain posture, and moves the body.

endosperm (EN-doe-sperm) Nutritive tissue that surrounds a flowering plant embryo and becomes food for the young seedling.

endospore A resting structure that forms around a copy of the chromosome and part of the cytoplasm of certain bacteria.

endosymbiosis In general, a mutually beneficial interdependence between two species, one of which resides permanently inside the other's body.

energy A capacity to do work.

energy carrier A molecule that delivers energy from one metabolic reaction site to another. ATP is the most common energy carrier in all cells.

energy flow pyramid A pyramid-shaped representation of an ecosystem's trophic structure, illustrating the energy losses at each transfer to a different trophic level.

enhancer A base sequence in DNA that is a binding site for an activator protein.

entropy (EN-trow-pee) A measure of the degree of disorder in a system (how much energy has become so disorganized and dispersed, usually as heat, that it is no longer readily available to do work). Any organized system tends toward entropy without energy inputs to make up for the flow of energy out of it.

enzyme (EN-zime) One of a class of proteins that enormously speed (catalyze) reactions between specific substances, usually at their functional groups.

eosinophil Fast-acting white blood cell; its enzyme secretions digest holes in parasitic worms during an inflammatory response.

epidermis The outermost tissue layer of a multicelled plant and of nearly all animals.

epinephrine (ep-ih-NEF-rin) Hormone of the adrenal medulla; raises blood levels of sugar and fatty acids; increases heart rate and the force of contraction.

epiglottis A flaplike structure at the start of the larynx, the position of which directs the movement of air into the trachea or of food into the esophagus.

epistasis (eh-PISS-tah-sis) An interaction between gene pairs. Two alleles of one gene mask expression of another gene's alleles, so expected phenotypes may not appear.

epithelium (EP-ih-THEE-lee-um) An animal tissue of one or more layers of adhering cells that covers the body's external surfaces and lines its internal cavities and tubes. It has one free surface; the opposite surface rests on a basement membrane between it and an underlying connective tissue. Epidermis is an example.

equilibrium, dynamic [Gk. *aequus*, equal, + *libra*, balance] The point at which a chemical reaction runs forward as fast as in reverse; the concentrations of reactant molecules and product molecules show no net change.

erosion The movement of land under the force of wind, running water, and ice.

erythrocyte (eh-RITH-row-site) [Gk. *erythros*, red, + *kytos*, vessel] Red blood cell.

esophagus (ee-SOF-uh-gus) Tubular portion of a digestive system that receives ingested food and leads to the stomach.

essential amino acid An amino acid that an organism cannot synthesize for itself and must obtain from a food source.

essential fatty acid A fatty acid that an organism cannot synthesize for itself and must obtain from food source.

estrogen (ESS-trow-jen) A sex hormone that helps oocytes mature, induces changes in the uterine lining during the menstrual cycle and pregnancy, and helps maintain secondary sexual traits; also influences bodily growth and development.

estrus (ESS-truss) [Gk. *oistrus*, frenzy] For mammals generally, the cyclic period of a female's sexual receptivity to the male.

estuary (EST-you-ehr-ee) A partly enclosed coastal region where seawater mixes with freshwater and runoff from the surrounding land, as by streams and rivers.

ethylene (ETH-il-een) Plant hormone that stimulates fruit ripening and abscission.

Eubacteria Kingdom of the most common species of bacterial cells.

eukaryotic cell (yoo-CARRY-oh-tic) [Gk. *eu*, good, + *karyon*, kernel] A cell having a "true nucleus" and other distinguishing membrane-bound organelles. *Compare* prokaryotic cell.

eutrophication Nutrient enrichment of a body of water, such as a lake, that typically results in reduced transparency and a phytoplankton-dominated community.

evaporation [L. *e-*, out, + *vapor*, steam] Heat energy converts a substance from the liquid to the gaseous state.

evolution, biological [L. *evolutio*, unrolling] Genetic change in a line of descent over time; brought about by microevolutionary processes (gene mutation, natural selection, genetic drift, and gene flow).

evolutionary systematics The branch of biology that applies evolutionary theory to the task of identifying patterns of diversity over time and in the environment.

evolutionary tree A treelike diagram in which the branches represent separate lines of descent from a common ancestor and branch points represent divergences.

excitatory postsynaptic potential (or EPSP) One of two competing signals at an input zone of a neuron; a graded potential that brings the neuron's plasma membrane closer to threshold.

excretion Any of several processes by which excess water, excess or harmful solutes, or waste materials leave the body by way of a urinary system or certain glands.

exergonic reaction (EX-ur-GONE-ik) A chemical reaction that shows a net loss in energy.

exocrine gland (EK-suh-krin) [Gk. *es*, out of, + *krinein*, to separate] Glandular structure that secretes products, usually through ducts or tubes, to a free epithelial surface.

exocytosis (EK-so-sigh-TOE-sis) Transport of a substance out of a cell by means of a vesicle, the membrane of which fuses with the plasma membrane, so that the vesicle's contents are released outside.

exodermis Layer of cells just inside the root epidermis of most flowering plants; helps control the uptake of water and solutes.

exon Any of the nucleotide sequences of a pre-mRNA molecule that become spliced together to form a mature mRNA transcript and ultimately get translated into protein.

exoskeleton [Gk. *exo*, out, + *skleros*, hard, stiff] An external skeleton, as in arthropods.

experiment A test of potentially falsifiable hypotheses about some aspect of nature. Its premise is that any aspect of the natural world has one or more underlying causes.

exponential growth (EX-po-NEN-shul) A pattern of population growth in which the population size expands by ever increasing increments during successive time intervals because the reproductive base becomes ever larger. The plot of population size against time has a characteristic J-shaped curve.

extinction, background A steady rate of species turnover that characterizes lineages through most of their histories.

extinction, mass An abrupt increase in the rate at which major taxa disappear, with several taxa being affected simultaneously.

extracellular fluid In animals generally, all the fluid not inside cells; includes plasma (the liquid portion of blood) and interstitial fluid (occupying the spaces between cells and tissues).

extracellular matrix A matrix that helps impart shape to many animal tissues; its ground substance contains fibrous proteins and other materials (mostly cell secretions).

FAD Flavin adenine dinucleotide, one of the nucleotide coenzymes that transfers electrons and unbound protons (H^+) from one reaction site to another. At such times it is abbreviated $FADH_2$.

fall overturn The vertical mixing of a body of water in autumn. Its upper layer cools, increases in density, and sinks; dissolved oxygen moves down and nutrients from bottom sediments move up.

family pedigree A chart of the genetic relationship of the individuals in a family through successive generations.

fat A lipid with a glycerol head and one, two, or three fatty acid tails. Tryglycerides (neutral fats) have three. Unsaturated tails have single covalent bonds in their carbon backbone; saturated tails also have one or more double bonds.

fate map A map of the surface of an animal embryo that shows the origin of each kind of differentiated cell in the adult.

fatty acid A molecule with a backbone of up to thirty-six carbon atoms, a carboxyl group (—COOH) at one end, and hydrogen atoms at most or all of the remaining bonding sites.

feedback inhibition Of cells or multicelled organisms, a control mechanism by which the output of a substance changes a specific condition or activity, which then triggers a decrease in further output of the substance or further activity.

fermentation [L. *fermentum*, yeast] A type of anaerobic pathway of ATP formation. It starts with glycolysis, ends with a transfer of electrons back to one of the breakdown products or intermediates, and regenerates NAD^+ required for the reaction. Its has a net yield of two ATP per glucose molecule.

fertilization [L. *fertilis*, to carry, to bear] The fusion of a sperm nucleus with the nucleus of an egg, which thus becomes a zygote.

fever A body temperature higher than a set point in the hypothalamic region that acts as the body's thermostat.

fibrous root system Of most monocots, all the lateral branchings of adventitious roots, which arose earlier from the young stem.

filter feeder An animal that filters food from a current of water that is directed through a body part, such as a sea squirt's pharynx.

filtration Of vertebrates, a process by which blood pressure forces water and solutes from capillaries into interstitial fluid. Filtration occurs in Bowman's capsule of a nephron.

fin Of fishes generally, an appendage that helps propel, stabilize, and guide the body through water.

first law of thermodynamics [Gk. *therme*, heat, + *dynamikos*, powerful] A law of nature stating the total amount of energy in the universe remains constant. Energy cannot be created from nothing and existing energy cannot be destroyed.

fish An aquatic animal of the most ancient, diverse vertebrate lineage, which includes jawless, cartilaginous, and bony fishes.

fitness An increase in adaptation to the environment, as brought about by genetic change.

fixation Of the individuals of a population, only one kind of allele remains at a specified locus; all individuals are homozygous for it.

fixed action pattern Of animals, a program of coordinated, stereotyped muscle activity that runs to completion independently of feedback from the environment.

flagellum (fluh-JELL-um), plural **flagella** [L. whip] Tail-like motile structure of many free-living eukaryotic cells; its core has a 9 + 2 array of microtubules.

flower A reproductive structure that distinguishes angiosperms from other seed plants and often attracts pollinators.

fluid mosaic model All cell membranes consist of a lipid bilayer and proteins. The lipids (phospholipids, mainly) impart basic structure, impermeability to water-soluble molecules, and (through packing variations and movements) fluidity. Diverse proteins spanning the bilayer or attached to one of its surfaces perform most of the membrane functions, including transport, enzyme activity, and reception of molecular signals or substances.

follicle (FOLL-ih-kul) A small sac, pit, or cavity, as around a hair; also a mammalian oocyte with its surrounding layer of cells.

food chain A straight-line sequence of who eats whom in an ecosystem.

food web A network of cross-connecting, interlinked food chains with some number of producers and consumers, as well as decomposers, detritivores, or both.

forebrain Most complex part of a vertebrate brain; it includes the cerebrum (and cerebral cortex), olfactory lobes, and hypothalamus.

forest A biome where tall trees grow together closely enough to form a fairly continuous canopy over a broad region.

fossil Recognizable, physical evidence of an organism that lived in the distant past.

fossil fuel Coal, petroleum, or natural gas; a nonrenewable source of energy formed in sediments by the compression of plant remains over hundreds of millions of years.

fossilization How fossils form. First an organism or traces of it becomes buried in sediments or volcanic ash. Water infiltrates the remains, which become infused with dissolved inorganic compounds. Sediments accumulate and exert pressure above the burial site. Over great spans of time, the pressure and chemical changes transform the remains to stony hardness.

founder effect A form of bottlenecking. By chance, allele frequencies of founders of a new population may not be the same as those of the original population. If there is no gene flow between the two, then natural selection will influence gene frequencies in drastically different ways through its interaction with genetic drift.

free radical A highly reactive, unbound molecular fragment that has the wrong number of electrons.

fruit [L. after *frui*, to enjoy] Of flowering plants, the expanded and ripened ovary of one or more carpels, some with accessory floral structures incorporated.

FSH Follicle-stimulating hormone; one of the hormones produced and secreted by the anterior lobe of the pituitary gland; serves reproductive roles in both sexes.

functional group An atom or group of atoms that is covalently bonded to the carbon backbone of an organic compound and that influences its chemical behavior.

functional-group transfer Donation of a functional group by one molecule to another.

Fungi The kingdom of fungi which, as a group, are major decomposers.

fungus A eukaryotic heterotroph that uses extracellular digestion and absorption; it secretes enzymes that break down an external food source into molecules small enough to be absorbed by its cells. Saprobes feed on nonliving organic matter, parasites feed on living organisms.

gall bladder Organ that stores bile secreted from the liver and that is connected by way of a duct to the small intestine.

gamete (GAM-eet) [Gk. *gametēs*, husband, and *gametē*, wife] A haploid cell, formed by meiosis and cytoplasmic division of a germ cell; required for sexual reproduction. Eggs and sperm are examples.

gamete formation Of animals, the first stage of development, in which sperm or eggs form and mature within reproductive tissues or organs of parents.

gametophyte (gam-EET-oh-fite) [Gk. *phyton*, plant] A haploid, multicelled, gamete-producing body that forms during the life cycle of most plants.

ganglion (GANG-lee-un), plural **ganglia** [Gk. *ganglion*, a swelling] A distinct clustering of cell bodies of neurons in regions other than the brain or spinal cord.

gastrulation (gas-tru-LAY-shun) The fourth stage of animal development; a time of major cellular reorganization when newly formed cells become arranged into two or three primary tissues, or germ layers.

gene [short for German *pangan*, after Gk. *pan*, all + *genes*, to be born] A unit of information about a heritable trait that is passed on from parents to offspring. Each gene has a specific location (locus) on a chromosome.

gene flow A microevolutionary process; the movement of alleles into and out of populations as a result of immigration and emigration.

gene frequency More precisely, allele frequency; the relative abundances of all the different alleles at a given gene locus that are carried by all individuals of a population.

gene locus The particular location of a gene along the length of a chromosome.

gene mutation [L. *mutatus*, a change] A heritable change in a DNA molecule by the deletion, addition, or substitution of one to several bases in its nucleotide sequence.

gene pair Of diploid cells, the two alleles at a particular gene locus (that is, on a pair of homologous chromosomes).

gene pool Sum total of all genotypes in a population.

gene therapy Generally, the transfer of one or more normal genes into an organism to correct or lessen the adverse effects of a genetic disorder.

genetic code [After L. *genesis*, to be born] The correspondence between the nucleotide triplets in DNA (then in RNA) and the specific sequences of amino acids in the resulting polypeptide chains; the basic language of protein synthesis in cells.

genetic disease An illness in which the expression of one or more genes increased the susceptibility of an individual to an infection or weakened its immune response.

genetic disorder An inherited condition that results in mild to severe medical problems.

genetic divergence Build-up of differences in gene pools of two or more populations of a species after a geographic barrier arises and separates them, because gene mutation, natural selection, and genetic drift are free to operate independently in each one.

genetic drift A random change in allele frequencies over the generations brought about by chance alone. The magnitude of its effect on genetic diversity and on the range of phenotypes relates to population size.

genetic engineering Altering the information content of DNA through use of recombinant DNA technology.

genetic equilibrium A state in which a population is not evolving. It occurs only if there is no mutation, if the population is large and isolated from other populations of the species, and if there is no natural selection (all members survive and reproduce equally by random mating).

genetic recombination A nonparental combination of some number of alleles in offspring; an outcome of gene mutation, crossing over, changes in chromosome structure or number, or recombinant DNA technology.

genome All the DNA in a haploid number of chromosomes of a given species.

genotype (JEEN-oh-type) Genetic constitution of an individual; a single gene pair or the sum total of an individual's genes. *Compare* phenotype.

genus, plural **genera** (JEEN-US, JEN-er-ah) [L. *genus*, race, origin] A taxon into which all species with phenotypic similarities and evolutionary relationship are grouped.

geographic dispersal Directional movement in which some residents of an established community leave their home range and take up residence elsewhere; they are considered to be exotic species in the new location.

geologic time scale A time scale for Earth history, the subdivisions of which are based on boundaries marked by episodes of mass extinction and which have been refined by radiometric dating.

germ cell Of animals, a cell of a lineage set aside for sexual reproduction; germ cells give rise to gametes. *Compare* somatic cell.

germ layer Of animal embryos, one of two or three primary tissue layers that form at gastrulation as forerunners of adult tissues. *Compare* ectoderm; endoderm; mesoderm.

germination (jur-min-AY-shun) Of resting spores and seeds, a resumption of activity following a period of arrested development.

gibberellin (JIB-er-ELL-un) A type of plant hormone that promotes stem elongation.

gill A respiratory organ, typically with a moist, thin vascularized layer of epidermis that functions in gas exchange.

gland A secretory cell or structure derived from epithelium and often connected to it.

glomerular capillaries Set of blood capillaries inside Bowman's capsule of the nephron.

glomerulus (glow-MARE-you-luss) [L. *glomus*, ball] First portion of the nephron, where water and solutes are filtered from blood.

glucagon (GLUE-kuh-gone) A hormone that stimulates cells to convert glycogen and amino acids to glucose; secreted by alpha cells of the pancreas when the blood level of glucose decreases.

glyceride (GLISS-er-eyed) One of the fats or oils; a molecule with one, two, or three fatty acid tails attached to a glycerol backbone.

glycerol (GLISS-er-oh) [Gk. *glykys*, sweet, + L. *oleum*, oil] A three-carbon molecule with three hydroxyl groups attached; one of the components of fats and oils.

glycogen (GLY-kuh-jen) A highly branched polysaccharide that is the main storage carbohydrate of animals; cleavage reactions at its many branchings yield an abundance of glucose monomers when required.

glycocalyx A sticky mesh of polysaccharides, polypeptides, or both around the cell wall of many bacteria.

glycolysis (gly-CALL-ih-sis) [Gk. *glykys*, sweet, + *lysis*, loosening or breaking apart] Initial energy-releasing reactions of aerobic and anaerobic pathways by which enzymes break down glucose (or some other organic compound) to pyruvate. It proceeds in the cytoplasm of all cells, it has a net yield of two ATP, and oxygen has no role in it.

glycoprotein A protein having linear or branched oligosaccharides covalently bonded to it. Nearly all surface proteins of animal cells and many proteins circulating in blood are glycoproteins.

gnetophyte Only gymnosperm known to have vessels in its xylem.

Golgi body (GOHL-gee) Organelle of lipid assembly, polypeptide chain modification, and packaging of both in vesicles for export or for transport to locations in the cytoplasm.

gonad (GO-nad) Primary reproductive organ in which animal gametes are produced.

graded potential Of neurons, a local signal that slightly alters the voltage difference across a patch of plasma membrane and that varies in magnitude according to the stimulus. With prolonged or intense stimulation, such signals may spread to a trigger zone of the membrane and initiate an action potential.

granum, plural **grana** In many chloroplasts, any of the stacks of flattened, membranous compartments with chlorophyll and other light-trapping pigments and reaction sites for ATP formation.

grassland A biome with flat or rolling land, 25-100 centimeters of annual rainfall, warm summers, grazing animals, and periodic fires that regenerate dominant species.

gravitropism (GRAV-ih-TROPE-izm) [L. *gravis*, heavy, + Gk. *trepein*, to turn] Tendency of a plant to grow directionally in response to the Earth's gravitational force.

gray matter Inside the brain and spinal cord, the unmyelinated axons, dendrites, and nerve cell bodies and neuroglial cells.

grazing food web A network of food chains in which energy flows from plants to an array of herbivores, then carnivores.

green alga An aquatic protistan with chlorophylls *a* and *b*; early members of its lineage may have given rise to plants.

green revolution In developing countries, the use of improved crop varieties, modern agricultural practices (including massive inputs of fertilizers and pesticides), and equipment to increase crop yields.

greenhouse effect Warming of the lower atmosphere as a result of the presence of increasing levels of greenhouse gases (such as carbon dioxide and methane).

ground meristem (MARE-ih-stem) [Gk. *meristos*, divisible] A primary meristem that produces the ground tissue system, hence the bulk of the plant body.

ground substance Of certain animal tissues, intercellular material made of cell secretions and other noncellular components.

ground tissue system Tissues making up the bulk of a plant body, the most common being parenchyma.

growth Of multicelled organisms, increases in the number, size, and volume of cells.

guanine A nitrogen-containing base; present in one of the four nucleotide building blocks of DNA and RNA.

guard cell Either of two adjacent cells with roles in the movement of carbon dioxide, oxygen, and water vapor across leaf or stem epidermis. An opening (stoma) forms when both swell with water and move apart; it closes when they lose water and collapse against one another.

gut A body region where food is digested and absorbed; of complete digestive systems, the gastrointestinal tract (the portions from the stomach onward).

gymnosperm (JIM-noe-sperm) [Gk. *gymnos*, naked, + *sperma*, seed] A vascular plant that bears seeds at exposed surfaces of reproductive structures, such as cone scales.

habitat [L. *habitare*, to live in] The type of place where an organism normally lives, as characterized by physical and chemical features and by its array of species.

hair cell A mechanoreceptor that may give rise to action potentials when bent or tilted.

half-life The time it takes for half of a given quantity of any radioisotope to decay into a different, less unstable daughter isotope.

halophile A type of archaebacterium that lives in extremely saline habitats.

haploidy (HAP-loyd) The presence of half the parental number of chromosomes in a gamete, as brought about by meiosis; the gamete has one of each pair of homologous chromosomes. *Compare* diploidy.

Hardy-Weinberg rule Allele frequencies will stay the same through the generations if there is no mutation, if the population is infinitely large and is isolated from other populations of the same species, if mating is random, and if all individuals survive and reproduce equally.

HCG Human chorionic gonadotropin. A hormone that helps maintain the lining of the uterus during the menstrual cycle and during the first trimester of pregnancy.

heart Muscular pump that keeps blood circulating through the animal body.

helper T cell A T lymphocyte which, when activated, produces and secretes chemicals that induce responsive T or B lymphocytes to divide and give rise to large populations of effector cells and memory cells.

hemoglobin (HEEM-oh-glow-bin) [Gk. *haima*, blood, + L. *globus*, ball] Iron-containing, oxygen-transporting protein that gives red blood cells their color.

hemostasis (HEE-mow-STAY-sis) [Gk. *haima*, blood, + *stasis*, standing] Stopping of blood loss from a damaged blood vessel through coagulation, blood vessel spasm, platelet plug formation, and other mechanisms.

herbivore [L. *herba*, grass, + *vovare*, to devour] Plant-eating animal.

hermaphrodite An individual with both male and female gonads; two individuals sexually reproduce by the mutual transfer of sperm.

heterocyst (HET-er-oh-sist) A self-modified cyanobacterial cell that makes a nitrogen-fixing enzyme when nitrogen is scarce.

heterotroph (HET-er-oh-trofe) [Gk. *heteros*, other, + *trophos*, feeder] Organism unable to synthesize its own organic compounds; it feeds on autotrophs, other heterotrophs, or organic wastes. *Compare* autotroph.

heterozygous condition (HET-er-oh-ZYE-guss) [Gk. *zygoun*, join together] Of a specified trait, having a pair of nonidentical alleles at a gene locus (on a pair of homologous chromosomes).

higher taxon (plural, **taxa**) One of the ever more inclusive groupings meant to reflect relationships among species. Family, order, class, phylum, and kingdom are examples.

hindbrain One of three divisions of the vertebrate brain; the medulla oblongata, cerebellum, and pons. Has reflex centers for respiration, blood circulation, and other basic functions; also coordinates motor responses and many complex reflexes.

histone Any of a class of proteins that are intimately associated with eukaryotic DNA and largely responsible for the organization of eukaryotic chromosomes.

homeostasis (HOE-me-oh-STAY-sis) [Gk. *homo*, same, + *stasis*, standing] Of animals, a physiological state in which physical and chemical aspects of the internal environment (blood and interstitial fluid) are maintained within ranges suitable for cell activities.

homeotic gene One of a class of master genes that specify the development of specific body part in animals.

hominid [L. *homo*, man] All species on or near the evolutionary road leading to modern humans.

hominoid Apes, humans, and their recent ancestors.

homologous chromosome (huh-MOLL-uh-gus) [Gk. *homologia*, correspondence] Of cells having a diploid chromosome number, one of a pair of chromosomes that are identical in size, shape, and gene sequence, and that interact during meiosis. Nonidentical sex chromosomes in a cell also interact during meiosis and are considered homologues also.

homology Similarity in one or more body parts in different species that is attributable to their descent from a common ancestor.

homologous structures The same body parts, modified in different ways, in different lines of descent from a common ancestor.

homozygous condition (HOE-moe-ZYE-guss) For a specified trait, having a pair of identical alleles at a gene locus (on a pair of homologous chromosomes).

homozygous dominant condition Having a pair of dominant alleles at a gene locus (on a pair of homologous chromosomes).

homozygous recessive condition Having a pair of recessive alleles at a gene locus (on a pair of homologous chromosomes).

hormone [Gk. *hormon*, to stir up, set in motion] Signaling molecule that stimulates or inhibits gene transcription in nonadjacent target cells (any cell having receptors for it).

horsetail A seedless vascular plant with photosynthetic stems that look like horsetails.

human genome project Worldwide basic research project to sequence the estimated 3 billion nucleotides present in the DNA of human chromosomes.

humus Decomposing organic matter in soil.

hydrogen bond A weak interaction between a small, highly electronegative atom of a molecule and a neighboring hydrogen atom that is taking part in a polar covalent bond.

hydrogen ion A free (unbound) proton; a hydrogen atom that has lost its electron and so bears a positive charge (H^+).

hydrologic cycle A biogeochemical cycle, driven by solar energy, in which water moves slowly through the atmosphere, on or through surface layers of land masses, to the ocean, and back again.

hydrolysis (high-DRAWL-ih-sis) [L. *hydro*, water, + Gk. *lysis*, loosening or breaking apart] Cleavage reaction in which covalent bonds break, splitting a molecule into two or more parts. Often H^+ and OH^- (derived from a water molecule) become attached to the exposed bonding sites.

hydrophilic substance [Gk. *philos*, loving] A polar substance that is attracted to the polar water molecule and dissolves easily in water. Sugars are examples.

hydrophobic substance [Gk. *phobos*, dreading] A nonpolar substance that is repelled by the polar water molecule and thus resists being dissolved in water. Oil is an example.

hydrosphere All liquid or frozen water on or near the Earth's surface.

hydrostatic pressure Any volume of fluid that exerts a force directed against a wall, membrane, or another structure enclosing that fluid. The greater its concentration of solutes, the greater will be the pressure that the fluid exerts.

hydrothermal vent ecosystem Ecosystem, near a fissure in the ocean floor, based on chemosynthetic bacteria that use dissolved minerals as their energy source.

hypha (HIGH-fuh), plural **hyphae** [Gk. *hyphe*, web] Of fungi, a filament with chitin-reinforced walls and, often, reinforcing cross-walls; component of the mycelium.

hypodermis A subcutaneous layer having stored fat that helps insulate the body; although not part of skin, it anchors skin and allows it some freedom of movement.

hypothalamus [Gk. *hypo*, under, + *thalamos*, inner chamber or possibly *tholos*, rotunda] Of the forebrain, a major center for homeostatic control of visceral activities (such as salt-water balance, temperature control, and reproduction), related forms of behavior (as in hunger, thirst, and sex), and emotional expression, such as sweating with fear.

hypothesis In science, a possible explanation of a specific phenomenon in nature, one that has the potential to be proved false by experimental tests.

immune response Events by which B and T lymphocytes recognize antigen, undergo cell divisions that form huge populations of lymphocytes, which differentiate into subpopulations of effector and memory cells. The effector cells destroy cells bearing antigen-MHC complexes. Memory cells are not activated until subsequent encounters with the same antigen.

immunization Various processes, including vaccination, that promote increased immunity against specific diseases.

immunoglobulin (Ig) One of five classes of antibodies, each with antigen-binding sites and other sites with specialized functions.

implantation Process by which a blastocyst adheres to the endometrium and establishes connections by which the mother and embryo will exchange substances during pregnancy.

imprinting A time-dependent form of learning that is triggered by exposure to sign stimuli and that usually occurs during a sensitive period when the animal is young.

inbreeding Nonrandom mating among close relatives, which have many identical alleles in common. Inbreeding is a form of genetic drift in a small group of relatives that are preferentially interbreeding.

incomplete dominance One allele of a pair is not fully dominant over its partner, so a heterozygous phenotype in between the two homozygous phenotypes emerges.

independent assortment, theory of By the end of meiosis in a germ cell, each pair of homologous chromosomes—hence the genes that they carry—have been sorted for shipment into gametes independently of how the other pairs were sorted out.

indirect selection A theory in evolutionary biology that self-sacrificing individuals can pass on their genes indirectly by helping relatives survive and reproduce.

induced-fit model A substrate induces change in the shape of an enzyme's active site when bound to it, the result being a more precise molecular fit between the two that promotes reactivity.

infection Invasion and multiplication of a pathogen in host cells or tissues. Disease follows if defenses cannot be mobilized fast enough to prevent the pathogen's activities from interfering with normal functions.

inflammation, acute Important aspect of nonspecific defenses and immune responses when cells of a local tissue are damaged or killed, as by infection; requires action of fast-acting phagocytes and plasma proteins, including complement proteins.

inheritance The transmission, from parents to offspring, of structural and functional patterns that have a genetic basis and are characteristic of their species.

inhibiting hormone A signaling molecule produced and secreted by the hypothalamus that suppresses a particular secretion by the anterior lobe of the pituitary gland.

inhibitor A substance that can bind with an enzyme and interfere with its functioning.

inhibitory postsynaptic potential (IPSP) Of neurons, one of two competing types of graded potentials at an input zone; it tends to drive the resting membrane potential away from threshold.

instinctive behavior A behavior performed without having been learned by experience in the environment. The nervous system of a newly born or hatched animal is prewired to recognize one or two sign stimuli (simple, well-defined cues in the environment) that can trigger a suitable response.

insulin Hormone secreted by beta cells of the pancreas that lowers the glucose level in blood by stimulating cells to take up glucose; also promotes protein and fat synthesis and inhibits protein conversion to glucose.

integration, neural [L. *integrare*, coordinate] Moment-by-moment summation of all the excitatory and inhibitory synapses acting on the neuron; it takes place at each level of synapsing in a nervous system.

integrator Of homeostatic systems, a control point such as a brain where information is pulled together in the selection of responses to stimuli.

integument Of animals, a protective body cover such as skin. Of seed-bearing plants, one or more layers around an ovule that harden, thicken, and form a seed coat.

integumentary exchange (in-teg-you-MEN-tuh-ree) A mode of respiration in which oxygen and carbon dioxide diffuse across a thin, moist, vascularized layer at the body surface of certain animals.

interleukin One of several chemical mediator molecules secreted by helper T cells that fan mitotic cell divisions and differentiation of responsive T and B cells.

intermediate A compound formed between the start and end of a metabolic pathway.

intermediate filament In different types of animal cells, a cytoskeletal element made of particular proteins.

interneuron Any neuron of the vertebrate brain and spinal cord.

internode In vascular plants, the stem region between two successive nodes.

interphase Of the cell cycle, an interval in between nuclear divisions when a cell increases in mass, roughly doubles the number of its cytoplasmic components, and then duplicates its chromosomes (replicates its DNA). The interval differs among species.

interspecific competition Individuals of different species compete with one another for a share of resources in their habitat.

interstitial fluid (IN-ter-STISH-ul) [L. *interstitus*, to stand in the middle of something] That portion of extracellular fluid occupying the spaces between cells and tissues of complex animals.

intertidal zone Generally, part of a rocky or sandy shoreline above the low water mark and below the high water mark; organisms that inhabit it are alternately submerged, then exposed, by tides.

intervertebral disk One of a number of disk-shaped structures containing cartilage that act as shock absorbers and flex points between bony segments of the vertebral column.

intraspecific competition All individuals of a population compete with one another for a share of resources in their habitat.

intron A noncoding portion of a newly formed mRNA molecule.

inversion A linear stretch of DNA within a chromosome that has become oriented in the reverse direction, with no molecular loss.

invertebrate Animal without a backbone.

in vitro fertilization Conception outside the body (literally, "in glass" petri dishes or test tubes).

ion, negatively charged (EYE-on) An atom or a compound that acquired an overall negative charge by gaining one or more electrons.

ion, positively charged An atom or a compound that acquired an overall positive charge by losing one or more electrons.

ionic bond Ions of opposite charge have attracted each other and are staying together.

isotonic condition Equality in the relative concentrations of solutes in two fluids; for two fluids separated by a cell membrane, there is no net osmotic (water) movement across the membrane.

isotope (EYE-so-tope) Of an element, an atom with more or fewer neutrons than the atoms having the most common number.

J-shaped curve The type of curve that emerges when population size is plotted against time; it represents unrestricted, exponential growth.

joint An area of contact or near-contact between bones.

juvenile Of many animals, a miniaturized form between the embryo and adult; it simply changes in size and proportion until it reaches sexual maturity.

karyotype (CARRY-oh-type) For an individual (or a species), a preparation of metaphase chromosomes sorted by length, centromere location, and other defining features.

keratin A tough, water-insoluble protein made by most epidermal cells that becomes concentrated in skin's outermost layers.

keratinization (care-AT-in-iz-AY-shun) Process by which keratin-producing epidermal cells die and accumulate as keratinized bags at the skin surface to form a barrier against dehydration, bacteria, many toxins, and ultraviolet radiation.

keystone species A species that dominates a community and dictates its structure.

key innovation A modified structure or function that allows a lineage to exploit the environment in more efficient or novel ways.

kidney One of a pair of vertebrate organs that filter mineral ions, organic wastes, and other substances from the blood; it controls the amounts returned to blood and thereby helps maintain the volume and solute levels of extracellular fluid.

kilocalorie 1,000 calories of heat energy; the amount of energy required to raise the temperature of 1 kilogram of water by 1°C; unit of measure for the caloric value of foods.

kinase One of a class of enzymes that catalyze phosphate-group transfers.

kinetochore Group of proteins and DNA at a chromosome's centromere where spindle microtubules become attached at mitosis or meiosis. Each chromatid of a duplicated chromosome has its own kinetochore.

Krebs cycle A cyclic pathway that occurs in mitochondria; together with a few preparatory steps, the stage of aerobic respiration in which pyruvate is completely broken down to carbon dioxide and water. Coenzymes accept unbound protons (H^+) and electrons stripped from intermediates and deliver them to the next stage.

lactate fermentation Anaerobic pathway of ATP formation; pyruvate from glycolysis is converted to the three-carbon compound lactate, and NAD^+ is regenerated. The net energy yield is two ATP.

lactation Milk production by hormone-primed mammary glands.

lake A body of standing freshwater with littoral, limnetic, and profundal zones.

large intestine Colon; a gut region that receives unabsorbed food residues from the small intestine and concentrates and stores feces until they are expelled from the body.

larva, plural **larvae** Of many animals, an immature developmental stage between the embryo and adult.

larynx (LARE-inks) A tubular airway to and from the lungs. In humans, it contains vocal cords.

lateral meristem A type of meristem in plants that show secondary growth; either vascular cambium or cork cambium.

lateral root Of taproot systems, a lateral branching from the first, primary root.

leaching The removal of some nutrients in soil as water percolates through it.

leaf For most vascular plants, a structure having chlorophyll-containing tissue that is the major region of photosynthesis.

learned behavior Variation or change in responses to stimuli after an animal has processed and integrated information gained from specific experiences.

lek Of some birds and other animals, a type of communal display ground occupied during times of courtship.

lethal mutation A mutation with drastic effects on phenotype that usually cause the individual's death.

LH Luteinizing hormone. Hormone secreted by the anterior lobe of the pituitary gland, with roles in male and female reproduction.

lichen (LY-kun) A symbiotic interaction between a fungus and a photoautotroph, such as a green alga.

life cycle A recurring pattern of genetically programmed events by which individuals grow, develop, maintain themselves, and reproduce.

life table Tabulation of age-specific patterns of birth and death for a population.

ligament A strap of dense connective tissue that bridges a joint.

light-dependent reactions The first stage of photosynthesis, in which sunlight energy is trapped and converted to the chemical energy of ATP alone (by a cyclic pathway) or ATP and NADPH (by a noncyclic pathway).

light-independent reactions The second stage of photosynthesis, in which ATP makes phosphate-group transfers required to build sugar phosphates. Often NADPH delivers electrons and hydrogen atoms for the synthesis reactions, which also require carbon from carbon dioxide. The sugar phosphates enter other reactions by which starch, cellulose, and other end products of photosynthesis are assembled.

lignification Of land plants, a process by which lignin is deposited in secondary cell walls. Lignin imparts strength and rigidity by anchoring cellulose strands in the walls, stabilizes and protects other components of the wall, and forms a waterproof barrier around the cellulose. A key factor in the evolution of vascular plants.

lignin A complex organic compound that strengthens and waterproofs cell walls in certain tissues of vascular plants.

limbic system Brain centers, located in the middle of the cerebral hemispheres, that interact to govern emotions and influence memory. Distantly related to the olfactory lobes; it still deals with the sense of smell.

limiting factor Any essential resource that, in short supply, limits population growth.

lineage (LIN-ee-age) A line of descent.

linkage group A quantitative measure of the relative positions of the genes along the length of a chromosome.

linkage mapping An investigative approach by which the relative linear positions of a chromosome's genes are deduced by tracking the percentage of recombinants in gametes.

lipid Mainly a hydrocarbon, greasy or oily, that strongly resists dissolving in water but quickly dissolves in nonpolar substances. In nearly all cells, certain lipids are the main reservoirs of stored energy. Other lipids are structural materials (as in membranes) and cell products (such as surface coatings).

lipid bilayer Structural basis of all cell membranes, with two layers of mostly phospholipid molecules. The hydrophobic tails of the molecules are sandwiched in between the hydrophilic heads, which are dissolved in the fluid that bathes them.

lipoprotein A molecule that forms when proteins in blood combine with cholesterol, triglycerides, and phospholipids that were absorbed from the gut, after a meal.

liver A large gland in vertebrates and many invertebrates that stores and interconverts organic compounds and helps maintain their blood concentrations. It additionally inactivates most hormone molecules that have served their functions as well as ammonia and other compounds that can be toxic at high concentrations.

local signaling molecules Secretions from cells of many local tissues that quickly alter chemical conditions in the vicinity, then are degraded before they can travel elsewhere.

logistic population growth (low-JISS-tik) A pattern of population growth in which a low-density population slowly increases in size, then enters a rapid growth phase that levels off once the carrying capacity has been reached.

loop of Henle The hairpin-shaped, tubular region of a nephron that functions in the reabsorption of water and solutes.

lung An internal respiratory surface in the shape of a cavity or sac that can be filled with air; a pair occur in a few fishes and in amphibians, birds, reptiles, and mammals.

lycophyte A seedless vascular plant of mostly wet or shaded habitats; requires free water to complete its life cycle.

lymph (LIMF) [L. *lympha*, water] Tissue fluid that has drained into the vessels of the lymphatic system.

lymph capillary A small-diameter vessel of the lymph vascular system that has no pronounced entrance; tissue fluid moves inward by passing between overlapping endothelial cells at the vessel's tip.

lymph node A lymphoid organ that is a battleground for immune responses; packed, organized arrays of lymphocytes inside cleanse lymph before it reaches the blood.

lymph vascular system [L. *lympha*, water, + *vasculum*, a small vessel] Of the lymphatic system, vessels that take up and transport excess tissue fluid and reclaimable solutes as well as fats absorbed from the digestive tract.

lymphatic system An organ system that supplements the circulatory system. Its vessels deliver fluid and solutes from interstitial fluid to the bloodstream; its lymphoid organs have roles in immunity.

lymphocyte Any of various T and B cells with roles in immune responses that follow recognition of antigen.

lysis [Gk. *lysis*, a loosening] Gross damage to a plasma membrane, cell wall, or both that allows the cytoplasm to leak out and thereby leads to cell death.

lysogenic pathway A common pathway in which a viral replication cycle enters a latent period and viral genes are integrated with a host chromosome. The recombinant molecule is a miniature time bomb that is passed on to all of the host cell's progeny.

lysosome (LYE-so-sohm) The main organelle of intracellular digestion, with enzymes that can break down nearly all polysaccharides, proteins, and nucleic acids, and some lipids.

lysozyme An infection-fighting enzyme in mucous membranes.

lytic pathway A common pathway in which a viral replication cycle proceeds rapidly and ends with lysis of the host cell.

macroevolution The large-scale patterns, trends, and rates of change among families and other more inclusive groups of species.

macrophage A leukocyte (white blood cell) that phagocytizes worn-out cells and tissue debris and anything bearing antigen. It may become an antigen-presenting cell and thus trigger immune responses.

mammal The only vertebrate whose females nourish offspring by milk, produced by and secreted from mammary glands.

mass extinction A sudden rise in rates of extinction above the background level; a catastrophic, global event during which a number of major taxa are wiped out simultaneously.

mass number The total number of protons and neutrons in an atom's nucleus.

mast cell In connective tissues, a basophil-like cell that releases histamine during an inflammatory response.

mechanoreceptor Sensory cell or nearby cell that detects mechanical energy (changes in pressure, position, or acceleration).

medulla oblongata Part of the hindbrain; it has reflex centers for vital tasks, such as respiration and circulation. It coordinates motor responses with complex reflexes, such as coughing. It influences brain regions concerned with sleep and arousal.

medusa (meh-DOO-sah) [Gk. *Medousa*, one of three sisters in Greek mythology having snake-entwined hair] Free-swimming, bell-shaped stage in cnidarian life cycles; oral arms and tentacles extend from the bell.

megaspore A haploid spore that forms in the ovary of seed-bearing plants; one of its cellular descendants develops into an egg.

meiosis (my-OH-sis) [Gk. *meioun*, to diminish] Two-stage nuclear division process that reduces the chromosome number of a germ cell by half, to the haploid number; each daughter nucleus receives one of each type of chromosome. It is the basis of gamete formation and (in plants) spore formation. *Compare* mitosis.

membrane excitability A special membrane property of any cell that can generate action potentials in response to stimulation.

memory The capacity of the brain to store and retrieve information about past sensory experience.

memory cell One of the subpopulations of B or T cells that form during an immune response to antigen but that do not act at once; it enters a resting phase, from which it is released during a secondary immune response.

menopause (MEN-uh-pozz) [L. *mensis*, month, + *pausa*, stop] End of the period of a human female's reproductive potential.

menstrual cycle Cyclic release of a secondary oocyte and priming of the endometrium to receive it, should it become fertilized; the complete cycle averages about twenty-eight days in female humans.

menstruation Cyclic sloughing of the blood-enriched endometrium (lining of the uterus) when pregnancy does not occur.

mesoderm (MEH-zoe-derm) [Gk. *mesos*, middle, + *derm*, skin] Of most animals, intermediate primary tissue layer that forms in the embryo and gives rise to muscle, most of the skeleton; circulatory, reproductive, and excretory organs; and connective tissue layers of the gut and integument.

mesophyll (MEH-zoe-fill) Of vascular plants, a tissue of photosynthetic parenchyma cells with an abundance of air spaces.

messenger RNA (mRNA) A single strand of ribonucleotides transcribed from DNA and translated into a polypeptide chain; the only RNA with protein-building instructions.

metabolic pathway (MEH-tuh-BALL-ik) One of many orderly sequences of enzyme-mediated reactions by which cells maintain, increase, or decrease the concentrations of substances. Different pathways are linear or circular, and one typically interconnects with others.

metabolism (meh-TAB-oh-lizm) [Gk. *meta*, change] All of the controlled, enzyme-mediated chemical reactions by which cells acquire and use energy to synthesize, store, degrade, and eliminate substances in ways that contribute to growth, survival, and reproduction.

metamorphosis (me-tuh-MOR-foe-sis) [Gk. *meta*, change, + *morphe*, form] Major change in body form during the transition from an embryo to the adult owing to hormonally controlled increases in size, reorganization of tissues, and remodeling of body parts.

metaphase Of meiosis I, the stage when all pairs of homologous chromosomes have become positioned midway between the spindle poles. Of mitosis or meiosis II, a stage when each duplicated chromosome has become positioned midway between the spindle poles.

metazoan A multicelled animal.

methanogen A type of archaebacterium that lives in oxygen-free habitats and produces methane gas as a metabolic by-product.

MHC marker Any of a variety of proteins that are self-markers. Some occur on all body cells of an individual; others are unique to the macrophages and lymphocytes.

micelle (my-SELL) Of fat digestion, a small droplet that is assembled from bile salts, fatty acids, and monoglycerides and that assists in fat absorption from the small intestine.

microevolution Change in allele frequencies resulting from mutation, genetic drift, gene flow, and natural selection.

microfilament [Gk. *mikros*, small, + L. *filum*, thread] Thin cytoskeletal element of two twisted polypeptide chains; it has roles in cell movement, especially at the cell surface, and in producing and maintaining cell shapes.

microorganism An organism, usually single-celled, that is far too small to be observed without the aid of a microscope.

microspore Of gymnosperms and flowering plants, a walled haploid spore that develops into a pollen grain.

microtubular spindle Of eukaryotic cells, a temporary bipolar structure composed of organized arrays of microtubules that forms during nuclear division and that moves the chromosomes to prescribed destinations.

microtubule (my-crow-TUBE-yool) A hollow, cylindrical cytoskeletal element, consisting mainly of tubulin, with roles in cell shape, motion, and growth and in the structure of cilia and flagella.

microtubule organizing center (MTOC) In the cytoplasm of eukaryotic cells, a small mass of proteins and other substances. The number, type, and location of MTOCs dictate the particular organization and orientation of microtubules inside the cell.

microvillus (MY-crow-VILL-us) [L. *villus*, shaggy hair] Of many cells, one of a number of very slender cylindrical extensions of the plasma membrane that serve in absorption or secretion.

midbrain Part of the vertebrate brain with coordination centers for reflex responses to visual and auditory input; also swiftly relays sensory signals to forebrain.

migration A recurring pattern of movement between two or more locations in response to environmental rhythms, such as circadian rhythms and seasonal changes in daylength. It requires activation or suppression of internal timing mechanisms that govern physiological and behavioral functions.

mimicry (MIM-ik-ree) In form, behavior, or both, a prey organism (the mimic) closely resembles a dangerous, unpalatable, or hard-to-catch species (its model).

mineral An element or inorganic compound formed by natural geologic processes and required for normal cell functioning.

mitochondrion (MY-toe-KON-dree-on), plural **mitochondria** Of eukaryotic cells, the organelle that specializes in formation of ATP; the second and third stages of aerobic respiration, an oxygen-requiring pathway, occur only in mitochondria.

mitosis (my-TOE-sis) [Gk. *mitos*, thread] Type of nuclear division that maintains the parental chromosome number for daughter cells; for eukaryotes, the basis of growth in size (and often of asexual reproduction).

mixture Two or more elements intermingled in proportions that can and usually do vary.

molar One of the cheek teeth; a tooth with a platform having cusps (surface bumps) that help crush, grind, and shear food.

molecular clock An accumulation of neutral mutations in a lineage that can be measured as a series of predictable "ticks" back through time; a way of calculating the time of origin of one lineage or species relative to others.

molecule A unit of matter in which chemical bonding holds together two or more atoms of the same or different elements.

mollusk An invertebrate with a tissue fold (mantle) draped around a soft, fleshy body; snails, clams, and squids are examples.

molting Shedding of hair, feathers, horns, epidermis, a shell, or cuticle in a process of increases in size or periodic renewal.

Monera In some traditional classification schemes, a kingdom that encompasses both archaebacteria and eubacteria.

monocot (MON-oh-kot) A monocotyledon; a flowering plant with seeds bearing one cotyledon, floral parts generally in threes (or multiples of three), and leaves typically parallel-veined. *Compare* dicot.

monohybrid cross [Gk. *monos*, alone] An experimental cross in which offspring inherit a pair of nonidentical alleles for a single trait being studied, so that they are heterozygous.

monomer Small molecule that is commonly a subunit of polymers, such as the sugar monomers of starch.

monosaccharide (MON-oh-SAK-ah-ride) [Gk. *monos*, alone, single, + *sakharon*, sugar] A simple carbohydrate, with only one sugar monomer. Glucose is an example.

monosomy A chromosome abnormality; the presence of a chromosome that has no homologue in a diploid cell.

morphogenesis (MORE-foe-JEN-ih-sis) [Gk. *morphe*, form, + *genesis*, origin] A program of orderly changes in an animal embryo's size, shape, and proportions, the outcome being specialized tissues and early organs. As part of the program, cells divide, grow, migrate, and change in size. Tissues expand and fold, and cells in some die in controlled ways at prescribed locations.

morphological convergence Lineages only remotely related evolved in response to similar environmental pressures, and they became similar in appearance, functions, or both. Analogous structures are evidence of this macroevolutionary pattern.

morphological divergence Of two or more lineages, change in appearance, functions, or both as they evolve from a shared ancestor. Homologous structures are evidence of this macroevolutionary pattern.

motor neuron A type of neuron that swiftly delivers signals from the brain or spinal cord to muscle cells, gland cells, or both.

multicelled organism An organism with differentiated cells arranged into tissues, organs, and often organ systems.

multiple allele system Three or more different molecular forms of the same gene (alleles) among individuals of a population.

muscle fatigue A decline in tension of a muscle kept in a state of tetanic contraction as a result of continuous, high-frequency stimulation.

muscle tension A mechanical force exerted by a contracting muscle; it resists opposing forces such as gravity and the weight of an object being lifted.

muscle tissue A tissue having cells able to contract under stimulation, then passively lengthen and return to the resting position.

mutagen (MEW-tuh-jen) Any environmental agent that can alter the molecular structure of DNA. Ultraviolet radiation and certain viruses are examples.

mutation [L. *mutatus*, a change, + *-ion*, result or a process or an act] A heritable change in the molecular structure of DNA. Mutations are the source of all alleles and, ultimately, of life's diversity. *See also* lethal mutation; neutral mutation.

mutation frequency The number of times a gene mutation has occurred in a population, as in 1 million gametes from which 500,000 individuals were produced.

mutation rate Of a gene, the probability of it undergoing spontaneous mutation during some specified interval, such as each DNA replication cycle.

mutualism [L. *mutuus*, reciprocal] A two-species, symbiotic ecological interaction that directly benefits both participants.

mycelium (my-SEE-lee-um), plural **mycelia** [Gk. *mykes*, fungus, mushroom, + *helos*, callus] A mesh of tiny, branching filaments (hyphae) that is the food-absorbing part of most fungi.

mycorrhiza (MY-coe-RIZE-uh) "Fungus-root"; a form of mutualism between the hyphae of a fungus and young roots of many plants. The fungus obtains carbohydrates from the plant and in turn releases dissolved mineral ions that plant roots can take up.

myelin sheath Of many sensory and motor neurons, an axonal sheath that enhances the propagation of action potentials; the plasma membranes of Schwann cells are wrapped repeatedly around the axon, and only a small node separates one from the other.

myofibril (MY-oh-FY-brill) One of the many threadlike structures inside a muscle cell; it is divided into many sarcomeres, the basic units of contraction.

myosin (MY-uh-sin) A motor protein with roles in contraction. In muscles, it interacts with actin to bring about contraction.

NAD⁺ Nicotinamide adenine dinucleotide; a nucleotide coenzyme. When carrying electrons and unbound protons (H^+) between reaction sites, it is abbreviated NADH.

NADP⁺ Nicotinamide adenine dinucleotide phosphate; a phosphorylated nucleotide coenzyme. When carrying electrons and unbound protons (H^+) between reaction sites, it is abbreviated $NADPH_2$.

natural killer cell A cytotoxic lymphocyte that reconnoiters for tumor cells and virus-infected cells, then touch-kills them.

natural selection A microevolutionary process; outcome of differences in survival and reproduction among individuals that show variation in heritable traits. Over the generations, it can lead to increased fitness (increased adaptation to the environment).

necrosis (neh-CROW-sis) Of multicelled organisms, the passive death of many cells that results from severe tissue damage.

negative feedback mechanism Homeostatic feedback mechanism in which an activity changes some condition in the internal environment; the altered condition triggers a response that reverses the change.

nematocyst (NEM-ad-uh-sist) [Gk. *nema*, thread, + *kystis*, pouch] A capsule formed only by cnidarians; it houses dischargeable, tubular threads that have prey-piercing barbs and that dispense toxins or sticky substances.

nephridium (neh-FRID-ee-um), plural **nephridia** Of earthworms and some other invertebrates, a system of regulating water and solute levels.

nephron (NEFF-ron) [Gk. *nephros*, kidney] Of the vertebrate kidney, a slender tubule into which water and solutes are filtered from blood, then selectively reabsorbed, and in which urine forms.

nerve Cordlike communication line of the peripheral nervous system, composed of axons of sensory neurons, motor neurons, or both bundled in connective tissue. In the brain and spinal cord, similar cord-like bundles are called nerve tracts.

nerve cord Of many animals, a cordlike communication line of axons of neurons.

nerve impulse *See* action potential.

nerve net A simple nervous system of a diffuse mesh of nerve cells that interact with contractile and often sensory cells in an epithelium.

nervous system An organ system in which nerve cells, such as neurons, are oriented relative to one another in signal-conducting and information-processing pathways. It detects and processes information about changes outside and inside the body and elicits responses from muscle and gland cells.

nervous tissue A type of connective tissue composed of neurons.

net energy Of energy resources available to the human population, the amount of energy that is left over after subtracting the energy used to locate, extract, transport, store, and deliver energy to consumers.

net population growth rate per individual (r) Of population growth equations, a single variable combining birth and death rates, which are assumed to remain constant.

neural tube Embryonic forerunner of the brain and spinal cord.

neuroglial cell (NUR-oh-GLEE-uhl) One of the cells that structurally and metabolically support neurons. Neuroglia represent about half the volume of the vertebrate nervous system.

neuromodulator A signaling molecule that can magnify or reduce the effects of a given neurotransmitter on neighboring or distant neurons.

neuromuscular junction A site of chemical synapsing between the axonal endings of a motor neuron and a muscle cell.

neuron (NUR-on) A nerve cell; the basic unit of communication in nervous systems. *See* motor neuron; interneuron; sensory neuron.

neurotransmitter Any of a class of signaling molecules that are secreted from neurons and act on immediately adjacent cells, then are rapidly degraded or recycled.

neutral mutation A mutation with little or no effect on phenotype; natural selection cannot increase or decrease its frequency in a population, for it cannot influence an individual's survival and reproduction. *Compare* molecular clock.

neutron A unit of matter, one or more of which occupies the atomic nucleus and has mass but no electric charge.

neutrophil The most abundant, fast-acting white blood cell; it phagocytizes bacteria during an inflammatory response.

niche (NITCH) [L. *nidas*, nest] The sum total of all activities and relationships in which individuals of a species engage as they secure and use the resources necessary to survive and reproduce.

nitrification (nye-trih-fih-KAY-shun) Of the nitrogen cycle, a process by which certain bacteria strip electrons from ammonia or ammonium in soil. The end product, nitrite (NO_2^-), is broken down to nitrate (NO_3^-) by other bacteria.

nitrogen cycle An atmospheric cycle; a biogeochemical cycle in which the largest nitrogen reservoir is the atmosphere.

nitrogen fixation Process by which a few kinds of bacteria convert gaseous nitrogen (N_2) to ammonia. This dissolves rapidly in their cytoplasm to form ammonium, which can be used in biosynthetic pathways.

nociceptor (NO-SEE-sep-tur) A pain receptor, such as a free nerve ending that detects tissue damage as by burns or distortions.

node A site where one or more leaves are attached to a plant stem.

noncyclic pathway of ATP formation (non-SIK-lik) [L. *non*, not, + Gk. *kylos*, circle] Photosynthetic pathway of ATP formation in which electrons from water molecules flow through two photosystems and two electron transport chains, and end up in NADPH.

nondisjunction The failure of two sister chromatids or of a pair of homologous chromosomes to separate at meiosis or mitosis, so that daughter cells end up with too many or too few chromosomes.

notochord (KNOW-toe-kord) Of chordates, a rod of stiffened tissue (not cartilage or bone) that serves as a supporting structure for the body.

nuclear envelope A double membrane (two lipid bilayers and associated proteins) that is the outermost portion of a cell nucleus.

nucleic acid (new-CLAY-ik) A single-stranded or double-stranded chain of four different kinds of nucleotides joined one after the other at their phosphate groups. They differ in which nucleotide base follows the next in sequence. DNA and RNA are examples.

nucleic acid hybridization Single strands of DNA or RNA from different species are recombined into double-stranded, hybrid molecules, which are then heated to break hydrogen bonds between them. The amount of heat required to separate the strands is a measure of biochemical similarity, which is greatest among closely related species and weakest among distantly related ones.

nucleoid (NEW-KLEE-oid) Of bacteria, the region of the cell in which the bacterial chromosome is physically organized (but not bounded by membrane, as with the nuclear envelope of eukaryotic cells).

nucleolus (new-KLEE-oh-lus) [L. *nucleolus*, a little kernel] In the nucleus of a nondividing cell, a site where the protein and RNA subunits of ribosomes are assembled.

nucleosome (NEW-KLEE-oh-sohm) One of the organizational units of the eukaryotic chromosome; a stretch of DNA looped twice around a "spool" of histone molecules.

nucleotide (NEW-KLEE-oh-tide) A small organic compound with a five-carbon sugar (deoxyribose), nitrogen-containing base, and phosphate group. Nucleotides are the structural units of adenosine phosphates, nucleotide coenzymes, and nucleic acids.

nucleotide coenzyme A protein that assists an enzyme by transporting electrons and hydrogen atoms released at one reaction site to another site.

nucleus (NEW-klee-us) [L. *nucleus*, a kernel] Of atoms, the central core of one or more positively charged protons and (in all but hydrogen) electrically neutral neutrons. In eukaryotic cells, a membranous organelle that physically isolates and organizes DNA out of the way of cytoplasmic machinery.

numerical taxonomy Practice by which traits of an unidentified organism are compared with those of a known group. The greater the total number of traits that it has in common with the known group, the closer is their inferred relatedness.

nutrient An element essential for growth and survival of a given organism; directly or indirectly, it has one or more roles in metabolism that no other element fulfills.

nutrition All of those processes by which food is selectively taken in, digested, absorbed, and later converted to the body's own organic compounds.

obesity An excess of fat in the body's adipose tissues, caused by imbalances between caloric intake and energy output.

oligosaccharide A carbohydrate consisting of a short chain of two or more covalently bonded sugar monomers. The type called a disaccharide has two sugar monomers. *Compare* monosaccharide; polysaccharide.

omnivore [L. *omnis*, all, + *vovare*, to devour] An animal able to obtain energy and carbon from more than one food source rather than being limited to one trophic level.

oncogene (ON-koe-jeen) Any gene having the potential to induce cancerous transformation.

oocyte An immature egg.

oogenesis (oo-oh-JEN-uh-sis) Formation of a female gamete, from a germ cell to a haploid ovum (mature egg).

operator A short base sequence intervening between a promoter and bacterial genes.

operon A promoter-operator sequence that services more than one bacterial gene.

orbital One of a number of volumes of space around the atomic nucleus in which electrons are likely to be at any instant.

organ A structure having definite form and function that consists of more than one tissue.

organ formation A stage of embryonic development in animals in which primary tissue layers split into subpopulations of cells, and different lines of cells become unique in structure and function; basis of growth and tissue specialization.

organ system Two or more organs that interact chemically, physically, or both in performing a common task.

organelle Of cells, an internal, membrane-bounded sac or compartment having one or more specific, specialized metabolic functions.

organic compound A molecule consisting of one or more elements covalently bonded to some number of carbon atoms.

osmoreceptor A sensory receptor that detects changes in water volume (hence solute concentration) in fluid that bathes it.

osmosis (oss-MOE-sis) [Gk. *osmos*, act of pushing] The diffusion of water in response to a water concentration gradient between two regions that are separated from each other by a selectively permeable membrane. The greater the number of molecules and ions dissolved in a solution, the lower its water concentration will be.

osmotic pressure At some point following the development of hydrostatic pressure in a cell, an amount of force that counters the inward diffusion of water and prevents any further increase in fluid volume.

ovary (oh-vuh-ree) A type of primary reproductive organ in which eggs form. In seed-bearing plants, the part of the carpel where eggs develop, fertilization takes place, and seeds mature; at maturity, it and sometimes other plant parts is a fruit.

oviduct (OH-vih-dukt) One of a pair of ducts through which eggs travel from an ovary to the uterus. Formerly called Fallopian tube.

ovulation (AHV-you-LAY-shun) The release of a secondary oocyte from an ovary during a menstrual cycle.

ovule (OHV-youl) [L. *ovum*, egg] Of seed-bearing plants, a female gametophyte with egg cell, surrounding tissue, and one or two protective layers (integuments). After fertilization, it matures into a seed.

ovum (OH-vum) Of vertebrates, the mature female gamete (egg).

oxidation-reduction reaction An electron transfer from one atom or molecule to another. Often hydrogen is also transferred along with an electron or electrons.

ozone thinning A pronounced seasonal thinning of the ozone layer in the lower stratosphere, especially above polar regions.

pancreas (PAN-cree-us) Gland that secretes enzymes and bicarbonate into the small intestine during digestion, and that secretes the hormones insulin and glucagon.

pancreatic islet Any of the 2 million clusters of endocrine cells in the pancreas, including alpha cells, beta cells, and delta cells.

parapatric speciation Neighbor populations become distinct species while maintaining contact along their common border.

parasite [Gk. *para*, alongside, + *sitos*, food] An organism that lives in or on other living organisms (hosts), feeds on specific tissues during part of its life cycle, and usually does not kill its host outright.

parasitism A two-species interaction in which one species lives in or on a host species and uses its tissues for nutrients. The parasite benefits; the host does not.

parasitoid An insect larva that grows and develops inside a host organism (usually another insect), eventually consuming the soft tissues and killing it.

parasympathetic nerve Of the autonomic nervous system, any of the nerves carrying signals that tend to slow down the body's activities and divert energy to basic tasks; such nerves work continually in opposition with sympathetic nerves to make small adjustments in internal organ activity.

parenchyma (par-ENG-kih-mah) A simple tissue that makes up the bulk of the plant body. Its cells take part in photosynthesis, storage, secretion, and other tasks.

parthenogenesis Development of an embryo from an unfertilized egg.

passive transport A transport protein that spans the lipid bilayer of a cell membrane passively allows a solute to diffuse through its interior, down its concentration gradient.

pathogen (PATH-oh-jen) [Gk. *pathos*, suffering, + *genes*, origin] A virus, bacterium, fungus, protozoan, parasitic worm, or some other infectious agent that can invade organisms, multiply in them, and cause disease.

pattern formation Of animals, a sculpting of nondescript clumps of embryonic cells into specialized tissues and organs according to an ordered, spatial pattern.

PCR Polymerase chain reaction; a method of enormously amplifying the quantity of DNA fragments cut by restriction enzymes.

peat bog An accumulation of the remains of peat mosses in an compressed, excessively moist, and highly acidic mat.

pedigree A chart of genetic connections among related individuals, constructed according to standardized methods.

pelagic province The full volume of ocean water; it is subdivided into a neritic zone (relatively shallow water over continental shelves) and an oceanic zone (water over ocean basins).

penis A male copulatory organ by which sperm can be deposited inside a female reproductive tract.

peptide hormone A protein hormone or some other water-soluble hormone unable to cross the lipid bilayer of a target cell. It enters by receptor-mediated endocytosis or activates cell surface receptors.

perennial [L. *per-*, throughout, + *annus*, year] A flowering plant that lives for more than two growing seasons.

pericycle (PARE-ih-sigh-kul) [Gk. *peri-*, around, + *kyklos*, circle] Of a root vascular cylinder, one or more layers just inside the endodermis that gives rise to lateral roots and contributes to secondary growth.

periderm Of plants that show extensive secondary growth, a protective covering that replaces epidermis.

peripheral nervous system (per-IF-ur-uhl) [Gk. *peripherein*, to carry around] The nerves leading into and out from a spinal cord and brain and the ganglia along those communication lines.

peristalsis (pare-ih-STAL-sis) Waves of contraction and relaxation of muscles in the wall of a tubular or saclike organ.

peritoneum (pare-ih-tah-NEE-um) A lining of the coelom that also covers and helps maintain the position of internal organs.

peritubular capillary One of the set of blood capillaries around the tubular parts of a nephron; it functions in reabsorption of water and solutes back into the body and in secretion of hydrogen ions and some other substances in the forming urine.

permafrost A permanently frozen, water-impenetrable layer beneath the soil surface in arctic tundra.

peroxisome Enzyme-filled vesicle in which fatty acids and amino acids are digested into hydrogen peroxide and converted to harmless products.

pest resurgence Insecticide resistance arises among a population of pests; the insecticide acts as an agent of directional selection.

PGA Phosphoglycerate (FOSS-foe-GLISS-er-ate) A key intermediate of glycolysis and of the Calvin-Benson cycle.

PGAL Phosphoglyceraldehyde. A key intermediate of glycolysis and of the Calvin-Benson cycle.

pH scale A measure of the concentration of free hydrogen ions in blood, water, and other solutions; pH 0 is the most acidic, 14 the most basic, and 7, neutral.

phagocyte (FAG-uh-sight) [Gk. *phagein*, to eat, + *kytos*, hollow vessel] A cell that obtains nutrients or destroys foreign cells and cellular debris by phagocytosis. The macrophages and amoeboid protozoans are examples.

phagocytosis (FAG-uh-sigh-TOE-sis) [Gk. *phagein*, to eat, + *kytos*, hollow vessel] The engulfment of foreign cells or substances by means of pseudopod formation and then endocytosis.

pharynx (FARE-inks) A muscularized tube by which food enters the gut. Also an organ of gas exchange in some chordates; in land vertebrates, a dual entrance for the tubular part of the digestive tract and windpipe.

phenotype (FEE-no-type) [Gk. *phainein*, to show, + *typos*, image] Observable trait or traits of an individual that arise from the interactions between genes, and between genes and the environment.

pheromone (FARE-oh-moan) [Gk. *phero*, to carry, + *-mone*, as in hormone] A nearly odorless, hormone-like, exocrine gland secretion that is a communication signal between individuals of the same species and that helps integrate social behavior.

phloem (FLOW-um) Of vascular plants, a tissue with living cells that interconnect and form the tubes through which sugars and other dissolved organic compounds are conducted.

phospholipid An organic compound with a glycerol backbone, two fatty acid tails, and a hydrophilic head (a phosphate group and another polar group). Phospholipids are the main structural materials of cell membrane.

phosphorus cycle Movement of phosphorus from rock or soil through organisms, then back to soil.

phosphorylation (FOSS-for-ih-LAY-shun) The attachment of unbound (inorganic) phosphate to a molecule; also the transfer of a phosphate group from one molecule to another, as when ATP phosphorylates glucose.

photoautotroph An organism able to synthesize all organic molecules it requires using carbon dioxide as the carbon source and sunlight as the energy source. All plants, some protistans, and a few bacteria are photoautotrophs.

photolysis (foe-TALL-ih-sis) [Gk. *photos*, light, + *-lysis*, breaking apart] A reaction sequence of the noncyclic pathway of photosynthesis, triggered by photon energy, in which water is split into oxygen, hydrogen, and electrons.

photoperiodism A biological response to a change in the relative length of daylight and darkness.

photoreceptor Light-sensitive sensory cell.

photosynthesis The trapping of energy from the sun and its conversion to chemical energy (ATP, NADPH, or both), followed by synthesis of sugar phosphates that become converted to sucrose, cellulose, starch, and other products. The main pathway by which energy and carbon enter the web of life.

photosystem One of the clusters of light-trapping pigments that are embedded in photosynthetic membranes and that donate electrons to transport systems that are involved in the formation of ATP, NADPH, or both. Examples are photosystems II and I of the thylakoid membrane of chloroplasts.

phototropism [Gk. *photos*, light, + *trope*, turning, direction] An adjustment in the direction and rate of plant growth in response to a source of light.

photovoltaic cell A device that can convert sunlight energy into electricity.

phycobilin (FIE-koe-BY-lins) One of the light-sensitive, accessory pigments that transfer energy they absorb to chlorophylls. The phycobilins are especially abundant in red algae and cyanobacteria.

phylogeny Evolutionary relationships among species, starting with most ancestral forms and including the branches leading to their descendants.

phytochrome Light-sensitive pigment, the activation and inactivation of which triggers plant hormone activities governing leaf expansion, stem branching, stem length and often seed germination and flowering.

phytoplankton (FIE-toe-PLANK-tun) [Gk. *phyton*, plant, + *planktos*, wandering] A community of floating or weakly swimming photoautotrophs in saltwater and freshwater habitats.

pigment A light-absorbing molecule.

pioneer species An opportunistic colonizer of barren or disturbed habitats, notable for its high dispersal rate and rapid growth.

pituitary gland An endocrine gland that interacts with the hypothalamus to control diverse physiological functions, including activity of many other endocrine glands. The posterior lobe stores and secretes hypothalamic hormones; the anterior lobe produces and secretes its own hormones.

placenta (play-SEN-tuh) A blood-engorged organ, consisting of endometrial tissue and extraembryonic membranes, that develops during pregnancy in the female placental mammal. It permits exchanges between the mother and her fetus without intermingling their bloodstreams, thus sustaining the new individual and allowing its blood vessels to develop apart from the mother's.

plankton [Gk. *planktos*, wandering] Any community of floating or weakly swimming organisms, mostly microscopic, living in freshwater and saltwater habitats. *See also* phytoplankton; zooplankton.

plant One of the eukaryotic organisms, nearly all of which are photoautotrophic, that have chlorophylls *a* and *b* and that generally have well-developed root and shoot systems.

Plantae The kingdom of plants.

plasma (PLAZ-muh) Liquid component of blood; water and dissolved ions, diverse proteins, sugars, gases, and other substances.

plasma membrane Of cells, the outermost membranous boundary between cytoplasm and external fluid bathing the cell. Its lipid bilayer is the basic structural part of the membrane; diverse proteins embedded in the bilayer or attached to its surfaces carry out most functions, such as transport and reception of extracellular signals.

plasmid Of many bacteria, a small, circular molecule of extra DNA that carries only a few genes and that replicates independently of the bacterial chromosome.

plasmodesma, plural **plasmodesmata** (PLAZ-moe-DEZ-muh) A plant cell junction; a membrane-lined channel across walls of adjacent cells that connects their cytoplasm.

plasmolysis An osmotically induced shrinkage of a cell's cytoplasm.

plate tectonics Theory that slabs, or plates, of the Earth's outer layer (lithosphere) float on a hot, plastic layer of the underlying mantle. All plates are in motion and they raft the continents to new positions over great spans of geologic time.

platelet (PLAYT-let) Any of the fragments of megakaryocytes that release substances necessary for clot formation.

pleiotropy (PLEE-oh-troe-pee) [Gk. *pleon*, more, + *trope*, direction] As a result of expression of alleles at a single gene locus, positive or negative effects on two or more traits. The effects may or may not be manifest at the same time in the individual.

polar body One of four cells that form during the meiotic cell division of an oocyte but that does not become the ovum.

pollen grain [L. *pollen*, fine dust] Depending on the species, the immature or mature, sperm-bearing male gametophyte of gymnosperms and angiosperms.

pollen sac In anthers of flowers, any of the chambers in which pollen grains develop.

pollen tube A tube formed after a pollen grain germinates; grows through carpel tissues, and carries sperm to the ovule.

pollination Of flowering plants, the arrival of a pollen grain on the landing platform (stigma) of a carpel.

pollutant Any substance with which an ecosystem has had no prior evolutionary experience, in terms of kinds or amounts, and that can accumulate to disruptive or harmful levels. Can be naturally occurring or synthetic.

polymer (POH-lih-mur) [Gk. *polus*, many, + *meris*, part] A large molecule with three to millions of subunits.

polymerase (puh-LIM-ur-aze) An enzyme that catalyzes a polymerization reaction. Examples are DNA polymerase (of DNA replication and repair), RNA polymerase (of gene transcription), and the enzyme that catalyzes cellulose synthesis.

polypeptide chain Organic compound with nitrogen atoms arrayed in a regular pattern in its backbone (-N-C-C-N-C-C-). A protein consists of one or more of these chains.

polymorphism (poly-MORE-fizz-um) [Gk. *polus*, many, + *morphe*, form] Of populations, the persistence of two or more distinctive forms of a trait (morphs).

polyp (POH-lip) Vase-shaped, sedentary stage of cnidarian life cycles.

polypeptide chain Three or more amino acids joined by peptide bonds.

polyploidy (POL-ee-PLOYD-ee) Having three or more of each type of chromosome in the interphase nucleus, compared to a diploid chromosome number.

polysaccharide [Gk. *polus*, many, + *sakcharon*, sugar] A straight or branched chain of many covalently linked sugar units, of the same or different kinds. The most common are cellulose, starch, and glycogen.

polysome A number of ribosomes translating the same mRNA molecule one after the other, at the same time, during protein synthesis.

pons Part of the hindbrain; a traffic center for signals passing between the cerebellum and integrating centers of the forebrain.

population Individuals of the same species occupying the same area.

population density A count of individuals of a population occupying a specified area or volume of a habitat.

population distribution The general pattern of dispersion of individuals of a population throughout their habitat.

population size The number of individuals that make up the gene pool of a population.

positive feedback mechanism Homeostatic mechanism by which a chain of events is set in motion that intensifies a change from an original condition; after a limited time, the intensification reverses the change.

predation An ecological interaction in which a predatory species eats (and so directly harms) a prey species.

predator [L. *prehendere*, to grasp, seize] An organism that feeds on other living organisms (its prey), that does not live in or on them, and that may or may not kill them.

prediction A statement about what you can expect to observe in nature if a theory or hypothesis is not false.

pressure flow theory Of vascular plants, a theory that organic compounds move through phloem because of gradients in solute concentrations and pressure between sources (such as photosynthetically active leaves) and sinks (such as growing parts).

primary growth Plant growth originating at root tips and shoot tips and resulting in increases in length.

primary immune response Defensive acts by white blood cells and their secretions, as elicited by a first-time encounter with an antigen; includes both antibody-mediated and cell-mediated activities.

primary productivity, gross Of ecosystems, the rate at which producers capture and store a given amount of energy in their cells and tissues during a specified interval.

primary productivity, net Of ecosystems, the rate of energy storage in the cells and tissues of producers in excess of their rate of aerobic respiration.

primate A mammalian lineage that includes the prosimians, tarsioids, and anthropoids (monkeys, apes, and humans).

primer A short nucleotide sequence that will base-pair with any complementary DNA sequence; later, DNA polymerases recognize them as start tags for replication.

prion A small protein linked with eight rare, fatal degenerative diseases of the nervous system. Prions are altered products of a gene that is present in normal as well as infected individuals.

probability The chance that each outcome of a given event will occur is proportional to the number of ways it can be reached.

procambium (pro-KAM-bee-um) Of vascular plants, a primary meristem that gives rise to primary vascular tissues.

producer Of ecosystems, an autotrophic (self-feeding) organism; it nourishes itself using sources of energy and carbon from its physical environment. Photoautorophs and chemoautotrophs are examples.

progesterone (pro-JESS-tuh-rown) A type of sex hormone secreted by the pair of ovaries in female mammals.

prokaryotic cell (pro-CARRY-oh-tic) [L. *pro*, before, + Gk. *karyon*, kernel] A bacterium; a single-celled organism, usually walled, that does not have the profusion of membrane-bound organelles seen in eukaryotic cells.

prokaryotic fission Division mechanism by which a bacterial cell reproduces.

promoter A short base sequence in DNA that signals the start of a gene; the site where RNA polymerase binds to start transcription.

prophase Of mitosis, the stage when each duplicated chromosome starts to condense, microtubules form a spindle apparatus, and the nuclear envelope starts to break up.

prophase I Of meiosis, the stage when the microtubular spindle starts to form, the nuclear envelope starts to break up, and each duplicated chromosome condenses and pairs with its homologous partner. Nonsister chromatids typically undergo crossing over and genetic recombination.

prophase II Brief first stage of meiosis II when chromosomes start moving toward spindle equator; each has already been separated from its homologue but still consists of two chromatids.

protein Organic compound of one or more chains of amino acids, which peptide bonds join in a linear sequence that is the basis of the protein's three-dimensional structure and chemical behavior.

Protista The kingdom of protistans.

protistan (pro-TISS-tun) [Gk. *prōtistos*, primal, very first] One of the staggeringly diverse eukaryotes, single-celled species of which may resemble the first eukaryotic cells. At present categorized in part by what they are *not* (not bacteria, fungi, plants, or animals).

proto-oncogene A gene sequence similar to an oncogene but that codes for a protein required for normal cell function. When a mutation alters its structure, it may trigger cancerous transformation.

proton Positively charged particle, one or more of which resides in the nucleus of each type of atom; when unbound, a proton is the same as a hydrogen ion (H^+).

protostome (PRO-toe-stome) [Gk. *proto*, first, + *stoma*, mouth] One of the lineage of coelomate, bilateral animals in which the first indentation to form in the embryo develops into a mouth. Includes mollusks, annelids, and anthropods.

protozoan A group of protistans, some predatory, others parasitic, that in certain respects resemble single-celled heterotrophs that presumably gave rise to animals. The amoeboid and animal-like protozoans, as well as sporozoans, are examples.

proximal tubule Of a nephron, the tubular portion into which water and solutes enter after being filtered from the blood at the preceding portion (Bowman's capsule).

pulmonary circuit Blood circulation route leading to and from the lungs.

Punnett-square method A way to predict the probable outcome of a genetic cross in simple diagrammatic form.

purine A type of nucleotide base with a double ring structure. Examples are adenine and guanine.

pyrimidine (phi-RIM-ih-deen) A type of nucleotide bases with a single ring structure. Examples are cytosine and thymine.

pyruvate (PIE-roo-vate) A small organic compound with a backbone of three carbon atoms. Glycolysis produces two molecules of pyruvate as end products.

r A variable used in population growth equations to signify net population growth rate; the assumption is that birth and death rates remain constant and therefore may be combined into this one variable.

radial symmetry Of animals, a body plan having four or more roughly equivalent parts arranged around a central axis.

radiometric dating A method of measuring the proportions of (1) a radioisotope in a mineral trapped long ago in a newly formed rock and (2) a daughter isotope that formed from it by radioactive decay in the same rock sample. Used to assign absolute dates to fossil-containing rocks and to a geologic time scale.

radioisotope An unstable atom, with an uneven number of protons and neutrons. It spontaneously emits particles and energy, and over a predictable time span becomes transformed (decays) into a different atom.

rain shadow A reduction in rainfall on the leeward side of high mountains, resulting in arid or semiarid conditions.

reabsorption In kidneys, diffusion or active transport of water and reclaimable solutes from a nephron into peritubular capillaries under the control of ADH and aldosterone. At a capillary bed, osmotic movement of some interstitial fluid into a capillary when there is a difference in the concentration of water between plasma and interstitial fluid.

rearrangement, molecular A conversion of one type of organic compound to another through a juggling of its internal bonds.

receptor, molecular Of cells, a membrane protein that binds a hormone or some other extracellular substance that triggers change in cell activities.

receptor, sensory Of nervous systems, a sensory cell or specialized cell adjacent to it that can detect a particular stimulus.

recessive allele [L. *recedere*, to recede] In heterozygotes, an allele whose expression is fully or partially masked by expression of its partner; only in the homozygous recessive condition is it fully expressed.

recognition protein Of mammalian cell membranes, a glycolipid or glycoprotein that functions as a molecular fingerprint; it identifies a cell as being of a specific type. Self-proteins of the body's own cells are an example.

recombinant technology Procedures by which DNA (genes) from different species may be isolated, cut, spliced together, and the new recombinant molecules multiplied in quantity, as by PCR or by a population of rapidly dividing bacterial cells.

red alga A protistan, most of which are multicelled, aquatic, and photosynthetic, with an abundance of phycobilins.

blood cell An erythrocyte; an oxygen-transporting cell in blood.

red marrow A site of blood cell formation in the spongy tissue of many bones.

reflex [L. *reflectere*, to bend back] A simple, stereotyped movement elicited directly by sensory stimulation.

reflex pathway [L. *reflectere*, to bend back] The simplest route by which stimulation of nerve cells leads to contractile or glandular action. In complex animals, sensory neurons directly stimulate or inhibit motor neurons, without intervention by interneurons.

refractory period Of neurons, the period following an action potential at a given patch of membrane when sodium gates are shut and potassium gates are open, so that the patch is insensitive to stimulation.

regulatory protein A protein involved in a positive or negative control system that operates during transcription or translation or after translation. The protein interacts with DNA, RNA, or a gene product.

releasing hormone A signaling molecule produced by the hypothalamus that enhances or slows down secretions from target cells in the anterior lobe of the pituitary gland.

repressor Regulatory protein that can block gene transcription.

reproduction Of nature, any of a number of processes by which a new generation of cells or multicelled individuals is produced. For eukaryotes, it may be sexual or it may be asexual (as by binary fission, budding, or vegetative propagation). For bacteria only, it occurs by prokaryotic fission.

reproductive isolating mechanism Any heritable feature of body form, functioning, or behavior that prevents interbreeding between one or more genetically divergent populations.

reproductive success The survival and production of offspring by an individual.

reptile A type of carnivorous vertebrate; its ancestors were the first vertebrates to escape dependency on standing water, largely by means of internal fertilization and amniote eggs. Examples are dinosaurs (extinct), crocodiles, lizards, and snakes.

resource partitioning Coexistence among competing species; they share the same resource differently or at different times.

respiration [L. *respirare*, to breathe] Of animals, the exchange of oxygen from the environment for carbon dioxide wastes from cells by way of integumentary exchange, a respiratory system, or both. *Compare* aerobic respiration.

respiratory surface An epithelium or some other surface tissue that functions in gas exchange in animals; it is thin enough to allow oxygen to diffuse into the body and carbon dioxide to diffuse out of it.

resting membrane potential Of a neurons and other excitable cells, a steady voltage difference across the plasma membrane in the absence of outside stimulation.

restoration ecology Attempts to reestablish biodiversity in abandoned farmlands and other large areas altered by agriculture, mining, and other severe disturbances.

restriction enzyme Of bacteria, one of a class of enzymes that can cut apart foreign DNA injected into the cell body, as by viruses; also used in recombinant DNA technology.

reticular formation A mesh of interneurons extending from the upper spinal cord, through the brain stem, and into the cerebral cortex; a low-level pathway through the vertebrate nervous system.

reverse transcriptase A viral enzyme that uses viral RNA as a template to synthesize either DNA or mRNA in a host cell.

reverse transcription A process by which reverse transcriptase assembles DNA on an RNA template. Used by RNA viruses as well as by technologists who use mature mRNA transcripts as the templates for synthesizing cDNA.

RFLP Short for restriction fragment length polymorphism. Such fragments are formed by using restriction enzymes to cut DNA; for each individual, they show slight but unique differences in their banding pattern.

Rh blood typing A laboratory method of characterizing red blood cells on the basis of a protein that serves as a self-marker at their surface; Rh^+ signifies its presence, and Rh^- its absence.

rhizoid A rootlike absorptive structure of some fungi and nonvascular plants.

ribosomal RNA (rRNA) A type of RNA molecule that combines with proteins to form ribosomes, on which the polypeptide chains of proteins are assembled.

ribosome In all cells, the structure at which amino acids are strung together in specified sequence to form the polypeptide chains of proteins. An intact ribosome consists of two subunits, each composed of ribosomal RNA and protein molecules.

RNA Ribonucleic acid. A general category of single-stranded nucleic acids that function in processes by which genetic instructions encoded in DNA are used to build proteins.

rod cell A vertebrate photoreceptor that is sensitive to very dim light and contributes to coarse perception of movement.

root hair Of vascular plants, an extension of a specialized root epidermal cell; root hairs collectively enhance the surface area available for absorbing water and solutes.

root nodule A localized swelling on roots of certain legumes and other plants that contain symbiotic, nitrogen-fixing bacteria.

root A part of a plant that typically grows belowground, absorbs water and dissolved nutrients, helps anchor aboveground parts, and often stores food.

RuBP Ribulose bisphosphate. A compound with a backbone of five carbon atoms; used for carbon fixation by C3 plants and also regenerated in the Calvin-Benson cycle of photosynthesis.

S-shaped curve A curve, obtained when population size is plotted against time, that is characteristic of logistic growth.

salination A salt buildup in soil as a result of evaporation, poor drainage, and heavy irrigation.

salivary gland A gland that secretes saliva, a fluid that initially mixes with food in the mouth and starts the breakdown of starch.

salt A compound that releases ions other than H^+ and OH^- in solution.

saltatory conduction Of myelinated neurons, a rapid form of action potential propagation by a hopping of excitation from one node to another between the jellyrolled membranes of cells making up the myelin sheath.

sampling error A rule of probability: The fewer times that a chance event occurs, the greater will be the variance from the expected outcome of that occurrence. Of experimental groups, large sampling tends to lessen the chance that differences among individuals will distort the results.

saprobe A heterotroph that obtains energy and carbon from nonliving organic matter and so causes its decay. Saprobic fungi are examples.

sarcomere (SAR-koe-meer) Of muscles, the basic unit of contraction; a region of myosin and actin filaments organized in parallel arrays between two Z lines of one of the many myofibrils inside a muscle cell.

sarcoplasmic reticulum (sar-koe-PLAZ-mik reh-TIK-you-lum) A membrane system around a muscle cell's myofibrils that takes up, stores, and releases the calcium ions required for cross-bridge formation in sarcomeres, hence for contraction.

Schwann cell One of a series of neuroglial cells wrapped like jellyrolls around an axon in a nerve and forming a myelin sheath.

sclerenchyma (skler-ENG-kih-mah) A simple plant tissue in which most cells have thick, lignin-impregnated walls. Sclerenchyma supports mature plant parts and commonly protects seeds.

sea-floor spreading Molten rock erupts from great, continuous ridges on the ocean floor, flows laterally in both directions, and hardens to form new crust. As the seafloor spreads out, it forces older crust down into great trenches elsewhere in the seafloor.

second law of thermodynamics A law of nature stating that the spontaneous direction of energy flow is from organized to less organized forms. Overall, the total amount of energy in the universe is spontaneously flowing from forms of higher to lower quality; with each conversion, some energy gets randomly dispersed in a form (usually heat) not as readily available to do work.

second messenger A molecule within a cell that mediates a hormonal signal by triggering a response to it.

secondary immune response An immune response to previously encountered antigen that is more rapid and prolonged than the first owing to the swift participation of memory cells.

secondary sexual trait A trait associated with maleness or femaleness but one with no direct role in reproduction.

secretion A product released across the plasma membrane of a cell that may act singly or as part of glandular tissue.

sedimentary cycle A biogeochemical cycle with no gaseous phase; the element moves from land to the seafloor, then returns only through long-term geological uplifting.

seed Of gymnosperms and angiosperms, a fully mature ovule (contains the plant embryo) with integuments that form the seed coat.

segmentation Of animals, a body plan with a repeating series of units that may or may not be similar to one another in appearance. Also, an oscillating movement created by rings of circular muscle in a tubular wall. In a small intestine, such movement constantly mixes and forces the contents of the lumen against the absorptive wall surface.

segregation, theory of [L. *se-*, apart, + *grex*, herd] A Mendelian theory that diploid organisms inherit pairs of genes for traits (on pairs of homologous chromosomes) and that the two genes segregate during meiosis and end up in separate gametes.

selective gene expression Of a multicelled organism, the activation and suppression of some fraction of the same inherited genes in different populations of cells; it leads to cell differentiation.

selective permeability Of a cell membrane, a capacity to allow some substances but not others to cross it at certain sites, at certain times, owing to its molecular structure.

selfish behavior A behavior by which an individual protects or increases its own chance of producing offspring, regardless of the consequences to the group to which it belongs.

selfish herd A society held together simply by reproductive self-interest.

semen (SEE-mun) [L. *serere*, to sow] Sperm-bearing fluid expelled from a penis during male orgasm.

semiconservative replication [Gk. *hēmi*, half, + L. *conservare*, to keep] How a DNA molecule duplicates itself. A DNA double helix unzips and a complementary strand is assembled on the exposed bases of each strand. Each conserved strand and its new partner are wound together into a double helix. Thus the outcome is two "half-old, half-new" molecules.

senescence (sen-ESS-cents) [L. *senescere*, to grow old] Sum of processes leading to the natural death of an organism or some of its parts.

sensation A conscious awareness of a stimulus. Not the same thing as perception, which is an understanding of the meaning of a sensation.

sensory neuron Any of the nerve cells or cells adjacent to them that detect specific stimuli (such as light energy) and relay signals to the brain and spinal cord.

sensory system Of animals, the "front door" of the nervous system; the components that detect external and internal stimuli and relay information about them into the central nervous system.

sessile animal (SESS-ihl) An animal that remains attached to a substrate during some stage (often the adult) of its life cycle.

sex chromosome Of animals and some plants, one of two types of chromosomes, the combinations of which govern gender. *Compare* autosome.

sexual dimorphism Of a species, having individuals with distinctively male or female phenotypes.

sexual reproduction Production of offspring by way of meiosis, gamete formation, and fertilization.

sexual selection A microevolutionary process; natural selection favors a trait that gives the individual a competitive edge in attracting and holding onto mates, hence in reproductive success.

shifting cultivation The cutting and burning of trees, followed by tilling of ashes into the soil; once called slash-and-burn agriculture.

shell model Model of electron distribution in atoms, in which all orbitals available to electrons occupy a nested series of shells.

shoot system The aboveground parts of a plant, such as stems, leaves, and flowers.

sieve-tube member Of flowering plants, a cellular component of the interconnecting sugar-conducting tubes in phloem.

sign stimulus A simple but important cue in the environment that triggers a suitable response to a stimulus that the nervous system is prewired to recognize.

sink region Of plants, any region using or stockpiling organic compounds for growth and development.

sister chromatid One of two DNA molecules (and associated proteins) of a duplicated chromosome that remain attached at their centromere region until separated at mitosis or meiosis; each ends up as a chromosome in a different daughter nucleus.

skeletal muscle An organ with hundreds to many thousands of muscle cells arrayed in bundles. Connective tissue surrounds the bundles and extends beyond the muscle, in the form of tendons that attach it to bone.

sliding filament model Model of muscle contraction in which each myosin filament in a sarcomere physically slides along and pulls actin filaments toward the center of the sarcomere, which shortens. The sliding requires ATP energy and formation of cross-bridges between the actin and myosin.

small intestine Of vertebrates, the portion of the digestive system where digestion is completed and most nutrients absorbed.

smog, industrial Polluted air, gray colored, that predominates in industrialized cities with cold, wet winters.

smog, photochemical Polluted air, brown and smelly, that occurs in large cities with many gas-burning vehicles and a warm climate.

social behavior Intraspecific interactions that require mixes of the instinctive and learned behaviors by which individuals send and respond to communication signals.

social parasite An animal that exploits the social behavior of another species to gain food, care for young, or some other factor necessary to survive, reproduce, or both.

sodium-potassium pump A membrane transport protein; when activated by ATP, it selectively transports potassium ions across the membrane, against its concentration gradient, and allows the crossing of sodium ions in the opposite direction.

soil A variable mixture of mineral particles and decomposing organic material; air and water occupy spaces between the particles.

solute (SOL-yoot) [L. *solvere*, to loosen] Any substance dissolved in a solution. In water, this means that hydration spheres surround the charged parts of individual ions or molecules and keep them dispersed.

solvent A fluid, such as water, in which one or more substances is dissolved.

somatic cell (so-MAT-ik) [Gk. *somā*, body] Of animals, any body cell that is not a germ cell (which gives rise to gametes).

somatic nervous system Nerves leading from the central nervous system to skeletal muscles.

somite One of many paired segments in a vertebrate embryo that will give rise to most bones, skeletal muscles of the head and trunk, and the overlying dermis.

source region Of plants, a site where organic compounds form by photosynthesis.

speciation (spee-cee-AY-shun) Evolutionary process by which a daughter species forms from a population or subpopulation of the parent species. The process can vary in its details and in the length of time it takes before the required reproductive isolation from each other is complete.

species (SPEE-sheez) [L. *species*, a kind] In general, one kind of organism. Of sexually reproducing organisms only, one or more populations of individuals that are interbreeding under natural conditions and producing fertile offspring, and that are reproductively isolated from other such groups.

sperm [Gk. *sperma*, seed] A type of mature male gamete.

spermatogenesis (sperm-AT-oh-JEN-ih-sis) Process by which mature sperm form from a germ cell in males.

sphere of hydration Through positive or negative interactions, a clustering of water molecules around individual molecules of a substance placed in water. *Compare* solute.

spinal cord The portion of the central nervous system that threads through a canal inside the vertebral column and that affords direct reflex connections between sensory and motor neurons, and that has tracts to and from the brain.

spleen One of the lymphoid organs; it is a filtering station for blood, a reservoir of red blood cells, and a reservoir of macrophages.

spore A single-celled reproductive structure or resistant, resting body; often walled, and often adapted to survive adverse conditions. An asexual spore may directly give rise to a new stage of the life cycle; a sexual spore must unite with another sexual spore to produce a new stage.

sporophyte [Gk. *phyton*, plant] Of plant life cycles, a vegetative body that grows by way of mitosis from a zygote and that produces spore-bearing structures.

spring overturn Of some bodies of water, movement during spring of dissolved oxygen from the surface layer to the depths and movement of nutrients from bottom sediments to the surface.

stabilizing selection A mode of natural selection by which intermediate phenotypes are favored, and extremes at both ends of the range of variation are eliminated.

stamen (STAY-mun) A male reproductive structure in a flower, commonly consisting of a pollen-bearing structure (anther) on a single stalk (filament).

start codon A base triplet in a strand of mRNA that serves as the start signal for the translation stage of protein synthesis.

stem cell A type of self-perpetuating cell that remains unspecialized. Some of its daughter cells also are self-perpetuating; others differentiate into cells of specialized structure and function. Stem cells in bone marrow that give rise to daughter stem cells and to blood cells are an example.

sterol (STAIR-all) A type of lipid with a rigid backbone of four fused carbon rings. Sterols differ in the number, position, and type of functional groups. They occur in eukaryotic cell membranes; cholesterol is the main type in animal tissues.

steroid hormone A lipid-soluble hormone, synthesized from cholesterol, that diffuses directly across the lipid bilayer of a target cell's plasma membrane and binds with a receptor inside the cell.

stigma Of many flowers, a sticky or hairy surface tissue on upper portion of ovary; it captures pollen grains and favor their germination.

stimulus [L. *stimulus*, goad] A specific form of energy, such as mechanical energy, light, and heat, that activates a sensory receptor having the capacity to detect it.

stoma (STOW-muh), plural **stomata** [Gk. *stoma*, mouth] A controllable gap between two guard cells in stems and leaves; any of the tiny passageways across the epidermis by which carbon dioxide moves into a plant and water vapor and oxygen move out.

stomach A muscular, stretchable sac that receives ingested food; of vertebrates, an organ between the esophagus and intestine in which much protein digestion occurs.

stop codon A base triplet in a strand of mRNA that serves as a stop signal during the translation stage of protein synthesis; it blocks further additions of amino acids to a new polypeptide chain.

strain One of two or more organisms with differences that are too minor to classify it as a separate species. An example is a strain of *Escherichia coli* or some other bacterium.

stratification Ancient layers of sedimentary rock resulting from the slow deposition of volanic ash, silt, and other materials, one above the other.

stream A flowing-water ecosystem that starts out as a freshwater spring or seep.

stroma [Gk. *strōma*, bed] The semifluid interior between a thylakoid membrane system and the two outer membranes of a chloroplast; the chloroplast zone where sucrose, starch, cellulose, and other end products of photosynthesis are assembled.

stromatolite A formation of sediments and matted remains of photosynthetic species that lived shallow seas and gradually accumulated in thin, cakelike layers.

substrate A reactant or precursor for a metabolic reaction; a specific molecule or molecules that an enzyme can chemically recognize, bind reversibly to itself, and modify rapidly in a predictable way.

substrate-level phosphorylation A direct, enzyme-mediated transfer of a phosphate group from a substrate of a reaction to a different molecule, as when an intermediate of glycolysis gives up a phosphate group to ADP, thus forming ATP.

succession, primary (suk-SESH-un) [L. *succedere*, to follow after] An ecological pattern by which a community develops in sequence, from the time pioneer species colonize a new, barren habitat to the climax community (an end array of species that remain in equilibrium over the region).

succession, secondary A pattern by which some disturbed area within a community recovers and moves back toward the climax state; typical of abandoned fields, burned forests, and storm-battered intertidal zones.

surface-to-volume ratio A mathematical relationship in which volume increases with the cube of the diameter, but surface area increases only with the square. Of growing cells, the volume of cytoplasm increases faster than the surface area of the plasma membrane that must service the cytoplasm. This constraint generally keeps cells small, elongated, or with membrane foldings.

survivorship curve A plot of age-specific survival of a group of individuals in a given environment, from the time of their birth until the last one dies.

swim bladder An adjustable flotation device that changes in volume as it exchanges gases with blood and that allows many fishes to maintain neutral buoyancy in water.

symbiosis (sim-by-OH-sis) [Gk. *sym*, together, + *bios*, life, mode of life] Literally, living together. For at least part of the life cycle, individuals of one species live near, in, or on individuals of another species. Commensalism, mutualism, and parasitism are examples of symbiotic interactions.

sympathetic nerve Of autonomic nervous systems, one of the nerves dealing mainly with increasing overall body activities at times of heightened awareness, excitement, or danger; also work continually in opposition with parasympathetic nerves to make minor adjustments in internal organ activity.

sympatric speciation [Gk. *sym*, together, + *patria*, native land] Species form within the home range of an existing species, in the absence of a physical barrier. It may happen instantaneously, as by polyploidy.

synaptic integration (sin-AP-tik) Of an individual neuron, the moment-by-moment combining of all excitatory and inhibitory signals arriving at its trigger zone.

syndrome A set of symptoms that may not individually be a telling clue but that collectively characterize a particular genetic disorder or disease.

systemic circuit (sis-TEM-ik) Of a closed circulatory system, a route by which oxygen-enriched blood flows from the lungs to the left half of the heart, through the rest of the body (where it gives up oxygen and takes up carbon dioxide), then back to the right half of the heart.

T lymphocyte One of a class of white blood cells that carry out immune responses. The helper T and cytotoxic T cells are examples.

taproot system A primary root together with its lateral branchings.

target cell Of a signaling molecule such as a hormone, any cell having the type of receptor that can bind with that molecule.

tectum The midbrain's roof. In fishes and amphibians, a center for coordinating most sensory input and initiating motor responses. In most vertebrates (not mammals), a reflex center that swiftly relays sensory input to higher integrative centers in the forebrain.

telophase (TEE-low-faze) Of meiosis I, the stage when one of each pair of homologous chromosomes has arrived at a spindle pole. Of mitosis and of meiosis II, the final stage when chromosomes decondense into threadlike structures and two daughter nuclei form.

temperature A measure of how rapidly the molecules or ions of a substance are moving.

tendon A cord or strap of dense connective tissue that attaches muscle to bones.

territory An area that one or more animals defend against competitors for food, water, mates, or suitable living space.

test A way to determine the accuracy of a prediction, as by conducting experimental or observational tests and by developing models. In science, such predictions must be based on potentially falsifiable hypotheses; that is, they must be tested in the natural world in ways that might disprove them.

testcross For an individual of unknown genotype that shows dominance for a trait, a type of experimental cross that may reveal whether the individual is homozygous dominant or heterozygous.

testis, plural **testes** Male gonad; a primary reproductive organ in which male gametes and sex hormones are produced.

testosterone (tess-TOSS-tuh-rown) Of male vertebrates, a sex hormone with key roles in the development and functioning of the male reproductive system.

tetanus Of muscles, a large contraction in which repeated stimulation of a motor unit causes muscle twitches to mechanically run together. In a disease by the same name, toxins block muscles from relaxing.

thalamus Of the forebrain, a coordinating center for sensory input and a relay station for signals to the cerebrum.

theory, scientific A testable explanation of a broad range of related phenomena, one that has been extensively tested and can be used with a high degree of confidence. A scientific theory remains open to tests, revision, and tentative acceptance or rejection.

thermal inversion The trapping of a layer of dense, cool air beneath a layer of warm air; can cause air pollutants to accumulate to high levels close to the ground.

thermophile A type of archaebacterium of unusually hot aquatic habitats, as in hot springs or near hydrothermal vents.

thermoreceptor A sensory cell or specialized cell by it that detects radiant energy (heat).

thigmotropism (thig-MOE-truh-pizm) [Gk. *thigm*, touch] An orientation of the direction of growth in response to physical contact with a solid object, as when a vine curls around a fencepost.

threshold Of a neuron and other excitable cells, the minimum amount by which the resting membrane potential must change to trigger an action potential.

thylakoid membrane system Of chloroplasts, an internal membrane system commonly folded into flattened channels and disks (grana) that have light-absorbing pigments and enzymes used in the formation of ATP, NADPH, or both during photosynthesis.

thymine A nitrogen-containing base of one of the nucleotides in DNA.

thymus gland A lymphoid organ that has endocrine functions; lymphocytes of the immune system multiply, differentiate, and mature in its tissues, and its hormone secretions affect their functioning.

thyroid gland An endocrine gland, the hormones of which affect overall metabolic rates, growth, and development.

tissue Of multicelled organisms, a group of cells and intercellular substances that function together in the performance of one or more specialized tasks.

tonicity The relative concentrations of solutes in two fluids, such as inside and outside a cell. When solute concentrations are isotonic (equal in both fluids), water shows no net osmotic movement in either direction. When one fluid is hypotonic (has less solutes), the other is hypertonic (has more solutes) and water tends to move into it.

toxin A normal metabolic product of one species with chemical effects that can harm or kill individuals of a different species that encounter it.

trace element Any element that represents less than 0.01 percent of body weight.

tracer A substance that has a radioisotope attached to it so that its pathway or destination in a cell, organism, ecosystem, or some other system can be tracked, as by scintillation counters that detect its emissions.

trachea (TRAY-kee-uh), plural **tracheae** An air-conducting tube of respiratory systems; of land vertebrates, the windpipe, which carries air between the larynx and bronchi.

tracheal respiration Of insects, spiders, and some other animals, a respiratory system consisting of finely branching tracheae that extend from openings in the integument and that dead-end in body tissues.

tracheid (TRAY-kid) Of flowering plants, one of two types of cells in xylem that conduct water and dissolved minerals.

tract A communication line inside the brain and spinal cord; comparable to a nerve of the peripheral nervous system.

transcription [L. *trans*, across, + *scribere*, to write] The first stage of protein synthesis, when an RNA strand is assembled on one of the two strands of a DNA double helix; the base sequence of the resulting transcript is complementary to the DNA template.

transfer RNA (tRNA) An RNA molecule that binds and delivers an amino acid to a ribosome and that pairs with an mRNA codon during the translation stage of protein synthesis.

translation The stage of protein synthesis when the encoded sequence of information in mRNA becomes converted to a sequence of particular amino acids, the outcome being a polypeptide chain; rRNA, tRNA, and mRNA interact to bring this about.

translocation Of cells, a stretch of DNA that physically moved to a different location in the same chromosome or in a different one, with no molecular loss. Of vascular plants, the process by which organic compounds are distributed by way of phloem.

transpiration Evaporative water loss from aboveground plant parts, leaves especially.

transport protein A membrane protein that allows water-soluble substances to move through their interior, which spans the lipid bilayer of a cell membrane.

transposable element A DNA region that moves spontaneously from one location to another in the same DNA molecule or a different one. Often such regions inactivate genes into which they become inserted and cause changes in phenotype.

triglyceride (neutral fat) A lipid having three fatty acid tails attached to a glycerol backbone. Triglycerides are the body's most abundant lipids and richest energy source.

trisomy (TRY-so-mee) Of cells, the presence of three chromosomes of a given type rather than the two characteristic of the parental diploid chromosome number.

trophic level (TROE-fik) [Gk. *trophos*, feeder] All of the organisms in an ecosystem that are the same number of transfer steps away from the energy input into the system.

tropical rain forest A biome where rainfall is regular and heavy, the annual mean temperature is 25°C, humidity is 80 percent or more, and biodiversity is spectacular.

tropism (TROE-pizm) Of plants, a directional growth response to an environmental factor, such as growth toward light.

true breeding Of a sexually reproducing species, a lineage in which the offspring of successive generations are just like the parents in one or more traits being studied.

tumor A tissue mass composed of cells that are dividing at an abnormally high rate.

turgor pressure (TUR-gore) [L. *turgere*, to swell] Internal fluid pressure applied to a cell wall when water moves into the cell by way of osmosis.

ultrafiltration Bulk flow of a small amount of protein-free plasma from a blood capillary when the outward-directed force of blood pressure is greater than the inward-directed osmotic force of interstitial fluid.

uniformity Various theories that mountain building, erosion, and other forces of nature have worked over the Earth's surface in the same repetitive ways through time.

upwelling An upward movement of deep, nutrient-rich water along coasts; it replaces surface waters that move away from shore when the direction of prevailing wind shifts.

uracil (YUR-uh-sill) Nitrogen-containing base of a nucleotide found in RNA molecules; like thymine, it can base-pair with adenine.

ureter A tubular channel for urine flow between the kidney and urinary bladder.

urethra A tubular channel for urine flow between the urinary bladder and an opening at the body surface.

urinary bladder A distensible sac in which urine is stored before being excreted.

urinary excretion A mechanism by which excess water and solutes are removed by way of a urinary system.

urinary system An organ system that adjusts the volume and composition of blood, and so helps maintain extracellular fluid.

urine A fluid formed in kidneys by filtration, reabsorption, and secretion; urine consists of wastes, excess water, and solutes.

uterus (YOU-tur-us) [L. *uterus*, womb] A chamber in which the developing embryo is contained and nurtured during pregnancy.

vaccination An immunization procedure against a specific pathogen.

vaccine An antigen-containing preparation, swallowed or injected, that increases immunity to certain diseases by inducing formation of armies of effector and memory B and T cells.

vagina Part of a reproductive system of mammalian females that receives sperm, forms part of the birth canal, and channels menstrual flow to the exterior.

variable Of an experimental test, a specific aspects of an object or event that may differ over time and among individuals.

vascular bundle Of vascular plants, the arrangement of primary xylem and phloem into multistranded, sheathed cords that extend lengthwise through the ground tissue system.

vascular cambium A lateral meristem that increases stem or root diameter (girth).

vascular cylinder Arrangement of vascular tissues as a central cylinder in roots.

vascular plant A plant with xylem and phloem, and well-developed roots, stems, and leaves.

vascular tissue system Xylem and phloem; conducting tissues that distribute water and solutes through the body of vascular plants.

vein Of the circulatory system, any of the large-diameter vessels that lead back to the heart; of leaves, one of the vascular bundles that thread through photosynthetic tissues.

ventricle (VEN-tri-kuhl) Of the vertebrate heart, one of two chambers from which blood is pumped out. Blood circulation is driven by ventricular contraction.

venule A small blood vessel that accepts blood from capillaries and delivers it to a vein.

vernalization Of flowering plants, stimulation of flowering by exposure to low temperatures.

vertebra, plural **vertebrae** One of a series of hard bones organized, with intervertebral disks, into a backbone.

vertebrate Animal with a backbone.

vesicle (VESS-ih-kul) [L. *vesicula*, little bladder] In the cytoplasm of cells, one of a variety of small membrane-bound sacs that function in the transport, storage, or digestion of substances or in some other activity.

vessel member A cell in xylem; although dead at maturity, its wall becomes part of a water-conducting pipeline.

villus (VIL-us), plural **villi** Any of several fingerlike absorptive structures projecting from the free surface of an epithelium.

viroid An infectious particle consisting only of very short, tightly folded strands or circles of RNA. Viroids may have evolved from introns, which they resemble.

virus A noncellular infectious agent that consists of DNA or RNA and a protein coat; it can replicate only after its genetic material enters a host cell and subverts the host's metabolic machinery.

vision Perception of visual stimuli; requires focusing of light precisely onto a layer of photoreceptive cells that is dense enough to sample details of a light stimulus, followed by image formation in a brain.

visual signal An observable action or cue that functions as a communication signal.

vitamin Any of more than a dozen organic substances that animals require in small amounts for metabolism but that they generally cannot synthesize for themselves.

vocal cord One of the thickened, muscular folds of the larynx that help produce sound waves for speech.

water potential Sum of two opposing forces (osmosis and turgor pressure) that can cause a directional movement of water into or out of a walled cell.

water table The upper limit at which the ground in a specified area has become fully saturated with water.

watershed Any specified region in which all precipitation drains into a single stream or river.

wavelength A wavelike form of energy in motion. The horizontal distance between two crests of every two successive waves.

wax A molecule with long-chain fatty acids packed together and linked to long-chain alcohols or to carbon rings. Waxes have a firm consistency and repel water.

white blood cell One of the eosinophils, neutrophils, macrophages, T and B cells, and other leukocytes which, together with their chemical weapons and mediators, defend the vertebrate body against attacks by pathogens and against tissue damage.

white matter Inside the brain and spinal cord, axons that have glistening white sheaths and that specialize in rapid signal transmission.

wild-type allele For a given gene locus, the allele that occurs normally or with the greatest frequency among individuals of a population.

wing A body part that functions in flight, as among birds, bats, and many insects. A bird wing is a forelimb with feathers, strong muscles, and extremely lightweight bones. An insect wing develops as a lateral fold of the exoskeleton.

X chromosome A sex chromosome with genes that, in humans, causes an embryo to develop into a female, provided that it inherits a pair of these.

X-linked gene Any gene located on an X chromosome.

X-linked recessive inheritance Recessive condition in which the responsible, mutated gene occurs on the X chromosome.

xylem (ZYE-lum) [Gk. *xylon*, wood] Of vascular plants, a tissue that transports water and solutes through the plant body.

Y chromosome A distinctive chromosome present in males or females of many species, but not both. In human males, one is paired with an X chromosome (XY); and in human females, the pairing is XX.

Y-linked gene One of the genes located on a Y chromosome.

yellow marrow A fatty tissue in the cavities of most mature bones that produces red blood cells when blood loss from the body is severe.

yolk sac Of land vertebrates, one of four extraembryonic membranes. In most shelled eggs, it holds nutritive yolk; in humans, part becomes a site of blood cell formation and some of its cells give rise to the forerunners of gametes.

zero population growth A population for which the number of births is balanced by the number of deaths over a specified period, assuming that immigration and emigration also are balanced.

zooplankton A freshwater or marine community of floating or weakly swimming heterotrophs, mostly microscopic, such as rotifers and copepods.

zygote (ZYE-goat) The first cell of a new individual, formed by the fusion of a sperm nucleus with the nucleus of an egg at fertilization; also called a fertilized egg.

CREDITS AND ACKNOWLEDGMENTS

CHAPTER 21 **21.1** Jeff Hester and Paul Scowen, Arizona State University, and NASA. **21.2** William K. Hartmann. **21.3** (a) Chesley Bonestell. **21.4** Art, Precision Graphics. **21.5** (a) Sidney W. Fox; (b) W. Hargreaves, D. Deamer. **21.7** Art, Precision Graphics. **21.8** Bill Bachman/ Photo Researchers. **21.9** (a) Stanley W. Awramik; (b-f) Andrew H. Knoll, Harvard University. **21.10** P. L. Walne and J. H. Arnott, *Planta*, 77:325-354, 1967. **21.11** Art, R. Ciemma. **21.12** Robert K. Trench. **21.13** (a) Neville Pledge/South Australian Museum; (b) Chip Clark. **21.14** (a,b,d) Art, R. Ciemma; (c) Patricia G. Gensel. **21.15** *Above*, Megan Rohn courtesy David Dilcher; *below* , Precision Graphics. **21.16** © John Gurche 1989. **21.17** (a) NASA Galileo Imaging Team. **21.18** Art, R. Ciemma. **21.19** (a,b) Paintings © Ely Kish; (c) Field Museum of Natural History, Chicago, and the artist Charles R. Knight (Neg. No. CK8T). **21.20** *Left*, Maps by Lloyd K. Townsend after A. M. Ziegler, C. R. Scotese, and S. F. Barrett, "Mesozoic and Cenozoic Paleogeographic Maps" and J. Krohn, J. Syndermann, "Paleotides Before the Permian" in F. Brosche and J. Syndermann (Eds.), *Tidal Friction and the Earth's Rotation II*, Springer-Verlag, 1983.

CHAPTER 22 **22.1** (a-c) Tony Brain and David Parker/SPL/Photo Researchers. **22.2** Lee D. Simon/Photo Researchers. **22.3** Art, R. Ciemma. **22.4** (a) Stanley Flegler/Visuals Unlimited; (b) CNRI/SPL/Photo Researchers. **22.5** L. J. LeBeau, University of Illinois Hospital/BPS. **22.6** L Santo; art, R. Ciemma. **22.9** (a) John D. Cunningham/Visuals Unlimited; (b) Tony Brain/SPL/Photo Researchers; (c) P. W. Johnson and J. McN. Sieburth, University of Rhode Island/BPS. **22.10** T. J. Beveridge, University of Guelph/BPS . **22.11** *Above*, Centers for Disease Control; *below*, Edward S. Ross. **22.12** (a) Richard Blakemore; (b) Hans Reichenbach, Gesellschaft fur Biotechnologische Forschung, Braunschweig, Germany. **22.13** (a) John Clegg/Ardea London; (b) Courtesy Jack Jones, *Archives of Microbiology*, Volume 136, 1983, pp. 254-261. Reprinted by permission of Springer-Verlag; (c) M. Abbey/Visuals Unlimited; (d) G. Shih and R. Kessel/Visuals Unlimited; (e) T. E. Adams/Visuals Unlimited. **22.15** (a) George Musil/Visuals Unlimited; (b) K. G. Murti/Visuals Unlimited; (c,d) Kenneth M. Corbett. **22.16, 22.17** Art, Palay/Beaubois and Precision Graphics. **22.18** R. Regnery, Centers fo Disease Control and Prevention, Atlanta. **22.19** Edmund Zottola.

CHAPTER 23 **23.1** (a) Ronald W. Hoham, Dept. of Biology, Colgate University; (b) Steven C. Wilson/Entheos; (c) Edward S. Ross; (d) Gary W. Grimes and Steven L'Hernault. **23.2** M. S. Fuller, *Zoosporic Fungi in Teaching and Research*, M. S. Fuller and A. Jaworski (eds.), 1987, Southeastern Publishing Company, Athens, GA. **23.3** (a) Heather Angel; (b) W. Merrill. **23.4** (a) Art, Leonard Morgan; (b) M. Claviez, G. Gerish, and R. Guggenheim; (c) London Scientific Films; (d-f) Carolina Biological Supply Company; (g) courtesy Robert R. Kay from R. R. Kay et al. *Development*, 1989 Supplement, pp. 81-90, © The Company of Biologists Ltd. 1989. **23.5** M. Abbey/Visuals Unlimited. **23.6** (a) Dr. Howard Spero; (b) John Clegg/Ardea, London; (c, e) redrawn from V. & J. Pearse and M. & R. Buchsbaum, *Living Invertebrates*, The Boxwood Press, 1987. Used by permission; (d) G. Shih and R. Kessel/Visuals Unlimited; (e) T. E. Adams/ Visuals Unlimited. **23.7** (a) Jerome Paulin/Visuals Unlimited; (b) David M. Phillips/Visuals Unlimited; (c) John D. Cunningham/Visuals Unlimited. **23.8** Steven L'Hernault; art, Leonard Morgan. **23.9** (a) Frieder Sauer/ Bruce Coleman Ltd.; (b) art, R. Ciemma redrawn from V. & J. Pearse and M. & R. Buchsbaum, *Living Invertebrates*, The Boxwood Press, 1987. Used by permission. (c) Art, R. Ciemma; (d) Sidney L. Tamm; (e) Ralph Buchsbaum, and sketch from V. & J. Pearse and M. & R. Buchsbaum, *Living Invertebrates*, The Boxwood Press, 1987, used by permission. **23.10** Art, R. Ciemma. **23.11** P. L. Walne and J. H. Arnott, *Planta*, 77:325-354, 1967; art, Palay/Beaubois. **23.12** (a) Greta Fryxell, University of Texas, Austin; (b-d) Ronald W. Hoham, Dept. of Biology, Colgate University. **23.13** (a) © S. Berry/Visuals Unlimited; (b) C. C. Lockwood; (c) Florida Department of Environmental Protection, Florida Marine Research Institute, St. Petersburg. **23.14** (a) D. P. Wilson/Eric & David Hosking; (b) Douglas Faulkner/Sally Faulkner Collection. **23.15** Art, R. Ciemma. **23.16** (a) J. R. Waaland, University of Washington/BPS; (b) Lewis Trusty/Animals Animals;*below*, Tom Garrison, *Oceanography: An Invitation to Marine Science*, Wadsworth, 1993. **23.17** (a) Hervé Chaumeton/Agence Nature; (b) © Linda Sims/Visuals Unlimited; (c) Brian Parker/Tom Stack and Associates; (d) Manfred Kage/Peter Arnold, Inc.; (e) Ronald W. Hoham, Colgate University. **23.18** D. J. Patterson/Seaphot Limited: Planet Earth Pictures; art, R. Ciemma. **23.19** Carolina Biological Supply Company.

CHAPTER 24 **24.1** Robert C. Simpson/Nature Stock. **24.2** (a), (b) Robert C. Simpson/Nature Stock. **24.3** Garry T. Cole, University of Texas, Austin/BPS; art, R. Ciemma. **24.4** (a) Jane Burton/Bruce Coleman Ltd.; (b) Thomas J. Duffy. **24.5** Ed Reschke; art, R. Ciemma. **24.6** (a) After T. Rost et al., *Botany*, Wiley, 1979; (b) M. Eichelberger/Visuals Unlimited; (c) Robert C. Simpson/Nature Stock; (d) G. L. Barron, University of Guelph; (e) Garry T. Cole, University of Texas, Austin/BPS. **24.7** N. Allin and G. L. Barron. **24.8** (a) After Raven, Evert, and Eichhorn, *Biology of Plants*, Fourth edition, Worth Publishers, New York, 1986; (b), (c) Gary Head; (d) Mark Mattock/ Planet Earth Pictures; (e) Edward S. Ross. **Page 392** J. D. Cunningham /Visuals Unlimited. **24.9** (a) © 1990 Gary Braasch; (b) F. B. Reeves. **24.10** (a) Dr. P. Marazzi/SPL/Photo Researchers; (b) Eric Crichton/Bruce Coleman Ltd. **24.11** John Hodgin.

CHAPTER 25 **25.1** (a) Pat & Tom Leeson/Photo Researchers; (b, c) Edward S. Ross. **25.2** Art, R. Ciemma after E.O. Dodson and Peter Dodson, *Evolution Process and Product, 3rd Edition*, p. 401, Prindle Weber and Schmidt. **25.4** Craig Wood/Visuals Unlimited; (b) Jane Burton/ Bruce Coleman Ltd.; art, R. Ciemma. **25.5** (a) Fred Bavendam/Peter Arnold; (b) John D. Cunningham/Visuals Unlimited. **25.6** (a,b) Kingsley R. Stern; (c) John D. Cunningham/ Visuals Unlimited. **25.7** (b) Kingsley R. Stern; (c) Ed Reschke/ Peter Arnold; (d) William Ferguson; (e) W. H. Hodge; (f) Kratz/ZEFA. **25.8** Art, R. Ciemma; *inset* photograph A. & E. Bomford/Ardea, London; Lee Casebere. **25.9** Brian Parker/Tom Stack & Associates; art, R. Ciemma; *below*, Field Museum of Natural History, Chicago; (Neg. #7500C). **25.10** (a) Kathleen B. Pigg, Arizona State University; (b) George J. Wilder/Visuals Unlimited. **25.11** (a) Stan Elems/ Visuals Unlimited. **25.12** (a) Jeff Gnass Photography; (b) R. J. Erwin/Photo Researchers; (c) Robert and Linda Mitchell; (d) Doug Sokell/Visuals Unlimited. **25.13** (a) Robert &Linda Mitchell; (b) Ed Reschke. **25.14** (a) Joyce Photographics/Photo Researchers; (b) Runk/Schoenberger/Grant Heilman, Inc.; (c) William Ferguson; (d) Kingsley R. Stern. **25.16** Edward S. Ross; art, R. Ciemma. **25.17** (a) David Hiser, Photographers/ Aspen, Inc.; (b) Terry Livingstone; (c) © 1993 Trygve Steen; (d) © 1994 Robert Glenn Ketchum. **25.18** (a) M.P.L. Fogden/Ardea London; (b) Heather Angel; (c) L. Mellichamp/ Visuals Unlimited; (d) Peter F. Zika/Visuals Unlimited; (e) Art, Jennifer Wardrip. **25.19** Art, R. Ciemma. **25.20** (a,b) © 1989, 1991 Clinton Webb.

CHAPTER 26 **26.1** (a) Courtesy of Department of Library Services, American Museum of Natural History (Neg. # K10273); (c) Lisa Starr, Jack Carey. **26.2** Art, D. & V. Hennings. **26.3** Art, R. Ciemma. **26.4** *Left*, after Laszlo Meszoly in L. Margulis, *Early Life*, Jones and Bartlett. **26.5** Bruce Hall. **26.6** (a-c) R. Ciemma; (d) *left*, Art, R. Ciemma, after Bayer and Owre, *The Free-Living Lower Invertebrates*, © 1968 Macmillan; *right*, Don W. Fawcett/Visuals Unlimited; (e) Marty Snyderman/ Planet Earth Pictures. **26.7** Art. R. Ciemma; (c) F. Schensky; (d) Kim Taylor/ Bruce Coleman Ltd. **26.8** R. Ciemma. **26.9** (a) Art, R. Ciemma; (b) Douglas Faulkner/Sally Faulkner Collection; (c) F. S. Westmorland/Tom Stack & Associates. **26.10** Art, Precision Graphics after T. Storer et al., *General Zoology*, Sixth edition, © 1979 McGraw- Hill. **26.11** (a) Christian DellaCorte; (b) Douglas Faulkner/ Photo Researchers; (c) Bill Wood/ Seaphot Limited: Planet Earth Pictures. **26.12** Andrew Mounter/ Seaphot Limited: Planet Earth Pictures. **26.13** R. Ciemma; (b) Kathie Atkinson/Oxford Scientific Films. **26.14** Kim Taylor/ Bruce Coleman Ltd.; art, R. Ciemma. **26.15** (a) Cath Ellis, University of Hull/SPL/ Photo Researchers; (b) Robert & Linda Mitchell. **26.16** Kjell B. Sandved. **26.17-26.18** R. Ciemma; **27.19** Art, R. Ciemma, micrograph Carolina Biological Supply Company. **26.20** (a) Lorus J. and Margery Milne; (b) Dianora Niccolini. **26.21** Art, R. Ciemma. **Page 430** Art, Palay/Beaubois. **26.22** (a) Gary Head; (b) Alex Kerstitch; (c) Jeff Foott/Tom Stack & Associates; (d) Frank Park/A.N.T. Photo Library; (e) J. Grossauer/ ZEFA. **26.23** (b) Kjell B. Sandved; (c) Hervé Chaumeton/Agence Nature. **26.24** (b) Hervé Chaumeton/Agence Nature. **26.25** (a) Art, R. Ciemma; (b) Bob Cranston; (c) © 1992 Sea World of California, All rights reserved. **26.26** (a) J.A.L. Cooke/Oxford Scientific Films; (b) © Cabisco/Visuals Unlimited; (c) Jon Kenfield/Bruce Coleman Ltd.; (d) *Left*, from Eugene N. Kozloff, *Invertebrates*, copyright © 1990 by Saunders College Publishing. Reproduced by permission of the publisher. *Right*, adapted from Rasmussen, *Ophelia*, Vol. 11 in Eugene N. Kozloff, *Invertebrates*, 1990. **26.27** Art, R. Ciemma. **26.28** Jane Burton/ Bruce Coleman Ltd. **26.29** *Above*, Angelo Giampiccolo/FPG; *below*, Jane Burton/Bruce Coleman Ltd. **26.30** (a) Redrawn from *Living Invertebrates*, V. & J. Pearse/M. & R. Buchsbaum, The Boxwood Press, 1987. Used by permission; (b) P. J. Bryant, University of California, Irvine/BPS; (c) Ken Lucas/ Seaphot Limited: Planet Earth Pictures; (d) John H. Gerard. **26.31** (a) Franz Lanting/ Bruce Coleman Ltd.; (b) Hervé Chaumeton; (d) Agence Nature; (f) Fred Bavendam/Peter Arnold, Inc. (e), (g) Art, R. Ciemma. **26.32** After David H. Milne, *Marine Life and the Sea*, Wadsworth 1995. **26.33** (a) Z. Leszczynski/ Animals Animals; (b) Steve Martin/Tom Stack & Associates. **26.34** Art, D. & V. Hennings. **26.35** From Georges Pasteur, "Jean Henri Fabre," *Scientific American*, July 1994. Copyright © 1994 by Scientific American, Inc. All rights reserved. **26.36** (a-d) Edward S. Ross; (e) C. P. Hickman, Jr.; (f-i) Edward S. Ross; (j) David Maitland/Seaphot Limited: Planet Earth Pictures. **26.37** (a) Chris Huss/The Wildlife Collection; (b) John Mason/Ardea London; (c) Kjell B. Sandved; (d) Ian Took/Biofotos. **26.38** (a) Art, Lewis Calver; (b-c) Hervé Chaumeton/Agence Nature. **Page 443** Jane Burton/Bruce Coleman Ltd. **26.39** Art, R. Ciemma. **26.40** Walter Deas/ Seaphot Limited: Planet Earth Pictures. **26.41** (a) J. Solliday/BPS; (b) Hervé Chaumeton/ Agence Nature.

CHAPTER 27 **27.1** (a) Jean Phillipe Varin/ Jacana/Photo Researchers; (b) Tom McHugh/ Photo Researchers. **27.3** (a) Rick M. Harbo; (b) Peter Parks/Oxford Scientific Films/ Animals Animals; (c) redrawn from *Living Invertebrates*, V. & J. Pearse and M. & R. Buchsbaum, The Boxwood Press, 1987. Used by permission. **27.4** (a) Art, R. Ciemma; (b) Runk & Schoenberger/ Grant Heilman, Inc. **27.6** Al Giddings/Images Unlimited; art, R. Ciemma, adapted from A. S. Romer and T. S. Parsons, *The Vertebrate Body*, Sixth edition, Saunders College Publishing, 1986. **27.7** (a, b) After David H. Milne, *Marine Life and the Sea*, Wadsworth, 1995; (c) Heather Angel. **27.8** (a) Erwin Christian/ZEFA; (b) Allan Power/Bruce Coleman Ltd.; (c) Tom McHugh/Photo Researchers. **27.9** (a) Bill Wood/Bruce Coleman Ltd.; (b) R. Ciemma. **27.10** (a) From Tom Garrison, *Oceanography: An Invitation to Marine Science*, Wadsworth, 1993; (b) Robert & Linda Mitchell; (c) Patrice Ceisel/© 1986 John G. Shedd Aquarium; (d) Peter Scoones/Seaphot Limited: Planet Earth Pictures. **27.11** (a) © Marianne Collins; (b, c) art, Laszlo Meszoly and D. & V. Hennings. **27.12** (a) Jerry W. Nagel; (b) Stephen Dalton/ Photo Researchers; (c) John Serraro/Visuals Unlimited; (d) Juan M. Renjifo/ Animals Animals. **Page 455** Art, Leonard Morgan adapted from A. S. Romer and T. S. Parsons, *The Vertebrate Body*, Sixth edition, Saunders College Publishing, 1986, and others. **27.13** (a) R. Ciemma; (b) © 1989 D. Braginetz; (c) Zig Leszczynski/ Animals Animals. **27.14** D. & V. Hennings. **27.15** (a) Kevin Schafer/Tom Stack & Associates; (b) D. Kaleth/ Image Bank; (c) Art, R. Ciemma; (d) Andrew Dennis/A.N.T. Photo Library; (e) Stephen Dalton/Photo Researchers; art, R. Ciemma; (f) Heather Angel. **27.16** (a) Gerard Lacz/A.N.T. Photo Library; (b) J.L.G. Grande/ Bruce Coleman Ltd.; (c) Rajesh Bedi; (d) Thomas D. Mangelsen/Images of Nature. **27.17** (b) Art, D. & V. Hennings. **27.18** (a) Sandy Roessler/ FPG; (b) Leonard Lee Rue III/FPG; (c) Art, R. Ciemma after M. Weiss and A. Mann, *Human Biology and Behavior*, Fifth edition, Harper Collins, 1990. **27.20** (a) D. & V. Blagden/A.N.T. Photo Library; (b) Jack Dermid; (c) Photo Researchers. **27.21** R. Ciemma. **27.22** (a) Clem Haagner/Ardea London; (b) J. Scott Altenbach, University of New Mexico; (c) Douglas Faulkner/ Photo Researchers; (d) Christopher Crowley; (e) Leonard Lee Rue III/FPG. **27.23** W. J. Weber/Visuals Unlimited.

CHAPTER 28 **28.1** *Left*, FPG; *right*, Douglas Mazonowicz/Gallery of Prehistoric Art. **28.2** *below*, Roger Burnard. **28.3** (a) Bruce Coleman Ltd.; (b) Tom McHugh/Photo Researchers;
(c) Larry Burrows/Aspect Picture Library. **28.4** D. & V. Hennings. **28.5** © Time Inc. 1965/Larry Burrows Collection. **28.8** Art, Precision Graphics after *National Geographic*, February 1997, page 82. **28.9** Art, D. & V. Hennings. **28.10** (a) Dr. Donald Johanson, Institute of Human Origins; (b) Louise M. Robbins; (c,d) Kenneth Garrett/ National Geographic Image Collection. **28.11** (a,b) Kenneth Garrett/National Geographic Image Collection; (c) Jean Paul Tibbles. **28.12** Photographs by John Reader ©1981. **28.13** Art, R. Ciemma. **Page 479** Gary Head.